Understanding Language

Man or Machine

FOUNDATIONS OF COMPUTER SCIENCE

Series Editor: Raymond E. Miller
Georgia Institute of Technology

PRINCIPLES OF DATA SECURITY
Ernst L. Leiss

UNDERSTANDING LANGUAGE: Man or Machine
John A. Moyne

Understanding Language

Man or Machine

John A. Moyne

Queens College and The Graduate School
The City University of New York
New York, New York

PLENUM PRESS • NEW YORK AND LONDON

Library of Congress Cataloging in Publication Data

Moyne, John A.
  Understanding language.

  (Foundations of computer science)
  Bibliography: p.
  Includes index.
  1. Psycholinguistics. 2. Linguistics—Data processing. 3. Comprehension. 4. Artificial intel-
ligence. 5. Grammar, Comparative and general. 6. Formal languages. I. Title. II. Series.
P37.M69   1985   401′.9   85-12341
ISBN 0-306-41970-X

© 1985 Plenum Press, New York
A Division of Plenum Publishing Corporation
233 Spring Street, New York, N.Y. 10013

Printed in the United States of America

To Claudia

Preface

This textbook is intended for graduate students in computer science and linguistics who are interested in developing expertise in natural language processing (NLP) and in those aspects of artificial intelligence which are concerned with computer models of language comprehension.

The text is somewhat different from a number of other excellent textbooks in that its foci are more on the linguistic and psycholinguistic prerequisites and on foundational issues concerning human linguistic behavior than on the description of the extant models and algorithms. The goal is to make the student, undertaking the enormous task of developing computer models for NLP, well aware of the major difficulties and unsolved problems, so that he or she will not begin the task (as it has often been done) with overoptimistic hopes or claims about the generalizability of models, when such hopes and claims are inconsistent either with some aspects of the formal theory or with known facts about human cognitive behavior. Thus, I try to enumerate and explain the variety of cognitive, linguistic, and pragmatic data which must be understood and formalized before they can be incorporated into a computer model.

In Chapter 7, I argue that artificial intelligence research must proceed along two parallel paths: (a) it must study and understand the nature, scope, and other aspects of human intelligence, and it must answer questions about biological endowment versus acquired knowledge as well as other relevant aspects of human behavior; (b) it must study and develop computing devices and programs which can perform tasks similar to those done by people, where intelligence and reasoning are required. The focus of the study of human language understanding in this book is more concerned with path (a) than path (b).

The chapters in this book are self-contained and need not necessarily be studied in the same sequence in each part. The instructor can omit or rearrange parts and chapters depending on the background of the students and on the scope of material to be covered in the class. No background need be assumed for the study of this book, other than the maturity and general knowledge of a graduate-level student. Thus, some students may be advised to consult some of the more basic references to gain more background, whereas other students may be able to omit or skim through the prerequisite chapters. In the text, a variety of algorithms, parsers, and instructions for computer projects have been provided for which I assume that the student already has the necessary programming skill, can obtain it from another course, or is being taught by the instructor as a supplementary part of the course (artificial intelligence courses sometimes start with instructions in LISP).

The author has found that the best results from the use of this text can be obtained by having students from computer science, linguistics, psychology, speech and hearing, and other areas in the same class, and by organizing small multidisciplinary groups of students who can complement each other and who can develop joint projects for study and implementation.

Many students and colleagues have contributed to this study. Substantial parts of Chapters 2, 3, and 7 were written by Ezra Black. Other contributors have included D. Terence Langendoen, Edwin Battistella, Thomas Maxfield, Constantine Kaniklidis, and Seyma Karmel. I am also grateful to L. S. Marchand and Barbara Verdi, editors at Plenum Press, for their support and skillful editorial improvement of this work. The original research for this study was supported in part by Grant IST-7923585 from the National Science Foundation.

The author wishes to thank the following authors and publishers for permission to reprint copyrighted material in this book.

From "Cognitive Psychology," by G. Bower. Copyright 1975. In *Handbook of Learning and Cognitive Processes*, Volume 1, W. Estes, ed., Hillsdale, N.J.: Lawrence Erlbaum. Adapted by permission of Lawrence Erlbaum Associates, Inc.

From "Levels of Processing: A Framework for Memory Research," by S. Craik and R. Lockhart. Copyright 1972. In *Journal of Verbal Learning and Verbal Behavior, 11*: 671–684. Reprinted by permission of Academic Press.

From "Retrieval Time from Semantic Memory," by A. Collins and A. Quillan. Copyright 1969. In *Journal of Verbal Learning and Verbal Behavior, 8*: 240–247. Adapted by permission of Academic Press.

From "Comprehension and Memory of Text," by W. Kintsch. Copyright 1978. In *Handbook of Learning and Cognitive Processes,* Volume 6, W. Estes, ed., Hillsdale, N.J.: Lawrence Erlbaum. Reprinted by permission of Lawrence Erlbaum Associates, Inc.

Contents

Introduction

Nothing is in the mind that did not pass through the senses.
Aristotle

This study is concerned with language comprehension as a phenomenon of human information-processing activity. Language comprehension, however, is one aspect of the general process of human perception and, therefore, language perception cannot be studied in isolation because other sensory processes such as visual, aural, and somesthetic perceptions are remarkably analogous. Thus, to construct an ideal model of human information processing for language comprehension, one must understand all the mechanisms of various perceptions and must be able to account for various other aspects of human mental processes and memory operations.

In Chapter 4, I touch on some philosophical aspects of these issues and give a survey of psycholinguistic views on language comprehension. Later in the study, I pose the challenging question—Is a *realistic* model of human language comprehension attainable? Computer models for natural language processing have been constructed, some with impressive achievements; but in retrospect, the builders are the first to admit that the models are far from adequate as realistic models. Furthermore, the inadequacies lie in some fundamental concepts and principles, rather than in the earlier popular notions about the size and speed of memory, vocabulary, data base, and other mechano-tactical considerations.

I have pursued this study on the thesis, borne out by the experience of the last three decades, that the understanding of human language use ("performance") should be a prerequisite for any attempt for constructing computer systems for natural language processing. Thus, the study has focused primarily on a theory of human language comprehension; and because this is ultimately linked with a theory of performance, the question arises as to what a theory of performance should include. In particular, a stand needs to be taken on the standard generative linguistic position that a theory of performance must include a theory of competence (Chomsky, 1965; Katz, 1977). According to Katz, in order to describe and explain how context and pragmatic factors can and do affect the meaning of utterances, we must first say what the meaning would be if it were not affected by such factors (this is Katz's "meaning in the null context"). Semantic theory, which is part of competence, will give us this meaning in the null context (i.e., the grammatical meaning); hence competence is needed as a part of, or is closely linked with, performance.

Is transformational grammar a model of competence? There are some recent proposals advocating that natural languages can be accounted for by context-free or even perhaps finite-state grammars. If this is true, then there is no need to have a grammar in the theory of performance that exceeds the power of a finite-state device. The matter is not, unfortunately, that simple, and in Part II I have discussed the details of the controversies. Thus, the study provides, in part, a catalogue of the complexities, controversies, and uncertainties associated with human language comprehension and with building of natural language-processing models.

I have not answered the question about the attainability of a *realistic* model, but I have worked out schemata (Chapter 9) for a model that approximates human comprehension to the extent that it has been possible to ascertain in this study. However, this model, given at the very end of this book, is only a skeletal proposal and has little bearing on the principal goals of this study. We are far from a complete understanding of human information processing and language comprehension and much research remains to be done for any model claiming emulation of human behavior (see Chapter 7). In a taxonomy of current models and their theoretical underlying concepts three approaches have been identified: linguistic, perceptual, and conceptual. Each of these approaches has its problems, controversies, and uncertainties. Our model purports, among other things, to integrate these seemingly different and at times contradictory approaches. Thus, in addition to its own complexities, the model tacitly inherits all the problems of the extant models. Problems aside, I recommend further research on such models.

There are obvious advantages in building natural language-processing models such as those developed in artificial intelligence (AI), over theoretical studies alone. In the former, one is forced to pay attention to details and to take a more global view of the processes; in the latter the tendency is to concentrate on narrow, tractable problems or subproblems. Moreover, computer models of language understanding can be useful tools for the study of perception and for testing hypotheses about language performance. Unfortunately, however, there have occasionally been implicit if not always explicit claims for generality and global utility of some of these models which are not founded or consistent with our present meager knowledge about languages and mental processes. There are also more serious issues concerning learning theory and theories of computability and decidability. These issues hit at the heart of the question of attainability. I touch on these issues in Part III.

In this study I try to define and explicate some of the crucial questions concerning language comprehension. I supply the necessary background in the disciplines involved and then discuss the central issues relevant to the goals of our study. Finally, I try to identify some of the characteristics that a realistic model of language comprehension must have, whether or not it will be (or can be) implemented on a computer. The study is then divided into three parts.

Part I consists of introductory chapters on automata and formal languages, linguistics, and psycholinguistics. The material in these chapters constitutes the necessary background and prerequisites for reading the rest of the study. Readers with sufficient background in these topics can skip this part. Of the three chapters in Part I, Chapter 1 is the most elementary and basic for the reading of the other two chapters.

Part II gives a critical survey of the literature and some of the more important models for language comprehension and natural language processing in the context of human information processing. I include the literature in linguistics, psychology, and artificial intelligence in this survey and try to spell out the immensity and complexity of linguistic structures, mental processes, memory structures and operations, brain functions, and other relevant topics. The problems are emphasized in order to demonstrate that the current models, though impressive and useful, are inadequate and inconsistent with the real world of human linguistic behavior.

The approaches discussed in this study must all deal with problems of representation in the memory of both linguistic structures and knowledge. Many questions about representations are unresolved and provide topics for future research. There are even controversies about the representation of the various modes of perception of the same object. For example, do the audio and visual perception of a horse have the same or different representations? Issues in representation are tied with data structures and storage and with semantics. Semantic theory is particularly rife with variations and controversies: truth conditional, sense structure, use theories, possible world, epistemic role, causal—to name some of the principal theories.

There are poorly understood and vexing problems concerning the processing devices for language perception. Here we are concerned with parsing systems, strategies, control systems, and the questions of hierarchical/serial versus parallel processing, on-line interactive versus autonomous processing, and the like. Limitations of the human information-processing capacity must be contrasted with the underlying Turing machine power of many of the computational models and linguistic proposals. The basic question is this: What sort of restrictions can be put on the perceptual models to bring them within the domain of primitive recursive capacity and yet preserve the breadth and richness of human perception?

In Part III the outline of a lexicon and a parser which the reader can further develop and implement on a computer has been provided. The parser is first presented as a simple device with backtracking capability. The parser is then revised in a couple of stages until it becomes a deterministic parser with no backtracking. The final version can potentially allow the device to "observe" the entire input sentence simultaneously and to do parallel processing of its elements.

PART I

Preliminaries and Prerequisites

CHAPTER 1

Formal Languages and Automata

Introduction

Study of formal languages and automata theory is a prerequisite for much of the material covered in this text. In this chapter, I will give a survey which is adequate for our purposes. The literature in the field is, however, extensive and the interested reader can gain a much deeper understanding of the topics discussed in this survey by studying some of the references cited at the end of this chapter, particularly Hopcroft and Ullman (1979) and Aho and Ullman (1972–73).

In this chapter, some of the theoretical aspects of the abstract underlying structures of languages, generative grammars, and abstract recognizers (acceptor automata) will be investigated. Within the present general theoretical approach, there is no difference in principle between natural languages and the artificial and formal languages of computers, mathematics, and automata theory. There are, of course, differences in the complexity of structures, and these differences, in turn, impose limitations on the models and on the understanding of the nature of some languages and of human linguistic behavior.

Given these premises, language L is defined as a finite or an infinite set of strings or *sentences*, constructed by concatenation out of a finite set of atomic elements called *alphabet* or *vocabulary*, in accordance with the rules of some grammar G. Of the strings over a vocabulary V, there is one with no symbols in it. We call this a null or *empty* string, denoted by the symbol ϕ. Then, if V is an alphabet, V^* is the set of all possible strings concatenated from the elements of V, including the empty string ϕ. We denote V^+ as the set of all strings over V except ϕ.

Symbols in a vocabulary can be concatenated in any arbitrary manner, but every string constructed in this way is not a sentence in a language L. We call the sentences of L *well-formed* sentences to distinguish them from the other strings made over the same vocabulary. An example in English would be

      (1)   A boy kicked the ball
      (2)   The kicked a boy ball

(1) and (2) are concatenated from the same vocabulary, but (1) is a sentence in English whereas (2) is not. In defining a language, we must, then, state some criteria for distinguishing between the strings that are well-formed sentences and those that are not. We cannot achieve this by merely listing all the sentences in L because in any interesting language system the number of sentences can be very large, possibly infinite. Furthermore, in many cases although the length of any given sentence is finite, there is no bound on sentence length. Consider the following well-known examples from English:

      (3a)   This is the cat
      (3b)   This is the cat that caught the rat
      (3c)   This is the cat that caught the rat that ate the cheese

It is clear that any number of *that* clauses can be added to (3a), each time forming a new and longer sentence. Any limit on the number of *that* clauses is imposed by human memory or patience, not by the structure of the language. Those familiar with recursive production rules will notice that the sentences in (3) could be described by two *rewrite* rules. Let X be an initial symbol, S be a sentence, and R a relative *that* clause; then

(4) $X \rightarrow S, S \rightarrow S \frown R$

where $\rightarrow$ is the rewrite or production symbol and $\frown$ is the concatenation symbol. Note that the two rules in (4) will generate an infinite number of sentences similar to those in (3).

    Listing of sentences aside, two other ways are open to us for specifying languages: (a) We can try to develop a device for testing sentencehood; that is, given a string, the device can somehow test it and decide whether or not the string is in L. This is sometimes done in a trivial and *ad hoc* way in compilers for programming languages of computers, but it has seldom been attempted in a systematic way. (b) We can try to construct a recursive procedure for enumerating the infinite list of sentences in L.

    The rules in (4) are trivial examples of the latter alternative, and further details of this approach will be developed in this chapter. The testing device in alternative (a) should not be confused with formal recognizers or acceptors, which will be discussed below. A recognizer R is a device which can receive a string σ as input and determine if σ is a well-formed string in L. But implicit in the definition of R is the existence of a grammar G which generates well-formed strings in L. A recognizer can therefore be thought of as an acceptor automaton, an abstract machine, which will accept a well-formed sentence of a grammar and halt, and which will reject a nonsentence.

A grammar should be finite because human beings and all our physical devices are finite organisms. It should, however, contain a finite set of rules which recursively specify (enumerate, generate) the potentially infinite number of sentences and exclude the nonsentences of a language. We denote as $L(G)$ the language L specified by grammar G, and we refer to G as the theory of L. However, we will see in the course of the subsequent sections that we would need more precise definitions of L, G, and R.

Synonymous with the notions of grammar and acceptor automata are *algorithm* and *procedure*. Algorithm is an old concept in mathematics. The term is derived from the name of the ninth century mathematician al-Khwarizmi, of the golden era of Islamic science.

An algorithm is a finite set of instructions for a sequence of operations which will result in solving a problem of a given type or in showing that the problem has no solution. The generality of an algorithm is guaranteed by its requirement to produce an answer for any problem in a given class. Conversely, in mathematics, a class of problems is considered solved if an algorithm can be discovered for it. The characteristic of an algorithm is that it *halts*; that is, an output is always produced and the processing stops. On the other hand, a set of instructions may never reach a halting state if the process cannot determine, for example, that there is no solution or result for a given problem—this is called a *procedure*. Thus, procedure is a broader term for algorithm; a procedure which always halts is an algorithm. There are many examples of procedures which do not halt, and here is an example:

Problem: Find the value of x such that $a + x = b$, where a and b are positive integers.

> Procedure Evaluate-x
> (1) Input values for a and b
> (2) Set $x \leftarrow 0$
> (3) Set $y \leftarrow a + x$

(4)   If $y = b$, print x and halt; else continue

(5)   Set $x \leftarrow x + 1$

(6)   Goto step (3)

Is this an algorithm? Does it meet the above specifications? Let $a = 5$ and $b = 15$. After running 11 cycles, the above procedure will print 10 for the value of x and will halt. Now, suppose we set $a = 5$ and $b = 3$; the procedure will not halt. Thus, this is not an algorithm. Another example which will demonstrate a somewhat different problem is:

(5) $$4x^2 - x + 2 = 0$$

The equation in (5) has no integer solution because it is clear that for any value of x, $4x^2 > x - 2$. Thus, there is no general solution or algorithm for the class of equations of this type with a number of unknowns. However, here we must not confuse problems with no solutions and undecidable problems. For one-variable diophantine equations, such as (5), there is in fact a decision algorithm for telling whether or not there exists a solution in integers and, if so, for finding it. The general problem was posed by David Hilbert in 1901 (among his famous challenging problems) for the multivariable system of polynomial equations. The Russian mathematician, J. V. Matijasevic, showed only recently that the general problem is unsolvable.

A simple example of a procedure (algorithm) which always halts is the Euclidean algorithm: Given two positive integers a and b, find their greatest common divisor. There is always at least one common divisor for any pair of positive integers—1. Hence, the following procedure will always terminate with a result. Let X and Y be two variables, respectively. We then reduce the division to a series of subtractions until $X = Y$. At this point, the value of X (or Y) is the answer.

(6)   Procedure Euclid
                      Set $X \leftarrow a$
                      Set $Y \leftarrow b$
                      AA: if $X = Y$, print X and halt
                          if $X > Y$ then
                                              $k \leftarrow X - Y$
                                              $X \leftarrow k$
                                              Goto AA
                                  else
                                              $k \leftarrow Y - X$
                                              $Y \leftarrow k$
                                              Goto AA
                      end

(Note: these procedures are not intended as examples of efficient computer programming!)

We will return to a more precise definition of algorithms and procedures later in this chapter. Turing (1936, 1937) gave one of the original exact definitions, and we owe much of the rigor and precision in the definition of languages and grammars to Chomsky (1956, 1959a, 1963).

Before turning to the discussion of grammars and languages, let us take a brief note of functions in the context of set theory. A function in this context is synonymous with a particular case of *mapping* or *relation*. Given two sets A and B, their Cartesian product is denoted by $A \times B$. This represents a collection of all ordered pairs of elements in A and B (a,b) such that $a \in A$ and $b \in B$. A relation R from set A to set B is a set; a subset of $A \times B$, is denoted as $R \subseteq A \times B$. If the pair (a,b) is in the set R, we write aRb or $R(a,b)$. Note that relation R indicates that for $a \in A$, there are corresponding elements in B, but the reverse is not necessarily true; that is, there may be elements in B with no corresponding elements in A. A relation R is a function if for every $a \in A$ there is exactly one $b \in B$ such that (a,b) is in R. Thus, we see that a function f is also a set. Consider the following examples: Let $A = \{a,b\}$ and $B = \{c,d,e\}$. Is R_1 a function?

$$R_1 = \{(a,d),(a,e),(b,e)\}$$

No, because $a \in A$ maps into two elements (d,e) in B. Are R_2 and R_3 functions?

$$R_2 = \{(a,c),(b,d)\}$$
$$R_3 = \{(a,c),(b,c)\}$$

Yes. For another example, assume that A is the set of all children and B is the set of all mothers.

In the above examples, set A is called the *domain* and set B the *range* of the function, and if f is a function from A to B, we write this as $f:A{\rightarrow}B$. In computer science, we often define a function by its name, its arguments, and its body, and we write it as a mathematical formula or as a series of instructions. Thus, $f(x) = x^3$ has the variable x as its argument and its body assigns to each real number its cube. The function $\text{sum}(a,b) = a + b$ has "sum" as its name, a and b as its arguments, and its body, $a + b$, returns the sum of the variables a and b.

   Recursive definition of a function is the definition permitting values of the function to be computed systematically in a finite number of steps, specifically by calling upon the function itself for computations in each cycle of the general solution procedure, namely, the recursive function. Let us consider some examples. We will develop a function to represent the expression "M raised to the power of N," where M and N are integers and N is positive. We can write this in function notation as power(m,n) where "power" is the name and m and n are the arguments. The entire formula can be written as

$$\text{power}(m,n) = m*m* \ldots \ldots *m \ (\text{for } n - 1 \text{ multiplications})$$

However, a less awkward notation would be:

$$\text{power}(m,n) = \begin{cases} m & \text{if } n = 1 \\ m*\text{power}(m,n - 1) & \text{if } n > 1 \end{cases}$$

The interpretation of this function is this: If $n = 1$, then the value of $m^n = m$, else $m^n = m*m^{n-1}$. This function is defined in terms of "itself"; that is, it has to be evaluated repeatedly to compute the value of m^{n-1} until $n = 1$. The accumulation of the values is then the value of the function. Let us assume we want to compute recursively the value of 4^3, which is $4*4*4 = 64$. To evaluate power $(4,3)$, we initialize m and n to 4 and 3, respectively, and follow the instructions of our recursive function:

$$\begin{aligned} &(1) \quad \text{power}(4,3) = 4*\text{power}(4,2) \\ &(2) \quad \text{power}(4,2) = 4*\text{power}(4,1) \\ &(3) \quad \text{power}(4,1) = 4 \end{aligned}$$

Now, substituting (3) in (2), we get

$$(2') \quad \text{power}(4,2) = 4*4 = 16$$

and substituting (2') in (1), we get

$$(1') \quad \text{power}(4,3) = 4*16 = 64$$

As an exercise, the reader can now show how n-factorial ($n!$) is computed by the following recursive function:

$$n! = \begin{cases} 1 & \text{if } n \leq 1 \\ n*(n - 1)! & \text{otherwise} \end{cases}$$

Note that in the above examples there is always a diminishing factor which systematically reduces the value of an argument in the process until it is brought to a final result and halt. This is a characteristic of a recursive function; it halts in a finite number of operations. Thus, a recursive function is an algorithm.

Returning to sets, we will be dealing with two types: (a) finite sets, containing a finite number of elements, and (b) countably infinite sets whose elements can be placed in a one-to-one correspondence with the set of positive integers. An example of a countably infinite set is the set of all even integers. An example of a set which is not countable is the set of real numbers; for given any arbitrary pair of real numbers, say 0.5 and 0.6, there are an infinite number of other real numbers which come between the two. There are grammars and languages which fall within all these categories of sets. Finally, there is a particular set of special interest. This is the set called pi, π, whose elements are {T,F} for true and false. A function whose range is π is called a *predicate*.

Grammars and Languages

The components of grammars under the present review are a vocabulary V and a set of rewrite rules or productions P. Both V and P are finite sets. The vocabulary has two parts or subsets: *nonterminal vocabulary N* and *terminal vocabulary T*. This can be more formally stated as

(7a) $$V = N \cup T$$
(7b) $$N \cap T = \emptyset$$

where (7b) denotes that N and T are disjoint sets, that is, no elements are shared between the two subsets of V. One special symbol in N is called initial, start, or distinguished symbol, and we will use S to represent this symbol.

The productions in P have the general form:

(8) $$\gamma\alpha\omega \rightarrow \gamma\beta\omega$$

where α, β, γ, and ω are strings over V and can be null, except α, which cannot be null. The rule in (8) can be read as "rewrite α as β in the context γ–ω." If the context is not specified in the rule, it will then apply in any context, including null. We will soon see that depending on the type of restrictions placed on the rule in (8), we get different types of grammars and languages. However, we can now give a general definition of a grammar as a quadruple:

(9) $G = (N,T,S,P)$

Let us take another example from English.

(10) $N = \{NP, VP, DET, ART, ADJ, N, V, S\}$
        $T = \{boy, ball, saw, kicked, tall, big, the, a\}$
        P: S → NP VP
              NP → (DET) N
              VP → V NP
              DET → ART (ADJ)
              N → boy, ball
              V → kicked, saw
              ART → the, a
              ADJ → tall, big

where S (the initial symbol) stands for sentence, NP = noun phrase, VP = verb phrase or predicate, DET = determiner, N = noun, V = verb, ART = article, ADJ = adjective. The symbols in parentheses in the P rules are to be taken as optional.

Of course, (10) is a fraction of a grammar of English, but it will nevertheless generate structures or analyses for sentences such as

        (11a)   A boy kicked the ball
        (11b)   The tall boy kicked the big ball
        (11c)   The boy saw the ball

The syntactic structure for (11b), according to grammar (10), is:

(12)   [s [NP [DET [ART the] [ADJ tall]] [N boy]] [VP [V kicked]
         [NP [DET [ART the] [ADj big]] [N ball]]]]]

(12) is a *labeled bracketing* representation of the sentence; the structure can also be represented as a syntactic tree which is usually referred to as a *phrase marker* (PM):

(13)

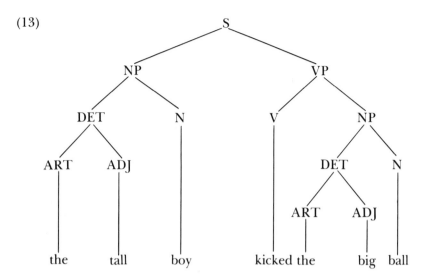

We will see in the course of the subsequent sections that the model of grammar in (10) is inadequate for natural languages. But even for the present fragmentary example, notice that there is nothing in the grammar to prevent formation of strings such as (an asterisk denotes that the string is not well-formed):

> (14a) *The tall ball kicked the big ball
> (14b) *The ball saw the boy

which are ungrammatical or anomalous at least from the semantic viewpoint, and should be blocked by any adequate grammar of English. For the time being, however, we will be concerned with formal languages for which this type of grammar seems to be adequate.

Let us consider a language L_1 whose sentences consist of a number of a's followed by the same number of b's and only these. We will represent strings of this kind as $a^n b^n$. The grammar G_1 for this language would be:

(15)
$$N = \{S\} \quad T = \{a,b\}$$
$$P: \quad S \to aSb$$
$$S \to ab$$

A slightly more complicated grammar would be one which can generate strings consisting of mirror image strings with a unique central element; for example, a string consisting of any concatenation of a's and b's, followed by a single c, followed by a mirror image of the first string: *abcba, bbacabb, ababacababa*. The grammar G_2 for this language L_2 is:

(16)
$$N = \{S\} \quad T = \{a,b,c\}$$
$$P: \quad S \rightarrow aSa$$
$$S \rightarrow bSb$$
$$S \rightarrow c$$

Notice that the grammar G_2 in (16) will generate the preceding strings, but it will also generate a string consisting of just c. If we want the grammar to generate the strings of the type $\alpha c \alpha^*$ (where α is any nonempty string over $\{a,b\}$, c is the central element, and α^* is the mirror image of α) and only these, we must sharpen the grammar by modifying the rules. Thus the following productions will satisfy the "only these" condition:

(17)
$$S \rightarrow aca \quad \text{(i)}$$
$$S \rightarrow bcb \quad \text{(ii)}$$
$$S \rightarrow aSa \quad \text{(iii)}$$
$$S \rightarrow bSb \quad \text{(iv)}$$

We can trace the *derivational history* of, for example, the string *abbaacaabba*, generated by the grammar in (17). The Roman numerals at the right of the following derivational lines refer to the corresponding number of the applicable rules in (17).

(18)
$$\begin{array}{ll} S & \\ aSa & \text{(iii)} \\ abSba & \text{(iv)} \\ abbSbba & \text{(iv)} \\ abbaSabba & \text{(iii)} \\ abbaacaabba & \text{(i)} \end{array}$$

In future examples a grammar will be represented only by its P rules, omitting the separate listing of N and T. As before, capital letters will be used for nonterminals or intermediate symbols, lower letters for terminal vocabulary, and Greek letters for strings.

If we call the grammar in (17) G_3, then the notation $S \xRightarrow[G_3]{*}$

abbaacaabba is read "S derives *abbaacaabba* in G_3" and, in general, if α_1, $\alpha_2, \ldots, \alpha_n$ are strings in V^*, and $\alpha_1 \underset{G}{\Longrightarrow} \alpha_2$, $\alpha_2 \underset{G}{\Longrightarrow} \alpha_3, \ldots, \alpha_{n-1} \underset{G}{\Longrightarrow} \alpha_n$, then we say $\alpha_1 \underset{G}{\overset{*}{\Longrightarrow}} \alpha_n$ (α_1 derives α_n in G). For any of the intermediate direct derivations, for example, $\alpha_2 \underset{G}{\Longrightarrow} \alpha_3$, we say "$\alpha_2$ directly derives α_3." Thus in example (18) S directly derives aSa and is written $S \underset{G_3}{\Longrightarrow} aSa$.

Types of Grammars

We defined the general format of the P rules in (8). In the following discussion the abbreviated version of that rule will be used with the context symbols γ–ω assumed but omitted. Let us, then, adopt $\alpha \rightarrow \beta$ as the most general type of rule. One can impose certain restrictions on this rule and obtain different types of grammars and different types of languages "generated" by these grammars. We will discuss four principal types, which are sometimes referred to as Chomsky languages (or Chomsky hierarchy) and were formalized first by Chomsky (1956).

The only common restriction in all our types of grammars is that α cannot be null. Now, if we construct a grammar with just this restriction, we will obtain what is called a *Type 0 grammar*. This is the most general or least restricted type of grammar. Recall that α and β are strings over V. If we require that the length of the string β (represented as $|\beta|$) in the rule $\alpha \rightarrow \beta$ be at least equal to or greater than the length of α, we will obtain a *context-sensitive* (CS) or *Type 1 grammar*. Next we define a *context-free* (CF) or *Type 2 grammar* as a grammar whose P rules have the general format $A \rightarrow \alpha$, where A is a single element in the nonterminal vocabulary. Finally, we have a *Type 3* or *finite-state* (FS) grammar, whose rules are of the type $A \rightarrow aQ$ or $A \rightarrow a$, where A and Q are symbols in the nonterminal vocabulary and a is a symbol in the terminal vocabulary. The last type is the most restricted grammar. It is sometimes referred to as *regular grammar*. The languages generated by these are also called Type 0, Type 1 (CS), Type 2 (CF), and Type 3 (FS) languages. Notice that the notion of CS grammar can be more intuitively understood if we write the rule as $\gamma A\omega \rightarrow \gamma\beta\omega$ and say that there must be at least one rule in the grammar in which the context γ–ω is not null. In CS grammars, β also cannot be null; otherwise, the length of β would become less than α.

We can now summarize the discussion of the types of grammars and languages. The following is given in the order of increasing restrictions (ϕ denotes a null string):

For all grammars: $G = (N,T,P,S)$, $\alpha \neq \phi$

Type 0: $\alpha \rightarrow \beta$

(19) Type 1 (CS): $\alpha \rightarrow \beta$; $|\beta| \geq |\alpha|$ (or $\gamma A\omega \rightarrow \gamma\beta\omega$)

Type 2 (CF): $A \rightarrow \alpha$; $A \in N$; $\alpha \in V$

Type 3 (FS): $A \rightarrow aQ$ (or $A \rightarrow x\ Q$; $x \in T^*$)

$A \rightarrow a$; $A, Q \in N$; $a \in T$

Finite-state and context-free grammars are the most highly developed and are discussed in connection with both natural languages and the artificial languages of computer programming. Context-sensitive grammars can deal with a number of inadequacies of CF and FS grammars for natural languages. But with the development of transformational grammars, CS grammars are rarely used for natural language descriptions. Type 0 grammars are too general and too powerful to be of any direct practical interest. We will therefore give just one example of a CS grammar but discuss the properties of FS and CF grammars at greater length.

Let us consider a language L which consists of strings of equal numbers of a, b, and c concatenated in that order. Thus $L = \{a^n b^n c^n\}$, where $n > 0$. The grammar for this language is:

(20)

$$
\begin{aligned}
S &\rightarrow aSBc &&\text{(i)} \\
S &\rightarrow abc &&\text{(ii)} \\
cB &\rightarrow Bc &&\text{(iii)} \\
bB &\rightarrow bb &&\text{(iv)}
\end{aligned}
$$

We can follow the derivational history of, for example, string *aaabbbccc* generated by this grammar as we did for the example in (18):

(21)

$$
\begin{aligned}
&aSBc &&\text{(i)} \\
&aaSBcBc &&\text{(i)} \\
&aaabcBcBc &&\text{(ii)} \\
&aaabBccBc &&\text{(iii)} \\
&aaabBcBcc &&\text{(iii)} \\
&aaabBBccc &&\text{(iii)} \\
&aaabbBccc &&\text{(iv)} \\
&aaabbbccc &&\text{(iv)}
\end{aligned}
$$

Finite-State Grammars

A finite-state grammar can be thought of as a generative device which can generate one terminal symbol at a time (or in each cycle). Associated with the generation of each symbol is a particular state of the device. Thus if the device is originally in some state, say q_i, and generates a symbol, it may go to state q_j, then it may generate another symbol and go to state q_k, and so forth. Schematically, this process can be represented as

(22)

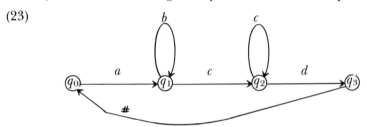

Note that the device can return to some previous state after generating some symbol. The following example illustrates several possibilities:

(23)

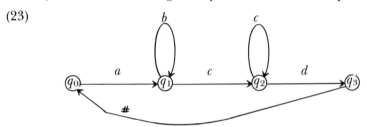

The example in (23) can generate strings such as *abbcccd#*, *abcd#*, *abcccccd#*, and so forth, where *#* marks the end of a string. We can now write the following grammar, which is equivalent to the *state diagram* in (23):

(24)

$$G = (N,T,P,S)$$
$$N = (q_0,q_1,q_2,q_3) \quad T = (a,b,c,d,\#) \quad S = q_0$$
$$P: \quad q_0 \rightarrow aq_1$$
$$q_1 \rightarrow bq_1$$
$$q_1 \rightarrow cq_2$$
$$q_2 \rightarrow cq_2$$
$$q_2 \rightarrow dq_3$$
$$q_3 \rightarrow \#q_0$$

Let us now consider an example for English:

(25)

$$S \rightarrow John\ X_1$$
$$X_1 \rightarrow saw\ X_2$$
$$X_2 \rightarrow Mary\ X_3$$
$$X_3 \rightarrow \#S$$

The grammar in (25) can generate *John saw Mary*, and with additional vocabulary and loops for articles and adjectives, comparable to the diagram in (23), it can generate sentences such as *The tall slender boy saw the short fat girl*. Perhaps the most elaborate system of this kind developed as a model for linguistic analysis is by Hockett (1955). However, the inadequacy of such models for natural languages is now well known (cf. Chomsky, 1957; Postal, 1964a). Furthermore, Chomsky (1956) has shown that FS grammars are inadequate to account for such simple languages as the example in (15) and even a simpler version of (16) given previously. We will see in Part III, however, that there are current attempts to develop FS and CF grammars and parsers for natural languages.

Context-Free Grammars

We have already defined a CF grammar in terms of its P rules, and the examples of grammars given in the discussion were CF grammars. As a further example, consider a language L whose sentences consist of a string of occurrences of one or more a, followed by a single occurrence of c, followed by a string of occurrences of b twice as long as the a string:

$$L = \{a^n c b^{2n}\}$$

$$n \geq 1$$

The grammar for this language will have these rules:

(26) $S \rightarrow aCbb \quad C \rightarrow aCbb \quad C \rightarrow c$

and, for example, for $S \overset{*}{\Longrightarrow} aacbbbb$ we have the PM or derivation tree:

(27)

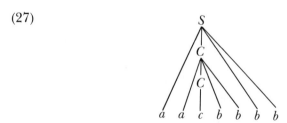

There are a number of ways for the "simplification" of CF grammars and for developing *normal forms* for them. These are often given as theorems with formal proofs. We will list here some of these properties of CF grammars and languages without giving the formal proofs. The reader can find the proofs in the literature (cf. for example, Harrison, 1978; Hopcroft and Ullman, 1979).

For every CF grammar there exists a CF language, which may be empty. Notice, however, that a language may be generated by more than one grammar. We call the grammars that can generate the same language *equivalent*. Now, given any CF grammar G generating a nonempty language, it is possible to show that there exists an equivalent grammar G_1 such that for every nonterminal symbol in G_1 there is a terminal string. Thus $A \underset{G_1}{\overset{*}{\Rightarrow}} \alpha$, where $A \in N$ and α is a string over T.

It is possible to show that there is a grammar G_2 with no P rules of the form $A \to B$; $A, B, \in N$. We can also show that there can be algorithms to determine whether a language generated by CF grammar G is empty, finite, or infinite. Furthermore, all these grammars with various properties are equivalent. These observations lead to the conclusion that there is a certain inherent property which allows one to write grammars with optimal efficiency, for example, by eliminating rules which do not generate terminal strings and other "wasteful" rules, including the generation of empty sentences.

Of special interest is a property known as *Chomsky normal form* (CNF). The theorem for this property asserts that any context-free language can be generated by a grammar whose rules are of the form $A \to BC$ or $A \to a$, where A, B, and C are nonterminal symbols and a is a terminal symbol. Notice that the CNF reduces the derivation of the phrase marker of any CF grammar to a binary form. For example, let us convert the grammar in (26) to its CNF. The method of converting is to take rules in the grammar which do not conform with the binary or unary rules of CNF and change them until they conform.

(28)
$$\begin{array}{lll} S \to aCbb & S \to ACBB & S \to DE \\ C \to aCbb & C \to ACBB & D \to AC \\ C \to c & & E \to BB \\ & & C \to DE \end{array}$$

Thus, the CNF for the grammar in (26) will have these rules:

(29)
$$\begin{array}{llll} S \to DE & D \to AC & E \to BB & C \to DE \\ A \to a & B \to b & C \to c \end{array}$$

Now observe that the derivation tree for the string *aacbbbb* will have this form (cf. [27]):

(30)

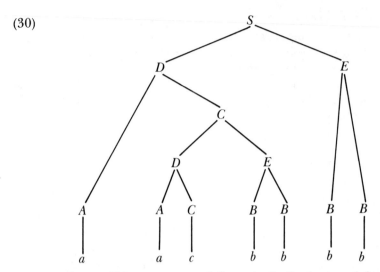

Another well-known normal form is *Greibach normal form* (GNF). Any context-free language can be generated by a grammar whose rules are of the form $A \rightarrow a\alpha$, where A is a nonterminal symbol, a is terminal, and α is a string (possibly null) of nonterminal symbols. We can again convert the grammar in (26) to GNF as an illustration. We use a simple example here for expositional purposes:

(31)
$$\begin{array}{ll} S \rightarrow aCbb & S \rightarrow aCBB \\ C \rightarrow aCbb & C \rightarrow aCBB \\ C \rightarrow c \end{array}$$

The final form of the grammar is then

(32)
$$\begin{array}{ll} S \rightarrow aCBB & C \rightarrow aCBB \\ C \rightarrow c & B \rightarrow b \end{array}$$

and the derivation tree for *aacbbbb* is:

(33)

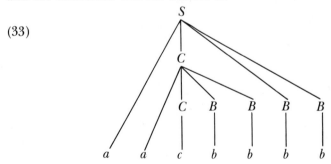

Of special interest in linguistics are *phrase structure* (PS) grammars, which are context free or context sensitive. We will discuss some further properties of these in Chapter 2.

Recognizers and Acceptor Automata

We saw in the previous section that grammars are finite devices for describing languages. There is yet another class of devices for describing languages: recognizers. Intuitively, a recognizer R is an algorithm which can process a string σ of a language L and determine whether the string is a well-formed sentence in a grammar G. Thus we can state, $R(G)$ terminates if $\sigma \in L(G)$. In this section we will discuss four such devices as acceptor automata, which are recognizers for the four types of languages just discussed.

Finite-State Automata

Finite-state automata (FSA) are acceptors for FS or Type 3 languages. They are the simplest and the most restricted automata of the four types we will discuss. A finite automaton can be formally defined as a quintuple $M = (K,\Sigma,\delta,q_0,F)$, where K is a finite, nonempty set of states, Σ is a finite alphabet (*input alphabet*), δ is a rule of M comparable to the P rules of grammars and consists of a mapping of $K \times \Sigma$ into K, q_0 is the *initial state* in K, and F is a set of *final states* (or *halting states*) in K and may be identical with q_0 ($F \subseteq K$).

The components of FSA are a finite control unit (CU), a tape, and a read head. The working of the device can be visualized in this manner:

(34)

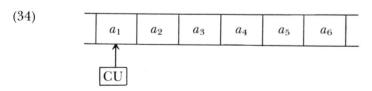

The tape is divided into squares, each of which can contain only one symbol from the input alphabet. The input string is recorded on the tape and the read head starts with the leftmost symbol and moves one square to the right in each cycle, scanning one input symbol in each move. The CU has a finite number of internal states. When M starts with the read head positioned on the first (leftmost) symbol, CU is in the initial state q_0. Every time a symbol is scanned and a move to the right is made CU can go into a new state. If M is in a final state when the reading of the input string is completed, then the string is said to have been *accepted* by the machine. The set of all strings or sentences accepted by an automaton of this kind is called a *regular set*.

Each cycle of M, consisting of a read, a move, and a change of state, can be formally represented by a δ-mapping, δ-function, or δ-rule, as they are variously called. The total behavior of a machine can be described by the set of its δ-functions. The general form of the mapping is $\delta(q_i,a_j) \rightarrow q_k$. As an example, let us describe an automaton M which will accept sentences consisting of occurrences of a and b with the further requirement that the number of both a and b occurrences be odd:

(35)
$$M = (K,\Sigma,\delta,q_0,F)$$
$$K = \{q_0,q_1,q_F\} \quad \Sigma = \{a,b\} \quad F = \{q_F\}$$
$$\delta(q_0,a) \rightarrow q_1 \quad \delta(q_1,a) \rightarrow q_0 \quad \delta(q_1,b) \rightarrow q_F \quad \delta(q_F,b) \rightarrow q_1$$

The following state diagram can further illustrate this example:

(36)

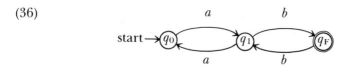

In the preceding and following diagrams a double circle is used to mark the final state or states and an arrow is used to mark the starting state.

A further example can be demonstrated by describing an FSA which will accept strings generated by the finite-state grammar in example (24):

(37)
$$K = \{q_0,q_1,q_2,q_3\} \quad \Sigma = \{a,b,c,d,\#\} \quad F = \{q_0\}$$
$$\delta(q_0,a) \rightarrow q_1 \quad \delta(q_1,b) \rightarrow q_1 \quad \delta(q_1,c) \rightarrow q_2$$
$$\delta(q_2,c) \rightarrow q_2 \quad \delta(q_2,d) \rightarrow q_3 \quad \delta(q_3,\#) \rightarrow q_0$$

The state diagram for this example is essentially the same as the diagram given in example (23). Because q_0 is both the initial and final state for this example, we need to put a double circle around the q_0 node in (23) to conform with the convention we have just adopted. Notice, incidentally, that the language accepted by the FSA in (37) is $L(M) = \{ab^nc^md\}*$ where $n \geqq 0$, $m \geqq 1$, and $*$ denotes any number of repetitions. Let us now define an automaton which will accept $L(M) = ab^*$; that is, strings consisting of an a followed by any number of occurrences of b or just an a:

(38) $\delta(q_0,a) \rightarrow \{q_1,q_2\}$ $\delta(q_0,b) \rightarrow \emptyset$ $\delta(q_1,a) \rightarrow \emptyset$
          $\delta(q_1,b) \rightarrow \{q_2\}$ $\delta(q_2,a) \rightarrow \emptyset$ $\delta(q_2,b) \rightarrow \{q_1\}$

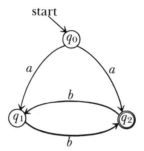

Note that example (38) differs from the previous ones in that if M is in state q_0 and scans an a, it has the choice of going to state q_1 or q_2. The previous examples were *deterministic* automata; the example in (38), where there is choice of states, is called *nondeterministic*. The output states for nondeterministic FSA are therefore sets, and they can be null; hence the second δ-function in (38).

It can be shown (Rabin and Scott, 1959) that for any language accepted by a nondeterministic FSA there exists a deterministic FSA which will accept the same language.

The examples of finite automata given so far assume that the read head can move only one way, from left to right. We can also describe a *two-way finite automaton* in which the read head can move in both directions; that is, it can rescan the symbols it has already read. We will use R for movement of one square to the right, L for one square to the left, and S for no move. If we choose D to represent these symbols ($D = \{R,L,S\}$), then the δ-mapping for a two-way automaton can be represented as $(q_i,a_j) \rightarrow (q_k,D)$. It can, however, be shown (Rabin and Scott, 1959; Shepherdson, 1959) that the class of languages or sets which are accepted by a two-way automaton is the same as those accepted by a one-way automaton.

Because two finite automata are equivalent if they accept the same language, we can conclude that the varieties of automata discussed so far are equivalent and that the basic automaton described at the beginning of this section has the same recognition (generative) power as the nondeterministic and two-way automata.

Pushdown Automata

The concept of *pushdown* or *stack* storage is well known to many computer programmers. In this section we will present a formalization of this concept in terms of automata theory.

The components of pushdown automata (PDA) are the following: A finite control unit (CU) with two heads, an input tape, and a storage tape. The input tape is divided into squares, each one of which can hold only one symbol from the alphabet. The tape is assumed to be infinitely long on both sides. The storage tape is also divided into squares which can hold only one symbol, but this tape is bounded on one side and infinitely extended on the other. The automata in this class derive their name from this storage tape, which is of the "last in, first out" type. This storage has often been compared with the stack of plates on a spring in many cafeterias, where the plate placed on the top is the first that one takes out, and every time a new plate is placed on the top, it pushes other plates further down the stack. Thus, the first plate placed on the stack comes out last and vice versa. The two heads consist of a read head, which can read one symbol at a time (or in each cycle) from the input tape, and a read-write-erase (RWE) head, which can read, write, or erase a symbol on the storage tape. As in the case of finite automata discussed in the previous section, the CU has a finite number of states, including an initial state and a set of final states. This device has two alphabets: an input alphabet and an output or storage alphabet. The input alphabet may be included in the output alphabet, and the two alphabets may be identical:

(39)   Input Tape

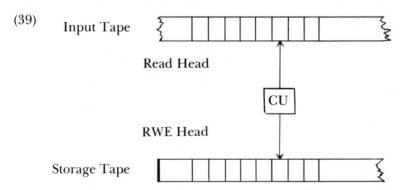

More formally, a PDA can be represented by this notation:

(40) $$M = (K, \Sigma, \Gamma, \delta, q_0, Z, F)$$

where K is a finite set of states; Σ is the finite input alphabet; Γ is the finite output or storage alphabet; δ is the mapping function as before; q_0 is the initial state of the device and is included in K; Z in Γ is the special starting symbol of the output alphabet; and $F \subseteq K$ is the set of final states. The mapping function has this general form:

$$(41) \qquad\qquad \delta(q_i,a,b) \to (q_j,c)$$

The interpretation of the "rule" in (41) is that the device in state q_i reads symbol a from the input tape and scans symbol b from the top of the storage tape; it will then go to state q_j and write symbol c on the top of the storage tape. We will call the triplet on the left of the arrow in (41) a *situation* of M, and each rule of the (41) type is a *configuration* of M.

There are two ways of visualizing the operation of the storage tape: (a) the tape moves two ways—if it moves forward, a symbol is written on it, if it moves backward, a symbol is erased, if it does not move, no change in the top symbol of the storage tape is made; (b) the device can write a new symbol on the top of the existing one, pushing the existing symbol down, erase the top symbol, popping the next symbol up, or do nothing. Methods (a) and (b) are equivalent.

An input string or sentence is considered *accepted* by a PDA if (a) when the reading of the string is completed the automaton is in a final state, or (b) when the reading of an input string is completed the storage tape is empty, except perhaps for the initial symbol Z. Again both these methods of recognition are equivalent. Notice that the device always starts in state q_0 and with Z on the top of the stack or the output tape.

Let us now introduce an identity symbol e in the mathematical sense such that if it occurs anywhere in the configuration of an automaton, it can be taken to represent any symbol and the rest of the configuration will apply. Thus, for example, $\delta(q_i,e,b) \to (q_j,c)$ denotes that if M is in state q_i, irrespective of what symbol it scans on the input tape, but reads a b on the storage tape, then it will go to state q_j and write a c on the top of the stack.

The device which we have described so far is *nondeterministic* since it has a finite number of choices of moves in each situation.

It has been proven (Chomsky, 1962b; Evey, 1963) that pushdown automata are acceptors for context-free or Type 2 languages. More precisely, it can be shown that if L is a context-free language, then there exists a PDA that will accept L, and, conversely, if L is a language accepted by a PDA, then L is a context-free language.

As an example, let us describe a PDA which will accept mirror-image strings with a central element (e.g., *abcba*, *aabbcbbaa*, etc.) in general, $\{\alpha c \alpha^* |\ \alpha \in \{a,b\}\}$. Recall that the context-free grammar given in example (17) generates strings of this type:

(42) $K = (q_0, q_1, q_2)$ $\Sigma = (a,b,c)$ $\Gamma = (a,b,c,Z,e)$ $F = \{q_0\}$

To reduce the number of δ-functions in our exposition, let us adopt a symbol α which ranges over $\{a,b,\}$. The identity symbol *e* may appear as an input symbol in our δ-rules, but, strictly speaking, it is not an input or output symbol. Its occurrence in the following rules only denotes that irrespective of whatever symbol is on the input or storage tapes, the configuration shall apply. Now the δ-functions for the preceding PDA are

(43) $\delta(\alpha, q_0, e) \rightarrow (q_1, \alpha)$ $\delta(\alpha, q_1, e) \rightarrow (q_1, \alpha)$
          $\delta(c, q_1, e) \rightarrow (q_2, e)$ $\delta(\alpha, q_2, \alpha) \rightarrow (q_2, \eta)$ $\delta(e, q_2, Z) \rightarrow (q_0, Z)$

The α symbol, as stated previously, is an abbreviation for *a* or *b*. Thus each of the preceding rules with α is an abbreviation for two rules. By convention, the α on both sides of an arrow in the same rule is the same. For example, the first line in (43) is an abbreviation for the two rules: $\delta(a, q_0, e) \rightarrow (q_1, a)$ and $\delta(b, q_0, e) \rightarrow (q_1, b)$. The η in the fourth rule in (43) is an erase instruction. Thus we are using the erase method and *empty-stack recognition* procedure in this example.

The operations of the automaton described in (42) and (43) can be illustrated by this state diagram:

(44)

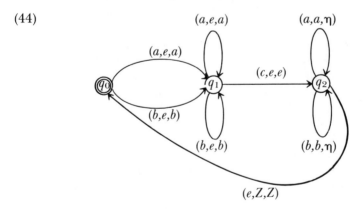

Notice that these operations can be represented as triplet situations (a_1, a_2, a_3), where a_1 is the symbol on the input tape, a_2 is the symbol on the top of the stack, and a_3 is the new symbol to be placed on the top of the stack.

As illustrated by the diagram in (44), the device starts in q_0 and reads a symbol a or b from the input tape, irrespective of what is on the top of the storage tape, and writes the same symbol (a or b) on the top of the storage tape. Then the device assumes state q_1 and repeats the operation, moving the input tape one square in each cycle, reading off a or b, and recording it on the storage. When c appears on the input tape (midway on the input string), the device reads the c, moves the tape to the next square but does not write anything on the storage tape, and changes its state to q_2. In state q_2 the device continues to read a or b on the input tape and to match it each time with a or b on the top of the storage tape. Every time a match is found, the symbol on the top of the storage tape is erased, causing the next symbol to pop up (the η instruction). If this process is continued with matches found in each cycle between the input and the storage symbols, the initial symbol Z will eventually appear on the top of the storage tape, signifying that the stack is empty and the input string has been "accepted" as well formed. The device then changes to state q_0 (resets itself) and halts.

Note that if we select mirror-image strings with no central element for our example, the nondeterministic nature of the device becomes more apparent. Mirror-image strings without a center element are aa, bb, $abbbba$, $babaaaabab$, and so forth, in general $\alpha\alpha^*$, with α again ranging over $\{a,b\}$. The context-free grammar which will generate such strings is:

(45) P: $S \rightarrow aSa$
                                      $S \rightarrow bSb$
                                      $S \rightarrow aa$
                                      $S \rightarrow bb$

The PDA recognizer for this grammar can be represented by these δ-functions:

(46)   (i) $\delta(\alpha,q_0,e) \rightarrow (q_1,\alpha)$
                 (ii) $\delta(\alpha,q_1,e) \rightarrow (q_1,\alpha)$
                 (iii) $\delta(\alpha,q_1,e) \rightarrow (q_2,\alpha)$
                 (iv) $\delta(\alpha,q_2,\alpha) \rightarrow (q_2,\eta)$
                 (v) $\delta(e,q_2,Z) \rightarrow (q_0,Z)$

Notice that configurations (ii) and (iii) in (46) have the same lefthand situation, indicating that the device has a choice of remaining in state q_1 or changing to state q_2. Because there is no symbol to mark the midpoint in the input string, the device has to "guess" when it has reached the focal point of the mirror and change states. If the guess is correct and the input string is well formed, it will be accepted by the empty-stack method as in the case of (43); otherwise, we have no sentence and the device will not halt. For diagram representation, this amounts to removing the (c,e,e) path in (44) and replacing it with two arrows:

(47)

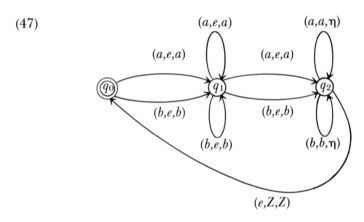

Turing Machines

In the first section we implicitly defined a procedure as a finite number of instructions which can be carried out by a person or perhaps a mechanical device. In his famous paper, Turing (1936) gave a precise characterization of this notion. He described a human "computer" which could carry computations in a prescribed manner, and then he defined a "machine" which could do the work of the human computer. This machine later became known as the Turing machine. Recall that when Turing wrote his paper in 1936 electronic computers had not been invented and his paper is regarded by many as the "invention" of computing machines. The so-called Turing thesis is that any process which could naturally be called a procedure can be realized by a Turing machine. (The general concept of this thesis is due to Church. Turing showed that his formulations were equivalent to the thesis of Church.) A procedure can then be regarded as a *program* (in the sense of electronic computer programs) for a Turing machine. An algorithm is a program which solves a class of problems. In automata theory a procedure is defined in terms of *computability* by a Turing machine. Some authors refer to an algorithm as an *effective procedure*.

We can formally define a Turing machine T as

(48) $$T = (K,\Sigma,\Gamma,\delta,q_0,F)$$

Note that, as in the case of PDA, K (the set of states), Σ (input alphabet), Γ (output or external alphabet), δ (the set of mapping functions, i.e., the program for T), and F (the set of final states in K) are finite, and q_0 in K is the unique initial state of T.

The components of T are the control unit CU, the read-write head RW, and an infinitely long tape divided into squares. The tape is bounded on the left: that is, there is a leftmost symbol, but it is infinitely extended to the right. An input sentence is a finite string on the tape with one symbol (or *word*) in each square. All other squares on the tape are assumed to contain blanks (#). As Turing pointed out, a "symbol," although finite, can be arbitrarily long; thus, for example, "an Arabic number such as 17 or 999999999999999 is normally treated as a single symbol." The tape can move in both directions, and RW can read a symbol, write a symbol over the symbol read, move the tape one square to right or left, or change states. Note that in all these machines we can have the tape being stationary and move the read/write head on the tape. The two methods are equivalent, but in writing the δ-functions one must be careful in marking directions, because the markings would be opposite depending on which method one chooses. A δ-mapping, representing a configuration of T, has the general form $\delta(q_i,a) \rightarrow (q_j,b,D)$, where q_i, q_j are possibly identical states of T, a and b are possibly identical symbols from the input and output alphabets, and $D = \{R,L,S\}$ represents the moves of the tape. Notice that erasure of a symbol in a square can be achieved if we assume that RW can also write a blank symbol (#).

The device which we have just defined is the simplest and most basic form of a Turing machine. We will subsequently mention some other, more complex models. It can be shown (Chomsky, 1962b) that all other models are equivalent to the preceding basic model.

Let us now consider a simple example. We will describe a Turing machine which will recognize strings $\{a^n b^n\}$ (e.g., *ab, aabb, aaabbb*, etc.):

(49) $$T = (K,\Sigma,\Gamma,\delta,q_0,F) \quad K = \{q_0,q_1,q_2,q_3,q_4,q_F\}$$

(50) $$\Sigma = \{a,b,\#\} \quad \Gamma = \{a,b,\#,X,Y\} \quad F = \{q_F\}$$

Table 1. A Turing Machine for Strings $a^n b^n$

Row	Input Q	Input Σ	Output Q	Output Γ	Output D	
(1)	0	a	1	X	L	
(2)	1	a	1	a	L	
(3)	1	b	2	Y	R	
(4)	1	Y	1	Y	L	
(5)	2	Y	2	Y	R	
(6)	2	a	4	a	R	
(7)	2	X	3	X	L	
(8)	3	Y	3	Y	L	
(9)	3	#	F	#	S	HALT
(10)	4	a	4	a	R	
(11)	4	X	0	X	L	

It is more convenient to write the δ-functions in the form of a matrix, and we will adopt this for our examples of Turing machines. In the matrix in Table 1 the columns under Q represent states of T, the symbols under Σ represent symbols read on the tape at various configurations of T, the symbols under Γ represent those written by RW at various configurations on the tape, and the symbols under D represent the movements of the tape.

Notice that T starts in state q_0 and scans the leftmost symbol on the tape. If this is an a, RW writes an X over it and moves the tape one square to the left. The state of T now changes to q_1. At this state the tape keeps moving to the left and scanning over any a (Row 2) until a b is scanned. T then assumes state q_2, a Y is written over the b, and the tape is moved to the right (Row 3). This process is continued, alternatively converting one a to X and one b to Y until a blank symbol is reached on the right-hand side of the input string. T then enters the final state q_F and halts. The input string is accepted as a well-formed sentence. For all other configurations δ is undefined. Notice, incidentally, now, that our T is an algorithm for the recognition of all $\{a^n b^n\}$ strings, where $n \geq 1$.

We will consider one further example, adopted from Minsky (1967). This is a "counter machine" which will count the number of all symbols on the input tape and write the total in a binary number. Note that this device can also be used for converting unary numbers to binary. For example, suppose that symbols on the tape are single strokes, |, with one stroke in each tape square:

(51)

The number of strokes (ten in this example) will represent a unary number, and the output from our Turing machine will be the binary representation of that number, that is, 1010.

The algorithm for this counter or converter can informally be stated as follows:

1. Starting from the left, convert every other stroke to an X.
2. Two situations may occur: (i) a stroke is left between the last Xed stroke and the right #; (ii) there is no stroke between the last X and the right #.
3. (a) If (i), record a zero at the left available #; (b) if (ii), record a one at the left available #.
4. If no strokes are left in the string, stop; else go to Step 1.

The string consisting of occurrences of one and zero thus formed at the beginning of the original string is the binary count of the strokes. In the following the reader can follow the counting of the strokes by the preceding algorithm:

(52)

```
# # # # # # I I I I I I I I I I # # #
# # # # # # X I X I X I X I X I # # #
# # # # # 0 X I X I X I X I X I # # #
# # # # # 0 X X X I X X X I X X # # #
# # # # 1 0 X X X I X X X I X X # # #
# # # # 1 0 X X X X X X X I X X # # #
# # # 0 1 0 X X X X X X X I X X # # #
# # # 0 1 0 X X X X X X X X X X # # #
# # 1 0 1 0 X X X X X X X X X X # # #
H A L T
```

The more formal Turing machine description for this problem is given in the matrix in Table 2. We can assume that the device will halt when there are no strokes left in the input string, although no provision for halting is included in Table 2.

Note that we can "program" a Turing machine to do the same problem in a different and more intuitive way. The algorithm (or Turing machine) in Table 3 will read each stroke from left to right and "X" it out, then the device will move to the left of the input string and

Table 2. A Turing Machine for a Counter

Q		Q		D
0	I	1	X	L
0	X	0	X	L
0	0	0	0	L
0	I	0	1	L
0	#	2	#	R
1	I	0	I	L
1	X	1	X	L
1	#	5	#	R
2	I	2	I	R
2	X	2	X	R
2	0	2	0	R
2	1	2	1	R
2	#	3	0	L
3	I	4	X	L
3	X	3	X	L
3	0	3	0	L
3	1	3	1	L
3	#	2	#	R
4	I	3	I	L
4	X	4	X	L
4	#	5	#	R
5	I	5	I	R
5	X	5	X	R
5	0	5	0	R
5	1	5	1	R
5	#	0	1	L

Table 3. Another Version of a Turing Counter Machine

Q		Q		D
0	I	1	X	R
0	X	0	X	L
0	0	0	0	L
0	1	0	1	L
0	#	F	#	S
1	X	1	X	R
1	0	0	1	L
1	1	1	0	R
1	#	0	1	L

do a simple binary addition for each stroke read. When the first blank on the right-hand side of the input string is reached the device will halt and the counting is completed.

There are two ways one can modify the basic Turing machine which we have described so far. One is in the direction of generalization, the other is in the opposite direction of further restrictions. As it was hinted before, it turns out that none of these modifications changes the computing power of the basic Turing machine. In the direction of generalization or removal of restrictions, one can dispense with the restriction that the input tape be bound on the left and have the tape extended infinitely on both sides. In fact we can go even further and have the tape extended infinitely on four sides or even on n directions, where n is any arbitrary finite number. A Turing machine can have more than one tape and more than one RW head. A Turing machine can be nondeterministic in the sense that in its description there might be more than one configuration whose left-hand sides are identical. For example, if T has the following δ-functions in its description,

(53)
$$\delta(q_1, a_1) \rightarrow (q_2, a_2, L)$$
$$\delta(q_1, a_1) \rightarrow (q_3, a_3, R)$$
$$\delta(q_1, a_1) \rightarrow (q_4, a_4, R)$$

it is clear that when T assumes state q_1 and reads symbol a_1, it has three choices, with three different results.

In the direction of restrictions, a Turing machine can have a read-only tape and one or more storage tapes. The movements of the tape and the number of states of T can be restricted, and the input symbols can be restricted to two: 0 and 1; or even further, to one starting symbol S on the tape and blanks. Any symbol can be *encoded* on this tape by representing it as a unique number which in turn can be represented on the tape by the number of blanks counted from the starting point S.

Several Turing machines can be combined as components of a more complex Turing machine. Similarly, Turing machines can be used as subroutines of other Turing machines. To use, say, T_2 as a subroutine of T_1, T_1 and T_2 must have disjoint states; then T_1 can "call" T_2 by simply entering one of its states.

A Turing machine of particular interest and importance is the following. Suppose one had to describe the operations of the T in Table 1 to a human "operator"; one would presumably say: "If T is in state q_0 and the read head scans the symbol a, change to state q_1, write an X on the tape, and move the tape one square to the left . . . ." Now suppose we were to encode these instructions on the tape of another Turing machine. Then, clearly the second machine could imitate the first one. More generally, we can describe a Turing machine U which can simulate any arbitrary Turing machine T, if we simply encode the configurations of T on the tape of U. U is called a *universal Turing machine*.

One apparent problem with U is that, like all other Turing machines, it must have a finite alphabet. But we can describe an infinite number of Turing machines, each with finite but different alphabets. How are we going to represent these infinite alphabets by the finite alphabet of U? We have already answered this question previously by mentioning the two-symbol machine. Suppose that U has only symbols 0 and 1 and possibly blank (#) in its external alphabet. We can surely represent any symbol in any other machine by some unique combination of zeros and ones.

One of the most important results concerned with Turing machines, and automata theory in general, is the following: Given any arbitrary Turing machine T, a finite input string on its tape, and some configuration C of T, there is no general algorithm which will decide whether or not T will eventually halt. This conclusion is known as the *halting problem* of Turing machines, and can be stated as a theorem: *The halting problem of Turing machines is undecidable.*

We cannot go into the formal proof of this important theorem here, but the line of argument for its proof is something like the following. Machine T enters configuration C; will it ever halt?

$$(54) \qquad\qquad T \Rightarrow C \overset{?}{\Rightarrow} \text{HALT}$$

We can reduce this to the *computability* problem by asking: Given a machine T and a configuration C, will it ever enter this configuration?

$$(55) \qquad\qquad T \overset{?}{\Rightarrow} C$$

Now, surely, if we can show that (55) is undecidable, it follows that (54) is also undecidable. The problem can be further narrowed by assuming a T whose tape contains a description of itself—a self-simulating machine. Now, it can be shown that the *self-computability* of T is undecidable, that is, (55) cannot be decided for a self-simulating machine, which is a particular case of the general problem. The way the latter assertion can be proven is to assume that self-computability is decidable and then show that this will lead to contradictions.

Finally, Chomsky (1959a) has shown that Turing machines are acceptors for Type 0 languages. More precisely, if a language L is generated by a Type 0 grammar, then L is recognized by a Turing machine; conversely, if a language L is recognized by a Turing machine, then L is a Type 0 language.

Linear Bounded Automata

Of the four types of grammars discussed so far, we have seen acceptor or recognizer devices for three of them:

Type 0: Turing machine
Type 2: Pushdown automata
Type 3: Finite-state automata

The acceptor for the remaining type (Type 1, context-sensitive) is a linear bounded automaton (LBA). That is, if a language L is context sensitive, then it is accepted by an LBA; conversely, if a language L is accepted by an LBA, then L is context sensitive (cf. Kuroda, 1964; Landweber, 1963).

An LBA is a Turing machine with the following restriction. The input string on the tape has a left and right boundary symbol, which could be #; the tape never passes these boundary limits in its movements to the left or right. Alternatively, we can say that the RW head never leaves the boundaries of the input string.

For an example we can describe an LBA which will accept strings

Table 4. A Turing Machine for Strings $a^n b^n c$

Q	Σ	Q	Γ	D	
0	a	1	X	L	
0	X	0	X	L	
0	#	F	#	S	Halt
1	a	1	a	L	
1	b	2	X	L	
1	X	1	X	L	
2	b	2	b	L	
2	c	3	X	R	
2	X	2	X	L	
3	X	3	X	R	
3	a	3	a	R	
3	b	3	b	R	
3	#	0	#	L	

Table 5

Name	Restrictions on P rules	Acceptor automata
Type 0	$\alpha \rightarrow \beta, \alpha \neq \phi$	Turing machine (T)
Type 1, context sensitive	$\alpha \rightarrow \beta, |\beta| \geq |\alpha|$	Linear bounded automata (LBA)
Type 2, context free	$A \rightarrow \alpha, A \in N, \alpha \in V^*$	Pushdown automata (PDA)
Type 3, finite state	$A \rightarrow aQ, A \rightarrow a$ $A,Q \in N, a \in T$	Finite-state automata (FSA)

generated by the grammar in example (20), that is, strings $\{a^n b^n c^n | n > 0\}$.

Informally stated, the algorithm for this acceptor is the following. Start in state q_0 and find an a, change that to an X ("X" it out), change to state q_1, and move the tape one square to the left. In state q_1 find a b, alter that to X, change to state q_2, and move the tape to the left. In state q_2 find a c, alter to X, change the state to q_3, and move the tape to the right. In state q_3 scan over all the symbols until the leftmost boundary (#) is reached, change to state q_0, and start over again. Notice that in state q_0 if the right boundary # is reached, the string has been accepted as well formed. In all other cases δ is undefined and the device will not halt.

The configuration matrix for this LBA is shown in Table 4.

We have now completed our exposition of the four types of grammars (languages) and their recognizers. The results can be summarized in Table 5 in order of decreasing generality.

Syntax-Directed Translation

Recall that a language L with respect to some grammar G is a subset of all possible strings made from the concatenation of the "alphabet" in the terminal vocabulary T. We have used the notation T^* for the set of all possible strings over T, including an empty string ϕ, and we have used T^+ to represent $T^* - \phi$. A language L, therefore, is a subset of T^+. More precisely we can state:

(56) $$L = \{x \mid S \overset{*}{\underset{G}{\Rightarrow}} x \;\&\; x \in T^+\}$$

Which can be read as: L is the set of any arbitrary string x such that x is derived from the initial symbol S by the application of zero or more rules of grammar G and such that x is a member of T^+. In the derivational sequence $S \Rightarrow \alpha_1 \Rightarrow \alpha_2 \Rightarrow \ldots \Rightarrow \alpha_n = x$, each α_i is called a *sentential form*. A *sentence* is a sentential form whose elements are drawn exclusively from the terminal vocabulary.

Now, given a language L_1 with its vocabulary T_1 and another language L_2 with its vocabulary T_2, we want to develop schemata for the translation or transformation of sentences in L_1 to sentences in L_2. In syntax-directed translation (SDT) we achieve this by attaching a translation element to each production. The productions will then have the dual function of inputting $x \in T_1^*$ and outputting $y \in T_2^*$, where x and y are sentences in L_1 and L_2, respectively. Recall that the general format of the rule for a context-free grammar was $A \to \alpha$. The revised production will be $A \to \alpha, \beta$.

The translation schema $\mathcal{T}$ is, then, a quintuple:

$$(57) \qquad \mathcal{T} = (N, T_1, T_2, R, S)$$

where N is the finite nonterminal vocabulary, T_1 is the finite terminal vocabulary of the input language L_1, T_2 is the finite terminal vocabulary of the output language L_2, R is the finite set of translation rules, and S is the starting symbol in N. The standard derivations also apply here, except that where in grammars we have $\alpha \underset{G}{\overset{*}{\Rightarrow}} \beta$, in SDT we have the pair $(\alpha, \beta) \underset{\mathcal{T}}{\overset{*}{\Rightarrow}} (\gamma, \omega)$.

As an example let us now consider a schema for translating standard (infix) arithmetic expressions into Polish notations. We will generate postfix or "reverse Polish" in our example. Thus, the transformations will be of this form:

$$(58) \qquad a + b * c \Rightarrow abc* +$$
$$a * b + c \Rightarrow ab*c +$$
$$a/(a + b) \Rightarrow aab + /$$

First, let us write a context-free grammar for generating sentences such as $a/(a + b)$:

$$(59) \qquad G = (N, T, S, P) = (\{S\}, \{a, b, /, (,), +\}, S, P)$$
$$P: \quad S \to (S)$$
$$S \to S + S$$
$$S \to S / S$$
$$S \to a$$
$$S \to b$$

Now, a translation schema $\mathcal{T}$ for achieving the transformation in (58) is:

(60) $\mathcal{T} = (N, T_1, T_2, R, S) = (\{S\}, \{a, b, /, \}, (, +\}, \{a, b, /, +\}, R, S)$

$$R: \quad S \to (S), S$$
$$S \to S + S, SS +$$
$$S \to S/S, SS/$$
$$S \to a, a$$
$$S \to b, b$$

Note that the device has the dual role of generating the expression $a/(a + b)$ and its Polish translation $aab + /$. The following *leftmost derivation* of these strings demonstrates this process:

$$(S, S) \Rightarrow (S/S, SS/) \Rightarrow (a/S, aS/) \Rightarrow (a/(S), aS/) \Rightarrow (a/(S + S),$$
$$aSS + /) \Rightarrow (a/(a + S), aaS + /) \Rightarrow (a/(a + b), aab + /)$$

The following phrase marker shows the syntactic structure for the expression $a/(a + b)$, as it would be generated by the grammar in (59), with the translations shown in boxes beside various nodes:

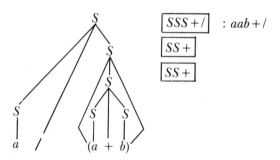

Translation Machines

The automata acceptors for the syntax-directed translation devices are called *transducers*. These automata must recognize the input strings but also must map them into output strings. We will give here examples of the simplest forms of finite and pushdown transducers which are acceptor automata for finite-state and context-free SDT schemata outlined in the previous section.

A finite transducer has the five components of a finite-state automaton plus an output alphabet. Thus, a finite transducer is a sextuple:

(61) $$\mathscr{T}_{\mathrm{fs}} = (K, \Sigma, \Delta, \delta, q_0, F)$$

where Σ is the finite input alphabet and Δ is the finite output alphabet. A general sketch of the device is:

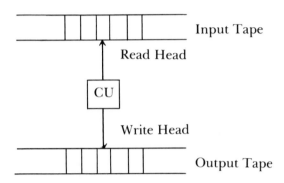

The δ-mapping is from input into output, more precisely: $K \times (\Sigma \cup \{\phi\}$, into $K \times (\Delta \cup \{\phi\})$. The use of the null string ϕ here indicates that there is not necessarily a one-to-one mapping between input and output strings and that the length of an input text T_1 and its translation T_2 may be different. For an input string x to be translated, it must first be accepted by the transducer, and the translation occurs from strings in Σ^* into strings in Δ^*. We can then formally define a translation process as:

(62) $\mathscr{T} = \{(x,y) \mid x \in \Sigma^*, y \in \Delta^*$ & as x is recognized, y is generated$\}$

It should also be pointed out that an input string x may have more than one output translation.

As an example, consider the language $L(G) = \{a^n b \mid n \geq 0\}$, with productions $S \to aS$, $S \to b$ in G. An SDT schema for translating this language to $L_1 = \{0^n 1\}$, where, for example, an input string $aaab$ will be translated to 0001, will have these rules:

(63) $R: \quad S \to aS, 0S$
                                    $S \to b, 1$

Now, to construct a finite transducer $\mathscr{T}_{\mathrm{fs}}$, we follow this general procedure:

(i) If $A \rightarrow aB, bB$ is in R, then obtain $\delta(q_i, a) = (q_j, b)$
for $a \in (\Sigma \cup \{\phi\})$ and $b \in (\Delta \cup \{\phi\})$
(ii) If $A \rightarrow a, b$ is in R, then obtain $\delta(q_i, a) = (q_F, b)$

q_i and q_j correspond with the nonterminals of the grammar underlying the SDT, and q_F is an additional final state which must be added for the transducer. Applying the above procedure to (63), we obtain:

(64)
$$\delta(q_0, a) = (q_0, 0)$$
$$(q_0, b) = (q_F, 1)$$

The design of the pushdown transducer is the same as the finite transducer with the addition of a stack or storage tape. A pushdown transducer $\mathcal{T}_{pd}$ has eight components:

(65)

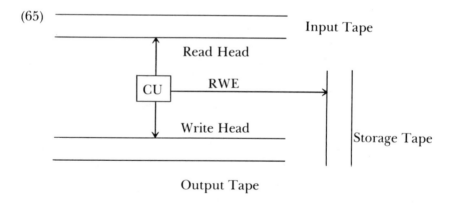

$$\mathcal{T}_{pd} = (K, \Sigma, \Delta, \Gamma, \delta, q_0, Z_0, F)$$

where Σ, Δ, and Γ are input, output, and stack alphabet, respectively, and Z_0 is the initial symbol in Γ.

Let us now write a transducer for encoding English messages to some numerical codes. We will arbitrarily assign some unique two-digit numbers to each letter and to some punctuations. We will use 0 in the output alphabet to represent a blank in the input alphabet and we will translate an end of message marker, #, in the input to 99 in the output. For decoding purposes, we want the number of words in the message to be recorded at the end of the output. We will use the storage tape

Table 6. Coding/Decoding Keys

A:11, B:22, C:31, D:12, E:21, F:92, G:13, H:32, I:62, J:33, K:93, L:85
M:34, N:45, O:16, P:74, Q:47, R:46, S:18, T:88, U:67, V:66, W:49, X:39
Y:17, Z:89, ,:29, ;:12, ::55, .:78, b:0, #:99

to accumulate a binary number representing the number of words in the input.

Let $\mathcal{T}_{pd} = (K,\Sigma,\Delta,\Gamma,\delta,q_0,Z_0,F)$ where $\Sigma = \{A,B,C, \ldots ,\#\}$, $\Delta = \{11,22,31, \ldots ,99\}$, and $\Gamma = \{0,1,Z_0\}$. A concise set of mapping functions for $\mathcal{T}_{pd}$ is given below, in which a_1 represents any symbol in Σ, a_2 represents the corresponding symbol in Δ, e is the identity symbol we had before which, when appearing in a triplet, can match with any symbol and indicates no action or changes on the relevant tapes, and $+$ represents the binary sum on the top of the stack.

(66)

    (i) $(q_0,a_1,Z_0) = (q_1,a_2,e)$

    (ii) $(q_1,a_1,e) = (q_1,q_2,e)$

    (iii) $(q_1,b,e) = (q_1,0,1)$

    (iv) $(q_1,\#,+) = (q_2,99,e)$

    (v) $(q_2,e,+) = (q_0,+,Z_0)$

In rule (i) $\mathcal{T}_{pd}$ starts in the initial state q_0, reads a letter from the input tape (a_1), and scans Z_0 on the top of the storage stack. It then changes to state q_1, writes a corresponding number from the output alphabet on the output tape, and does nothing with the storage stack. In (ii) in state q_1 the device keeps reading letters from the input and printing corresponding numbers on the output, disregarding the storage. In (iii) if a blank b is read, a 0 is printed on the output and a 1 is written on the stack. Here it is assumed that a binary addition takes place (for example, by calling the "counter machine" we developed in (52) as a subroutine) and the current sum is placed on the top of the stack. We represent this sum as $+$. In (iv) if the end marker $\#$ is encountered in the input with $+$ on the top of the stack, number 99 is written on the output tape and the state is changed to q_2. Finally, in (v) in state q_2, when $+$ is scanned on the stack, it is written at the end of the output, and the state and stack are reset to q_0 and Z_0 and the process terminates.

$\mathcal{T}_{pd}$ will translate, for example, the following input to the output code:

>Input: Jack and Jill went up the hill.
>Output: 33113193011451203362858504921458806774088322103262858578990111

Parsing Algorithms

We will be concerned with some details of parsing algorithms and procedures for natural languages in Parts II and III. In this section, I will give some introductory remarks about parsing of formal and artificial languages. Given a string x in some language L, an acceptor automaton analyzes the string and makes a binary decision: whether or not x is well formed. A parsing algorithm, in addition to arriving at the well-formedness decision, must provide the structure of x; that is, it must determine the sequence of productions which was used to derive x. If x is ambiguous, there would be more than one parse or sequence of productions, unless the ambiguity can be resolved by the *environment* in which x occurs.

In general there are two approaches to parsing. In *top-down* parsing, one begins with the initial symbol S in the grammar and attempts to find the sequence of productions in P which will derive x. This will result in the *leftmost* derivation of x. The other approach is to start with x and apply the applicable productions in P to it in cycles until the initial symbol S is reached. This is called *bottom-up* parsing and results in the *rightmost* derivation of x. The following example will demonstrate these two approaches:

Consider grammar $G = (N,T,S,P)$ with production $S \rightarrow aSb$ and $S \rightarrow ab$. We know that this grammar generates the language $L = \{a^n b^n \mid n \geq 1\}$. In (67) we convert the productions of this grammar to Chomsky normal form for expositional purposes:

(67)
1. $S \rightarrow AB$ 4. $A \rightarrow a$
2. $S \rightarrow AC$ 5. $B \rightarrow b$
3. $C \rightarrow SB$

The sentence $x = aaabbb$ is generated by this grammar. The following leftmost derivation of x represents the top-down parsing:

$$S \Rightarrow AC \Rightarrow aC \Rightarrow aSB \Rightarrow aACB \Rightarrow aaCB \Rightarrow aaSBB \Rightarrow$$
$$aaABBB \Rightarrow aaaBBB \Rightarrow aaabBB \Rightarrow aaabbB \Rightarrow aaabbb$$

The following rightmost derivation represents bottom-up parsing:

$$S \Rightarrow AC \Rightarrow ASB \Rightarrow ASb \Rightarrow AACb \Rightarrow AASBb \Rightarrow AASbb \Rightarrow$$
$$AAABbb \Rightarrow AAAbbb \Rightarrow AAabbb \Rightarrow Aaabbb \Rightarrow aaabbb$$

The reader can follow these derivations by examining the following phrase marker for x:

(68)

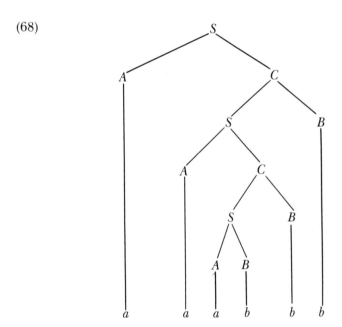

Let us now consider some parsing algorithms. In the first example, top-down parsing is illustrated with pushdown stacks. The parser, P, is sketched in (69). It has two stacks S1 and S2, and, for the convenience of exposition, we write strings in linear form in the stacks but with the understanding that the rightmost symbol in S1 and the leftmost symbol in S2 are on the top of their respective stacks. Thus, for example, in Step 3 (69), A is on the top of S1 and a is on the top of S2. Z_0 is the stack symbol (end marker) for S1 and $\dashv$ is the marker for S2.

(69)

Step	S1	S2	Rule #
1	Z_0	aaabbb⊣	
2	Z_0 S	aaabbb⊣	2
3	Z_0 CA	aaabbb⊣	4
4	Z_0 Ca	aaabbb⊣	pop a
5	Z_0 C	aabbb⊣	3
6	Z_0 BS	aabbb⊣	2
7	Z_0 BCA	aabbb⊣	4
8	Z_0 BCa	aabbb⊣	pop a
9	Z_0 BC	abbb⊣	3
10	Z_0 BBS	abbb⊣	1
11	Z_0 BBBA	abbb⊣	4
12	Z_0 BBBa	abbb⊣	pop a
13	Z_0 BBB	bbb⊣	5
14	Z_0 BBb	bbb⊣	pop b
15	Z_0 BB	bb⊣	5
16	Z_0 Bb	bb⊣	pop b
17	Z_0 B	b⊣	5
18	Z_0 b	b⊣	pop b
19	Z_0	⊣	

The parser in (69) uses the grammar in (67) to parse the string x = *aaabbb*. It works in the following way: we start with the stack symbol Z_0 on the top of S1 and the string *aaabbb* on the top of S2 (more precisely, letter *a* on the top of S2). We now enter the starting symbol of the grammar, S, in S1 and *a* remains on the top of S2 (Step 2). Rule 2 in (67) applies to S and rewrites it as AC. Thus, AC is pushed into S1 in reverse order, so that A appears on the top of S1 (Step 3). Rule 4 in (67) applies to A and rewrites it as *a* (Step 4). Now the *a* on the top of S1 matches the *a* on the top of S2. Therefore, the *a* in S2 is popped. In Step 5, we have C on the top of S1 and *a* on the top of S2 (the string *aabbb* is in the stack). Rule 3 applies to C in S1 and rewrites it as SB. In Step 6, we have S on the top of S1 and *a* on the top of S2. Once again Rule 2 applies to S, and the process continues to end, where in Step 19 we are left with Z_0 on the top of S1 and ⊣ on the top of S2. The string has been parsed and the sequence collected in S1 represents the derivational history of x. Note that in Steps 2 and 6 Rule 2 was applied to S, and in Step 10 Rule 1 was applied. Because there are two rules for the expansion of S (Rules 1 and 2), in each case we could have applied the other rule. The sequence in (69), however, is the correct sequence for the parsing of x. If we do otherwise; for example, if in step 2 we apply Rule 1, x will not parse, and we need to *backtrack* and try the alternative rule. This backtracking is a characteristic of parsers for nondeterministic grammars. There are, however, procedures for eliminating backtracking; some of these will be discussed when we examine parsers for natural languages in Parts II and III.

Cocke-Younger-Kasami Parsing Algorithm. This algorithm is in effect a bottom-up parsing algorithm, and it requires a context-free grammar in Chomsky normal form. We can then use the grammar in (67) again with the input string $x = aaabbb$. We need to set up a pair of axes x, y with indices i, j referring to the positions on them. The process builds up a triangular table from the x axis (cf. the figure in [70]). We refer to each location in the table as t_{ij}. Thus, t_{11} refers to the leftmost location in the bottom (Row 1) of the table. Now, for string $x = aaabbb$ we observe that Rules 4 and 5 in (67) apply in reverse from left to right to generate the preterminal nodes $AAABBB$. This sequence is entered in Row 1 of t. At Row 2, we take a pair of symbols and see if any rules in (67) apply. We note that the only pair subject to a rule is AB (Rule 1: $S \rightarrow AB$). Thus in location t_{32} we have S, but in all other locations in Row 2 we have $\emptyset$ indicating no applicable rule. At the next level (row), we take triplets, then four, and so on until all combinations have been tried and the table is completed.

(70)

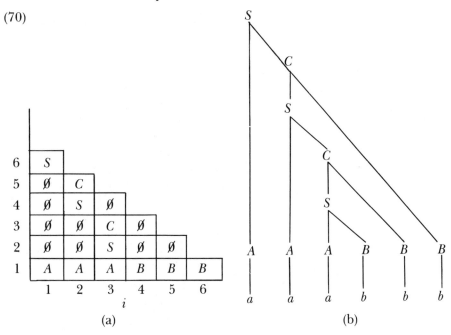

(a)   (b)

Let us now write the algorithm in a more formal manner: We label the elements in the input string as a_i for $1 \leq i \leq n$ (in the case of string x, it would be $a_1a_2a_3a_4a_5a_6$).

Step 1: Set $j = 1$ and compute t_{i1} for $1 \leq i \leq 6$.
For $a_1 = a$, we have $t_{11} = \{A\}$, since we have $A \rightarrow a$
For $a_2 = a$, we have $t_{21} = \{A\}$
For $a_3 = a$, we have $t_{31} = \{A\}$
For $a_4 = b$, we have $t_{41} = \{B\}$, since we have $B \rightarrow b$
For $a_5 = b$, we have $t_{51} = \{B\}$
For $a_6 = b$, we have $t_{61} = \{B\}$

Step 2: Set $j = 2$ and compute t_{i2} for $1 \leq i \leq 5$
For $a_1a_2 = aa$, we have $t_{12} = \emptyset$, since there are no rules such as $S \rightarrow AB$, $A \rightarrow a$, $B \rightarrow a$.
For $a_2a_3 = aa$, we have $t_{22} = \emptyset$
For $a_3a_4 = ab$, we have $t_{32} = \{S\}$, since we have $S \rightarrow AB$, $A \rightarrow a$, $B \rightarrow b$.
For $a_4a_5 = bb$, we have $t_{42} = \emptyset$
For $a_5a_6 = bb$, we have $t_{52} = \emptyset$

Step 3: Set $j = 3$ and compute t_{i3} for $1 \leq i \leq 4$.
For $a_1a_2a_3 = aaa$, we have $t_{13} = \emptyset$, since no combinations of rules in (67) apply to derive aaa.
For $a_2a_3a_4 = aab$, we have $t_{23} = \emptyset$
For $a_3a_4a_5 = abb$, we have $t_{33} = \{C\}$, since $C \Rightarrow SB \Rightarrow Sb$ $\Rightarrow ABb \Rightarrow Abb \Rightarrow abb$
For $a_4a_5a_6 = bbb$, we have $t_{43} = \emptyset$

Step 4: Set $j = 4$ and compute t_{i4} for $1 \leq i \leq 3$.
For $a_1a_2a_3a_4 = aaab$, we have $t_{14} = \emptyset$
For $a_2a_3a_4a_5 = aabb$, we have $t_{24} = \{S\}$
For $a_3a_4a_5a_6 = abbb$, we have $t_{34} = \emptyset$

Step 5: Set $j = 5$ and compute t_{i5} for $1 \leq i \leq 2$.
For $a_1a_2a_3a_4a_5 = aaabb$, we have $t_{15} = \emptyset$
For $a_2a_3a_4a_5a_6 = aabbb$, we have $t_{25} = \{C\}$, since $C \Rightarrow SB$ $\Rightarrow Sb \Rightarrow ACb \Rightarrow ASBb \Rightarrow ASbb \Rightarrow AABbb \overset{+}{\Rightarrow} aabbb$

Step 6: Set $j = 6$ and compute t_6.
For $a_1a_2a_3a_4a_5a_6 = aaabbb$, we have $t_{16} = \{S\}$.

Step 7: Halt, the table is complete.

Note that the table constructed in (70a) by the above algorithm is a representation of the tree structure in (70b), each of which gives a parse of $x = aaabbb$. Also observe that in the assignments for t in the above algorithm we have indicated sets (e.g., $\{S\}$, $\{A\}$, $\{B\}$); this is because the results could potentially be sets (see Exercise 3).

Exercises

1. Grammar $G_1 = (\{A,B,C\},\{a,b,c\},A,P)$, with productions $A \rightarrow aBC$, $B \rightarrow aBB$, $C \rightarrow CC$, $B \rightarrow b$, $C \rightarrow c$, can generate the string $x = a^n b^n c^n$. (a) Is G_1 a context-free grammar? (b) If so, how do you explain the generation of x?

2. Write an automaton acceptor for the language generated by G_1 in Exercise 1.

3. Given grammar $G_2 = (\{S,A\},\{a,b\},S,P)$; $S \rightarrow AA$, $S \rightarrow AS$, $S \rightarrow b$, $A \rightarrow SA$, $A \rightarrow AS$, $A \rightarrow a$, and string $w = abaab$, write a table showing the Cocke-Younger-Kasami parse of w. *Hint:* the nodes in the table may contain more than one element.

4. Write a context-free grammar G_3 to generate simple declarative English sentences such as $y = $ *the old men visit the nursing home*. Give a leftmost derivation of y in G_3.

5. Write a translation schema and a transducer to read as input decimal numbers and output their equivalent binary numbers. Use the following table for BCD conversion:

0	0000	5	0101
1	0001	6	0110
2	0010	7	0111
3	0011	8	1000
4	0100	9	1001

Example: $924 = 100100100100$.

6. Write a grammar in GNF to generate the language $L = \{ab^n c^n \mid n \geq 1\}$. Write a top-down parser for L.

7. We say a grammar G is ambiguous if there is more than one derivation for a string w in G. Show that the following grammar is ambiguous. $S \rightarrow abC$, $S \rightarrow aB$, $B \rightarrow bc$, $bC \rightarrow bc$.

8. Prove that for the following finite-state grammar G there exists a finite automaton M such that G and M are equivalent. $G = (\{SAB\},\{abc\},P,S)$; $S \rightarrow aS$, $S \rightarrow aB$, $B \rightarrow bB$, $B \rightarrow cA$, $A \rightarrow cA$, $A \rightarrow c$. *Hint:* prove by construction.

9. Let $L(G) = \{xx \mid x \in \{a,b\}^*\}$. That is, the sentences of L, represented here as xx, consist of two parts; the first part is a string x consisting of a's and b's, and the second half is exactly the same as the first part. Examples: *abab, aabaaaba*. Write the grammar to generate L.

10. A grammar G is said to be *right-linear* if each of its productions is of the form $A \to wB$ or $A \to w$, where A and B are members of the nonterminal vocabulary N and w is a string in T^*. A *left-linear* grammar has productions of the type $A \to Bw$ or $A \to w$. By definition every right-linear grammar is also a context-free grammar. Under what condition is a right-linear grammar a regular grammar?

11. The following productions represent an important grammar for certain arithmetic expressions.

$$E \to E + T \quad E \to T$$
$$T \to T * F \quad T \to F$$
$$F \to (E) \quad F \to a$$

Examples of expressions generated by this grammar are: $a + a * a$ and $a * (a + a + a)$. Write a set of syntax-directed rules R for converting the strings generated by this grammar into Polish (not reverse Polish!)

References

For more detailed study of formal languages and automata theory, see Aho and Ullman (1972–73), Vol. I; Hopcroft and Ullman (1979); and Harrison (1978). Minsky (1967) contains lively discussions of computations by and models of Turing machines. Original contributions to Chomsky Hierarchy (languages) and Turing machines are in Chomsky (1956,1959a,1962b,1963) and Turing (1936,1937).

A good source for further study of algorithms and procedures is Knuth (1968). See also Machtey and Young (1978) and Cutland (1980) for theories of algorithms and recursive functions.

We will include more discussions on translation procedures and parsing techniques in later chapters, but Aho and Ullman (1972–73), Volume 1, contains a thorough treatment of these topics as well as other important topics which we have not included in this chapter.

For mathematical foundations of the topics discussed in this chapter, consult Knuth (1968) and Birkhoff and Bartee (1970).

CHAPTER 2

Linguistics

Introduction

Language is perhaps the oldest subject matter of inquiry and speculation from the biblical accounts of divinity to the amazingly precise and rule-governed grammar of Panini, to Aristotle, the medieval philosophers, the 17th century rationalists, the 19th century linguists, neogrammarians and behaviorists, the structuralists in the 20th century, and the revolutionary revival of mentalism in the 1950s. Because of the nature and role of language throughout the known history of humanity, linguistics has remained a dynamic and vital science or art, depending on one's disposition.

At present, linguistics is frequently treated as a science, a natural science, inasmuch as it deals with formal, abstract objects, and with the empirical observation of their properties and behavior. It also deals with the psychology of cognition and learning. It could equally be (and has been) regarded as a discipline in arts and humanities because of the inevitable interconnection of language with other human activities, including artistic, communication, media, literary, and other cultural creations. The place of linguistics as a social science is also well attested: its close affinity with anthropology and its perhaps not so obvious impingement on education, history, philosophy, sociology, and other fields of study. The role of linguistics in computer science is well established (Moyne, 1975) and is gaining increasing significance with advances in artificial intelligence, knowledge systems, and natural language processing.

Modern Linguistics

Ferdinand de Saussure is generally credited as the founder of modern linguistics. His lecture notes, posthumously published by his students in 1915, laid the foundations of modern *descriptive* or *structural linguistics*. This is to be contrasted with the so-called *traditional grammar*, whose origins are to be found in Greece and India, and *historical* or *comparative linguistics*, which arose and flourished in the last century. According to de Saussure (1916/1955), language (*langage*) comprises a mentally stored system of signs (*langue*) and the use of this system in actual speech (*parole*). A sign (*signe*) is an unbreakable bond between a concept and a sound image. Thus, the sign "tree" links the concept of a tree and the mental representation of a particular sequence of speech sounds. To grasp even a single sign, one must understand the whole system to which it belongs. At first blush, the English *sheep* is equivalent to the French *mouton*, but to represent the meat of this animal, there is a shift to a different sign in only one of the two sign systems: *mutton* in English. Hence, *sheep* and *mouton* are said to have different values (*valeur*). Similarly, it will not be sufficient to learn the signs corresponding to different colors in one language and apply them to all languages, because it would not seem possible to predict how many segments the color spectrum is divided in any given language. A major claim of de Saussure was that such divisions were *arbitrary* in the sense that there are no preestablished partitions within the system of concepts: each speech community has a completely free hand in developing its own system. The same is true in two ways for the acoustic-image side of the sign. First, there is no intrinsic connection between a particular sound representation and some concept, that is, there is nothing more "doglike" about the acoustic image of *dog* than about *perro* or *chien*. Second, systems of sound units can differ from one language to another just as freely as systems of concepts. In an idiom with only three vocalic (vowel) sound units, say, those of m*ee*t, m*o*tt, and m*oo*t, American English pronunciation of the words *meet*, *met*, and *mate* could be identical sound images.

De Saussure distinguished *synchronic* from *diachronic linguistics*. The latter studies language change; the former investigates sign systems at a particular moment in time, abstracting from linguistic evolution. *Langue*, not *parole*, is the focus of synchronic research. A complete synchronic account of *langue* for a given language is an exhaustive classification, or *taxonomy*, of its signs and of the elements of these: sound units, word units, and coherent sub- and superword units.

Sentence formation is included in *parole*; as such it is not a concern of Saussurian language study, strictly defined.

During the first half of the present century various schools of Saussurian linguistics developed (Martinet, 1953). The American school—Leonard Bloomfield (Bloomfield, 1933, Hockett, 1970) and his followers—was highly selective as to which tenets of Saussure's doctrine it accepted (Levin, 1965). On the one hand, the followers of this school emphasized the notion of the arbitrariness of the linguistic sign, downplaying any common features among the world's languages, affirming rather their propensity to differ from each other in limitless ways. On the other hand, as behaviorists, the Bloomfieldians eschewed "mentalism"; that is, they excluded as unscientific the postulation of objects or events located in the "mind." Clearly this ruled out de Saussure's signs as objects of linguistic study. But the latter had defined the discipline as the analysis of the linguistic sign systems. With *langue* rejected, the American school turned to *parole*, the actual speech record. Sound units became not acoustic images, as they had been for de Saussure, but either the "convenient fictions" of the instrumentalist or some sort of pastiche of the physical properties of the unit's variants. Curiously, members of this school still considered themselves to be studying *langue*; they had redefined the concept perhaps without being fully aware of it in each case. The American structuralists kept Saussure's taxonomic orientation; but with the basis for the theory having evaporated, linguistics was reconceived as a methodology, a set of procedures for obtaining "tabulations of the noises produced in speech" (Katz, 1981, p. 2).

The great systematizer of the American structuralist methodology was Zellig Harris at the University of Pennsylvania. Harris's *Methods in Structural Linguistics* (1951) provided step-by-step instructions for the processing of linguistic data from the collection of a *corpus* of *utterances* to the final presentation of a set of sentence formation rules. An utterance is a stretch of speech that is preceded and followed by silence (pause), and is minimal in the sense that one utterance cannot be contained within another. A corpus is a record of one or more (usually many) utterances. It can be obtained by capturing, mechanically or via manual phonetic transcription, a conversation or a narrative, or by asking questions or prompting a native speaker (*informant*) who works with a linguist studying a particular language. The important point about Harris's *Methods* is that it is an attempt to define Bloomfieldian linguistic concepts operationally, that is, in terms of the manipulation of intersubjectively observable data. Thus, Harris provided a truly scientific procedure for segmentation and classification or taxonomy of the contents of utterances, which was the major goal of linguistics of the era.

The sophistication and rigor of Harris's procedures contributed to the perception, already widespread, that linguistics was a discipline nearing consummation. Soon, it was believed that the operations of the *Methods*, with slight modifications, could be performed automatically by digital computers that were just beginning to appear. Moreover, the role of linguistics as a model for other social sciences was consecrated with the publication of Harris's book. Margaret Mead, for example, in reviewing it, called for a reshaping of anthropology along the lines of Harrisonian structuralism (Mead, 1952).

Soon after *Methods*, Harris's studies led to another epoch in linguistics. Harris postulated that languages, for example English, had a small number of underlying *kernel* sentence structures and that all the sentences of a language can be derived from the set of kernels by the application of various types of *transformational* rules. The transformations, in turn, were divided into a number of primary or elementary transformations and complex transformations which were constructed from the elementary ones (Harris, 1957). This then laid the foundations of *transformational grammars* (TG).

Transformational-Generative Linguistics

Transformational theory as it is known today was developed by Noam Chomsky, a student of Harris, and by followers of Chomsky during the past two decades. Chomsky developed transformational grammar as a generative grammar for natural languages along the general line of formal languages and generative grammars that we have seen in Chapter 1. The grammar is intended to explicate all and only sentences of a language, and the underlying theory is purported to give adequate descriptive and explanatory accounts of language. However, the impact of Chomsky and his theories has gone beyond linguistics and has affected science, philosophy, and psychology, among other disciplines. From the viewpoint of perception, no modern theory of language comprehension can ignore TG linguistics, and many of the models constructed within this theory have been based directly or indirectly on TG linguistics. Thus, for our study, some familiarity with TG is necessary.

Let us begin then with two fundamental terms, *competence* and *performance*. As a first step toward defining these terms, we will consider a simple analogy: imagine an individual *x* who knows how to do arithmetic. Now if in a particular instance x makes an error in calculation, or momentarily forgets how to write a certain number, or gets angry and tears up the piece of paper on which he or she is writing, we do not necessarily say that x does not know the laws of arithmetic. Rather we might say that x's human limitations have prevailed on an instance and he or she has failed to "apply" his or her knowledge. Here then we recognize an underlying mental reality, that is, x's knowledge of arithmetic, which is distinct from his or her behavior or use of that knowledge; for example, the instance of x's botched attempts at calculating. Within the limitations of this rather crude analogy, it is possible to say that competence is the mental reality underlying one's use of language; that is, it is the knowledge of language underlying its performance.

If we look more closely at this analogy, we will perceive its strong and weak points. One of the former is that although there is no principled limit to the length of a calculation within the ambit of one's "arithmetical competence," there is a very real limit to the length of the calculations one can actually produce or understand. The identical statement is true of linguistic competence if we substitute "sentence" for "calculation." For it never becomes intrinsically "un-English," or ungrammatical, to extend a sentence such as "I see the man who punched the salesman who sold Bob the book that describes the man who . . . ." It simply becomes tedious, or foolish. Considerations of performance, not those of competence, impose the limitations. All this is the same as saying that the laws of arithmetic, and those of a language, do not operate in real time; people do.

A second strength of our analogy has to do with science's relative lack of understanding of the functioning of the human brain. When we speak of a deep-seated mental reality such as competence, we do so without a nuts-and-bolts understanding of the corresponding brain structures and activities. Any accounting we can presently hope to give of competence involves a radical abstraction from physical reality. This is true both for language and for arithmetic. Of course, we do have a formal account of arithmetical competence; it can be cast as a formula or as a recursive procedure. Linguistic science aims at developing similar formal accounts of linguistic competence.

Last of the strong points is the (not completely uncontroversial) observation that no lower animals, and practically all humans, possess both types of competence.

The only weakness in our analogy, though others exist, concerns the acquisition of competence. The analogy does not answer the objection of the behaviorists who would have language (and arithmetic) acquisition be the same as any other learning phenomenon, say swimming, by instructions and experience.

Setting our analogy aside for the moment, we note a further aspect of language competence. According to Chomsky (1965, p. 3):

> Linguistic theory is concerned primarily with an ideal speaker–listener, in a completely homogeneous speech-community, who knows its language perfectly and is unaffected by such grammatically irrelevant conditions as memory limitations, distractions, shifts of attention and interest, and errors (random or characteristic) in applying his knowledge of the language in actual performance.

In other words, it is only by abstracting from accidental factors, such as unfamiliarity with a certain vocabulary word or the influence of, say, the knowledge of some foreign language, that we can get at the competence of an arbitrary speaker–listener of a language.

To summarize, competence, the speaker–hearer's knowledge of his language, is abstract and mental; performance, the actual use of language in concrete situations, is tangible and physical. Linguistics attempts to construct complete theories of both performance and competence, and to demonstrate the relationship between the two.

To revert to our analogy, notice that it is possible to construe arithmetical competence as a system taking as input an ordered sequence of numbers (or variables) and operators, and yielding as output two entities. One is a representation of the input sequence in which all significant relations among numbers and operations have been made explicit. The other is a numerical value (the result). For instance, given

$$a + b * c$$

such a system returns both some answer (call it $\mathcal{A}$) and a structural representation either of the form:

$$[_\beta\ a\ +\ [_\alpha\ b\ *\ c_\alpha]_\beta]$$

or of the form:

The system returns, with an error message, any input in improper form, such as + 6/.

We can be sure that both types of the output are necessary, because getting the right answer absolutely depends on properly organizing the given elements.

In the light of our analogy, we expect a system representing linguistic competence to mirror the above arrangement. Input should consist of a sequence of abstract linguistic elements, output of a structural account of the sequence and of either the value *grammatical* or the value *ungrammatical*. Grammatical sequences would be those which meet the requirements of the system, that is, sentences of the language in question. Ungrammatical sequences would be everything else, that is, nonsentences of the language.

For example, if we input two separate sound sequences, represented at this point as (1)/ahyseemáwris/ and (2)/zhuhvwahmawrées/, a system embodying the competence of a speaker of English should output, very roughly:

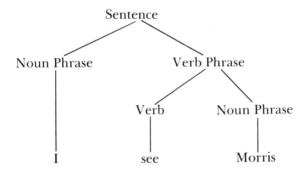

as grammatical, and (2) zhuhvwahmawrées, as ungrammatical. Sequence (2) is rejected because it is a nonsentence of English. (That it happens to be a sentence of French—*Je vois Maurice*—ought to make no difference to a system of English-language competence.)

We now have some expectations concerning the input and output of a linguistic competence system, but what is the internal structure of this system? One could perhaps say that it is simply a list of all the grammatical sentences of a particular language. But this suggestion will not work, because, as we have already seen, there is no longest sentence of English, even if we could have a bound on the number of sentences. An alternative approach would be to represent competence as a set of *rules*. In this case we would need only to store the rules and the individual elements manipulated by them. This seems like a plausible version of the internal structure of a competence system; a moment's reflection reveals that we would have to describe arithmetical competence in the same way.

Of course, a linguistic rule system would not physically produce any sentences—that is the job of performance. The rules, taken together, would simply *imply* the existence of all grammatical sentences of the language represented and fail to imply the existence of any nonsentences.

Let us assume then that the linguistic competence of a speaker–listener of a given language is to be represented by means of a *rule system* that *generates* the set of all grammatical sentences of that language, each with its associated *structural description(s)*. This system, moreover, *fails to generate* either any nonsentence of the language in question, or any *incorrect structural descriptions* of grammatical sentences.

Such a system, formally a type of *axiomatic system*, is called a *generative grammar*. In Chapter 1 we presented such grammars as abstract formal systems, including the *Chomsky hierarchies* of languages and grammars. We can now ask about the relevance of this hierarchy to linguistic competence. One of the goals of linguistic theory is to develop a formal system by which one could identify or measure the adequacy of grammars for natural languages. Chomsky (1956) and Postal (1964a), among others, have claimed that Types 2 and 3 grammars (discussed in Chapter 1) are not adequate for natural languages (see Part III for some current views on this issue). Type 1, context-sensitive grammar, was considered by some as an adequate generative grammar for natural languages, but with the development of TG no serious attempts were made to develop a CS grammar for a natural language.

The adequacy of a grammar can also be viewed from its *generative power*. Grammars with *weak generative capacity* are those that can simply generate strings or sentences belonging to a category of languages. On the other hand, a grammar is said to have *strong generative capacity* if for every sentence that it generates it also produces a structural description of that sentence.

Grammars for Natural Languages

Viewing grammar as the theory of language and assuming for the present that some restricted form of Type 1 grammar is the minimum capacity requirement for natural languages, we can ask what further requirements must be put upon a proposed grammar of a natural language?

First, it must display *observational adequacy*: it must accurately reflect the *judgments* of the speaker or speakers whose competence it represents, as to the *grammatical well-formedness* of all sentences and parts of sentences it generates, and of all nonsentences (which, of course, it should fail to generate). Second, it has to possess *descriptive adequacy*, in the sense that it correctly states all requisite rules for exhibiting the speaker's *structural knowledge* of his or her idiom. Third, it must follow from a general theory of competence which evinces *explanatory adequacy*, which accounts for the judgments and knowledge just referred to, as well as for the formal characteristics of the language data under description, by means of *universally* or at least generally valid generalizations about the form of human language as such.

The overall aim of this general theory of human language is to describe formally what is posited to be the *innate language faculty* that exists potentially in all humans and begins to develop with the birth of the organism. Linguistics, in the broadest sense, is an attempt to explain how it is that the human infant begins some process, which culminates roughly by the time he or she is six years old, at the end of which he or she is substantially in command of a system—a human language—which it has eluded the best efforts of the keenest scholars to describe. This is to be accomplished partially by means of an account of linguistic *universals*—the common denominators of all human languages which recur in some form or other from language to language. Linguists distinguish *formal* universals from *substantive* universals. The former include, but are not limited to, *operators* such as the rewrite symbol, →, *abbreviatory conventions* on linguistic rules, such as we will see later, the *general formal description* of linguistic rules, as well as *constraints* on the scope of these. Substantive universals consist of universal vocabularies of symbols or units from which the structures of language may be constructed.

Another way of stating the goals of linguistic theory is to say that its aim is to describe two different devices which are assumed to be part of our biological endowment as a species. The first device, called the *language acquisition device* (LAD) is used most obviously by the child as he learns to communicate. Its input is whatever utterances the child hears around him or her; its output is a generative grammar of the language spoken in his or her environment; and its internal structure, it is assumed, includes the general theory of language referred to above, along with an *evaluation measure* which enables the child to reject inappropriate hypotheses as to the grammar under construction and to come up ultimately with a grammar that is adequate in all the above respects, and probably of optimal concision.

raw linguistic data ———→ | LAD | ——→ optimal generative grammar

The second device is a perceptual model which, for a given utterance, assigns a structural description within a particular language and makes a judgment as to the utterance's grammaticality. In this device, a grammar plays only one part; other major roles are assumed by performance strategies of various types.

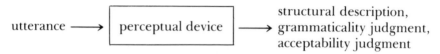

utterance ——→ | perceptual device | ——→ structural description,
                                           grammaticality judgment,
                                           acceptability judgment

Although research within the TG framework has focused on the competence involved in the perceptual device above, this emphasis has a theoretical justification in that, as the device is described, it is clear that it cannot be adequately accounted for without a prior understanding of competence. Some of the efforts which have been made to characterize this device by those working in the field of TG linguistics will be taken up later. At this point, however, we can note the distinction between *grammaticality* and *acceptability*. The former, a phenomenon of competence, consists in the conformity of an abstract string of linguistic elements to the rules of a particular generative grammar. The latter, part of the conceptual apparatus of performance, applies to an utterance in some natural language, that is, what we can consider a *concrete representation* of an abstract underlying linguistic structure of competence. An utterance is acceptable if it is relatively easily interpretable by speakers of the language in question, and not considered particularly "funny," "weird," or "strange" by them. It is important to note that a string or utterance can be more or less grammatical or acceptable, respectively, and that a given utterance can be relatively acceptable although its underlying representation is relatively ungrammatical, and vice versa. For example, an acceptable utterance whose underlying linguistic representation is ungrammatical, in English, is: "I aahh believe that aah you know they said that aahh aahh we're supposed to meet th... aahh they'll be there around five o'clock." On the other hand, a string such as the following is considered to be grammatical but unacceptable (+ signifies "concatenated with"): *he + is + the + man + the + butler + the + private + detective + the + police + work + with + questioned + incriminated.* (This means the same as, *He is the man who was incriminated by the butler who was questioned by the private detective with whom the police work.)*

Before we go further, let us stop to recollect what we have learned so far about TG theory. Competence is the unconscious knowledge of the structure of a natural language on the part of a speaker of it, whereas performance is the use of this knowledge, along with other systems, in concrete situations. Competence is described via a *generative grammar*, a rule system that generates, or implies, the set of all grammatical sentences of the language being represented, along with a set of *structural descriptions* for each such sentence.

A grammar weakly generates a set of terminal strings and strongly generates a corresponding set of structural descriptions. To the extent that a grammar is correct, it is adequate on three levels: it is *observationally adequate* in that it accounts correctly for the strings or sentences of the language; it is *descriptively adequate* in that it accurately sets forth the structural rules of the language; and it is part of a general theory of competence that is *explanatorily adequate* in that it accounts by means of significant generalizations for the fact that the data and grammatical judgments of the language are as they are and not some other way. A *theory of language* seeks to describe and explain linguistic universals, both formal and substantive, and attempts to characterize both the LAD and the perceptual device. In the interest of describing the LAD it incorporates an *evaluation measure*. Finally, a string conforming to the rules of some grammar G is said to be grammatical with respect to G, whereas an utterance understood with little difficulty by speakers of a language and not questioned or rejected by them as bizarre, is called an acceptable utterance with respect to the language of those speakers.

The Standard Theory

Aspects of the Theory of Syntax (Chomsky, 1965) presented a theory which differs in major respects from the one put forth in Chomsky (1955), which until 1965 had been the fullest statement of TG theory. Widely accepted for years by linguists, and still considered correct in most respects by a number of scholars, the *Aspects* version of TG theory has come to be known as the *standard theory* (Chomsky, 1972).

In order to obtain an overview of the standard theory, we consider the following *first approximation* of it. At the heart of a TG grammar of some natural language is a Type 1 grammar whose terminal vocabulary consists of abstract mental objects called *formatives* and whose nonterminal vocabulary is made of other abstract objects denominated *syntactic category symbols*. This Type 1 grammar is included, along with other systems, in what we will call the base of the principal component of the grammar, namely the *syntactic component*. The structures generated by the base, called *deep structures*, are modified terminal strings of the language implied by the above-mentioned Type 1 grammar. Deep structures are fed into an integrated set of rules within the syntactic component—the *transformational subcomponent*—and emerge as *surface structures*. The two remaining components of a generative grammar, the *semantic* and *phonological components*, are parasitic on the syntactic component in the sense that they are barren unless supplied with syntactic structures to process. They are the two *interpretive components*. The semantic component interprets deep structures and outputs their *semantic representations*; the phonological component interprets surface structures and yields their *phonetic representations*. In brief, then, the two immediately apparent sets of structures of a linguistic system—the set of meanings and the set of sound sequences—are creations of the covert set of syntactic structures.

The following diagram represents our first approximation of the structure of a TG grammar within the framework of the standard theory:

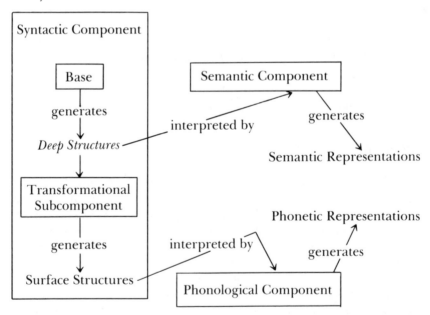

We can arrive at a somewhat more detailed understanding of the standard model by discussing each of the components of a grammar, as defined by this theory, in turn.

The syntactic component as stated above consists of the base and the transformational subcomponents. Taking the former subcomponent first, we can analyze the base into two sets. The first is the set of *categorial* rules, the second, the *lexicon*. Categorial rules are actually productions of a Type 1 grammar whose nonterminal symbols represent, theoretically, the universally valid (i.e., identical for all language) *syntactic categories*, and whose terminals are *formatives*, that is, minimal syntactic elements of some language, and place-holders for formatives. Some syntactic categories are: S (Sentence), NP (Noun Phrase), VP (Verb Phrase), PP (Prepositional Phrase), N (Noun), V (Verb), A (Adjective), P (Preposition). Of these, three are *lexical categories*: Noun, Verb, and Adjective. It is the latter that are rewritten within the categorial component as place-holders for formatives. In particular, one of the productions of the categorial system is $\mathcal{A} \to \Delta$, where $\mathcal{A}$ stands for a lexical category (N, V or A) and Δ is a place-holder, or *dummy symbol*. Those nonterminals which are not exponents of lexical categories are rewritten as formatives. These latter formatives are called *grammatical formatives*, whereas those whose place is now being held by dummy symbols are the *lexical formatives*.

For an example we can consider this greatly simplified segment of a grammar for English, where $\mathcal{A}$ = N, V, A.

Categorial rules:

$$S \to NP \; AUX \; VP$$
$$NP \to DET \; N$$
$$AUX \to TENSE$$
$$TENSE \to \left\{ \begin{matrix} present \\ past \end{matrix} \right\}$$
$$VP \to V \; (NP)$$
$$\mathcal{A} \to \Delta$$

We will assume that N = {S, NP, VP, N, AUX, V, A, DET, TENSE} and that T = { present, past, Δ} for this segment of English grammar. Note that *present* and *past* are grammatical formatives and that Δ is the dummy symbol. Two abbreviatory conventions are in use here: curly braces { }, which mean "rewrite as any of the elements listed, but not as more than one of these," and parentheses, which mean "include optionally the element(s) within parentheses." A derivation with respect to the above section of the English grammar is

S $\Rightarrow$ NP AUX VP $\Rightarrow$ DET N AUX VP $\Rightarrow$ the N AUX VP $\Rightarrow$ the Δ_N
AUX VP $\Rightarrow$ the Δ_N TENSE VP $\Rightarrow$ the Δ_N past VP $\Rightarrow$ the Δ_N past V
NP $\overset{*}{\Rightarrow}$ the Δ_N past Δ_V DET N $\Rightarrow$ the Δ_N past Δ_V the N $\Rightarrow$ the Δ_N past
Δ_V the Δ_N

(Although no rule was given to indicate that DET [*Determiner*] is
rewritten as *the*, we will assume this here; it is immaterial to the
discussion at hand.) We can construct a tree diagram for the above
derivation according to methods already indicated. We will call tree
diagrams *phrase markers* (P-markers) in the present context, because they
function to specify the componential structure of *phrases*, or strings of
words organized on linguistic principles. Further, the string generated
via the above derivation, namely, *the Δ_N past Δ_V the Δ_N*, will be called
a *preterminal string* rather than a terminal string, as an additional
operation is required to convert it into a terminal structure of the base.
The P-marker for our preterminal string, then, is

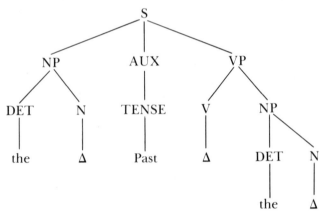

Now this structure represents a class of English sentences—a very large,
and perhaps infinite, class. It is crucial to observe that it also stands for
an equally large class of ungrammatical strings—strings which do not
underlie any English sentences, strings which are grammatically *ill-
formed*. Two of the grammatically well-formed strings (i.e., sentences
represented by our preterminal P-marker are

(1)   the + aardvark + Past + chase + the + cockroach
(2)   the + boy + Past + kill + the + bug

On the other hand, a pair of ill-formed strings which fit the same
structure are (by convention, the symbol * signifies an ill-formed string)

(3)* the + aardvark + Past + sleep + the + cockroach
(4)* the + boy + Past + put + the + bug.

The utterances these strings give rise to are, of course, (1) *The aardvark chased the cockroach*, (2) *The boy killed the bug*, (3)* *The aardvark slept the cockroach*, (4)* *The boy put the bug.*

There seems to be a problem here—we do not want our grammar to generate ill-formed strings. It is plain, however, that the categorial rules, left to their own devices, do exactly this. But as we saw above, there is a second portion of the base subcomponent of a grammar's syntactic component, namely the lexicon. The lexicon, along with the process of lexical insertion, will remedy our present difficulties.

Before actually turning to the lexicon, however, we must define a certain number of terms. First, a P-marker is composed of *branches* or *edges* which connect *nodes* (i.e., the endpoints, or vertices, of the branches). Each node is *labeled* by a particular symbol within the vocabulary of the grammar, that is, either by a category symbol such as S, NP, VP, N, V, A, DET, etc., or by a grammatical (e.g., past, present) or lexical formative (so far we have seen only the universal place-holder for lexical formatives—Δ). Higher nodes are said to *dominate* lower nodes. When there is no third node intervening between a dominating and a dominated node, the former is said to *immediately dominate* the latter. For instance, AUX immediately dominates TENSE in our P-marker, and V immediately dominates Δ. Given the above two definitions, it is possible to define what are referred to as *grammatical functions*, with respect to P-markers. A *deep-structure subject* is the NP immediately dominated by S, and the *deep-structure object* is the NP immediately dominated by a VP which, itself, is immediately dominated by S. The subject function is represented [NP, S], the object function [NP, VP, S]. Other such relations can be defined quite clearly, but none are of immediate interest to us.

The lexicon features two parts, one basically a vocabulary list, the other a set of *lexical insertion rules*. The first portion of the lexicon, then, is an unordered set of *lexical entries*, each of which is itself a set of *features* (defining characteristics). A lexical entry possesses phonological, semantic, and syntactical features; the first two classes will be taken up below. Suffice it to say here that such features specify the sound and meaning content of the lexical item in abstract terms. The syntactic features are subdivided into *inherent* and *contextual* features. Inherent features include *category features* such as [+N], [+V], and the like, which simply signify that the particular item is a member of the category in question. In addition, the class of inherent features includes features

such as [± Common], [± Abstract], [± Animate], and [± Human]. All
of the features just cited apply to Nouns: *book* is [+ Common], *John*
[− Common], *truth* [+ Abstract], *bottle* [− Abstract], *goat* [+ Animate],
rock [− Animate], and so on. Actually, each of the above lexical items
is a complex of such features, for instance,

$$
\overset{book}{\begin{bmatrix} +\,N \\ +\,Common \\ -\,Abstract \\ -\,Human \\ -\,Animate \end{bmatrix}}
\overset{truth}{\begin{bmatrix} +\,N \\ +\,Common \\ +\,Abstract \\ -\,Human \\ -\,Animate \end{bmatrix}}
\overset{goat}{\begin{bmatrix} +\,N \\ +\,Common \\ -\,Abstract \\ -\,Human \\ +\,Animate \end{bmatrix}}
$$

The second class of syntactic features, that of contextual features,
is limited to *strict subcategorization features*. We can define these by
example. The reason that (3) is grammatically ill-formed is that in the
process of *lexical insertion* (i.e., replacing dummy symbols with lexical
items to yield a deep structure; this is elucidated below), the strict
subcategorization features of the verb *sleep* were violated. That is, part
of the lexical entry for the Verb *sleep* is the *complex symbol* [−−−NP].
The presence of this symbol has the result that *sleep* is prohibited from
being inserted into a P-marker under a VP node, just in case it would
be followed, within that VP node, by an NP node. Now recall the
definition above of the concept *deep structure object* ([NP, VP, S]). This
is precisely the configuration into which *sleep* must *not* be introduced
according to its strict subcategorization feature. So the effect of this
particular feature is to ensure that *sleep* is always treated as an intransitive
verb (to employ traditional terminology), that is, as a verb without a
deep structure object. In our P-marker above, then, the strict subcategorization feature [−−−NP] has the function of preventing lexical
insertion in the following context:

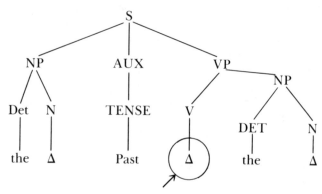

The consequence of ignoring strict subcategorization features in the process of lexical insertion is ungrammaticality. On the other hand, one of the reasons that (1) is grammatically well formed is that its verb, *chase*, has been inserted into a context which *is* in accord with its strict subcategorization feature, namely [+——NP (Adv$_{loc}$)]. This particular complex symbol specifies that *chase* is to appear in a VP in which it is followed by an NP, and optionally by an adverbial expression of location as well. The first of these two situations is that of our P-marker, which is one reason why the insertion of *chase* under the V node there preserves grammaticality.

Strings (2) and (4) above present a slightly different contrast with respect to strict subcategorization from that of (1) and (3). There is no problem with the insertion of *kill* into our P-marker because it is subcategorized by means of the feature [+——NP (Adv$_{manner}$)]. The corresponding feature of the verb *put*, however, requires the presence of an element not to be found in the P-marker above, namely an adverbial expression of location, that is, it is subcategorized by the complex symbol [+——NP Adv$_{loc}$]. (Note that Adv$_{loc}$ is *not* enclosed in parentheses, so that it is a necessary element of the feature as opposed to an optional one.) Because (4) violates the strict subcategorization of *put* it is, therefore, ill-formed.

To summarize, then, strict subcategorization features are complex symbols that specify the categorial restrictions on the insertion of particular lexical items into preterminal structures. More broadly, the set of syntactic lexical features is divided first into those which are inherent and those which are contextual. The latter class is equivalent to the ensemble of strict subcategorization features; the former is again split into category and noncategory features.

The phonological specification of a lexical entry is a concatenated sequence of + or − values for phonetic features. That is to say, from the point of view of sound sequences, the output of the syntactic component, once lexical items have been inserted into preterminal P-markers, is in the form of ordered sets of what we can informally refer to as elements of the phonetic vocabulary of the language being described. For instance, because *kill* is a lexical formative with the grammar of English, its lexical entry includes what we might informally designate as the sequence /k/ + /i/ + /l/, where /k/,/i/ and /l/ are "elements of the phonetic vocabulary" of English. We call such elements *systematic phonemic segments*. In actuality, these segments are merely informal symbolic representations for sets of the feature values referred to at the beginning of this paragraph. Each systematic phonetic segment of each natural language, then, is represented as a series of values for *distinctive phonetic features*. Those features which are distinctive for a given language, that is, which form the basis for the speaker's distinction between different words of the language, are drawn from a universal set of phonetic features that represents "the phonetic capabilities of man" (Chomsky and Halle, 1968, p. 299). So, to return to the phonological portion of the lexical entry, for instance the lexical item *kill*, in English, is a string of distinctive feature values: $[+\text{Cons}_1] + [-\text{Syll}_1] + [-\text{Son}_1] + [-\text{Nasal}_1] + [+\text{Lateral}_1] + [+\text{High}_1] + [+\text{Back}_1] + [-\text{Voiced}_1] + [-\text{Cons}_2] + [+\text{Syll}_2] + [+\text{Son}_2] + [-\text{Nasal}_2] + [-\text{Lateral}_2] + [+\text{High}_2] + [-\text{Back}_2] + [+\text{Voiced}_2] + [+\text{Cons}_3] + [+\text{Syll}_3] + [+\text{Son}_3] + [-\text{Nasal}_3] + [+\text{Lateral}_3] + [-\text{High}_3] + [-\text{Back}_3] + [+\text{Voiced}_3]$. Here Cons (Consonantal), Syll (Syllabic), Son (Sonorant), Nasal, and so forth are distinctive phonetic features, and $\alpha\, F_n$ refers to the value ($\alpha = +$ or $-$) of feature F within the set n of features, that is, within segment n. Phonological specifications of formative strings will be discussed, albeit briefly, later when we come to the phonological component of a generative grammar. It is to be noted, however, that for the purposes of the present study a real familiarity with phonological matters is unnecessary.

Semantic features of a lexical entry, as in the case of syntactic features, fall into two categories—lexical and contextual features. And this parallelism extends to the division of one of these two categories (that of lexical features in this case) into two subparts. Lexical semantic features, then, are of two kinds: *semantic markers* and *semantic distinguishers*. At this point, the difference between semantic markers and distinguishers can be said to exist in the fact that the former reflect whatever systematic relations hold between the particular lexical entry a semantic marker is being used to characterize and the rest of the vocabulary of the language whereas the latter reflect what is idiosyncratic about the meaning of the entry. The lexical semantic features of a given lexical item are organized, in the lexical entry, into *readings*, each reading corresponding to a different sense of the item. For the English noun *bachelor*, for instance, four readings are included in the item's lexical entry (cf. Katz and Fodor, 1963):

Reading 1. (Human) (Male) [who has never married]
Reading 2. (Human) (Male) [young knight serving under the standard of another knight]
Reading 3. (Human) [who had the first or lowest academic degree]
Reading 4. (Animal) (Male) [young fur seal when without a mate during the breeding season]

Parenthesized terms here are semantic markers, whereas bracketed expressions are distinguishers. The net effect of the representation of the meanings of lexical items via semantic markers and distinguishers is the analysis of the semantic side of lexical items into *atomic concepts*— indivisible, perhaps universal, units of meaning.

Contextual semantic features are limited to *selectional restrictions*. These are formal statements, in terms of grammatical functions, of lexical *cooccurrence restrictions*, that is, of allowable and proscribed combinations of lexical items within a sentence or within some other context. Lexical cooccurrence restrictions permit, for example, the collocations *the green lamp, the old lamp, the expensive lamp*, but block *the pedantic lamp*. By considering the following strings we can come to understand selection restrictions:

(5)   the + reporter + Past + interview + the + politician
(6)   the + repairman + Past + repair + the + fan
(7)*the + baboon + Past + interview + the + politician
(8)*the + repairman + Past + repair + the + mayor

Strings (7) and (8) are disallowed because they violate the selection restrictions on the verbs *interview* and *repair*, respectively. The selection restriction on *interview* is something like this:

[(Human) AUX—DET (Human)]

This means that *interview* requires a (Human) subject as well as a (Human) object, as these terms are structually defined above. However, (Human) is, of course, not among the semantic lexical features of the noun *baboon*, so selection restrictions are violated in (7) and the string is *semantically ill formed*. This problem fails to arise in (5) because *reporter* is (Human). To take one more example, the verb *repair* has selection restrictions such as:

[(Human) AUX—DET (Nonhuman) (Inanimate) (Object)]

Clearly this is the context into which *repair* is inserted in (6), but not in (8), *mayor* being (Human) rather than (Nonhuman) and (Animate) instead of (Inanimate).

The semantic portion of a lexical entry, then, consists of one or more readings, each composed of semantic markers and perhaps semantic distinguishers, as well as a statement of the selectional restrictions for each reading.

Two points concerning selectional restrictions are of interest here. First, we promised a fuller definition of semantic markers and distinguishers. We can now say that semantic markers are expressions of those atomic concepts that appear in the selection restrictions of at least one lexical item, and that semantic distinguishers are those which do not. This is a more formal way of opposing the systematic value of markers to the idiosyncratic value of distinguishers. Second, it is worth noting that violations of selection restrictions, although semantically ill-formed, are often quite acceptable when manifested in utterances. Numerous metaphors and jokes turn on such semantic ill-formedness.

To sum up our presentation of the lexicon, we can recall that it is an unordered set of lexical entries, each representing a different lexical formative. Each entry is itself a set and contains syntactic, phonological, and semantic features. The following chart may prove helpful in learning the organization of a lexical entry:

Syntactic Features	Phonological Features	Semantic Features
Inherent Features	$F_{1_1} + F_{1_2} + F_{1_3} \ldots +$	Lexical Features
Category Features	$F_{1_n} + F_{2_1} + F_{2_2} +$	Semantic Markers
Non-Category	$F_{2_3} \ldots + F_{m_n}$	Semantic
Features	where $F_{x_y} = +$ or $-$	Distinguishers
Contextual Features	value, on the part of	Contextual Features
(= Strict	the xth set of distinctive	(= Selection
Subcategorization	features, for the yth	Restrictions)
Features)	member of the	
	ordered set of	
	distinctive features	

Associated with the set of lexical entries in the lexicon is a set of *lexical insertion rules*. The mechanism of lexical insertion is not of prime importance for our purposes. However, the function of lexical insertion rules is necessary to understand. These rules function to insert lexical entries, in their entirety, into preterminal P-markers, thereby replacing the dummy symbols Δ. An entry is inserted only if the structure into which it is to be introduced meets the requirements of the item's strict subcategorization feature. Lexical insertion, however, disregards selectional restrictions. A string such as (7) or (8), then, would be grammatically well formed; its semantic ill-formedness would be recognized by the semantic component of the grammar.

P-markers that have undergone lexical insertion are called *deep structures*. A representation is given of the deep structure of the string *the + butcher + Past + hit + the + dog*, in simplified form:

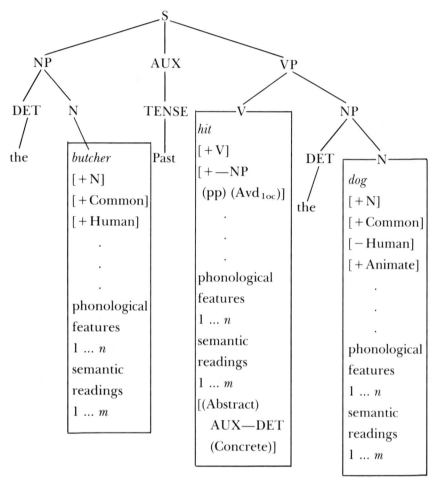

At this point it is well to observe that we do *not* have, in the above deep structure representation, a picture of either the sound sequence or the meaning structure of the sentence *the butcher hit the dog*. This follows from the nature of the syntactic component. The function of syntax is to generate an infinite set of abstract structures relating meaningful elements. However, syntax is mute. It does not generate *interpretations* of its output configurations, nor does it specify actual sound sequences to symbolize the sentences it produces.

We have described the base subcomponent of the syntactic component of a generative grammar according to the standard theory. In order to complete the account of this component, the transformational subcomponent remains to be treated.

The transformational subcomponent, then, is a set of *transformational rules*, (or simply transformations). In order to make sense of this question-begging definition, we need to define the notion transformational rule. This is a key concept of TG. In discussing the adequacy of Type 0–3 grammars as vehicles for the formal representation of natural language competence, we concluded that Type 1 grammars could do the job *with outside help*. This assistance is provided by transformations. Why is there a need for help anyway, and what role do transformations play in providing it?

One of the problems with phrase structure grammars such as the one included in the base subcomponent of a generative grammar, concerns groups of sentences such as:

(9a)   the girl is reading the book
(9b)   is the girl reading the book
(9c)   the girl is not reading the book
(9d)   what the girl is doing is reading the book
(9e)   the book is being read by the girl

The linguistic intuition of a speaker of English, which is based on his or her competence, includes the knowledge that (9a)–(9e) are closely related. Yet the sentences, in many cases, are assigned radically different structural descriptions by a surface grammar. There would seem to be no principled way to account for this aspect of competence by means of a phrase structure grammar alone.

A phrase structure grammar encounters even more difficulty when confronted with sentences such as

(10a)   Morty is eager to please
(10b)   Morty is easy to please

Although it may take him or her some reflection to call it to consciousness, a speaker of English clearly knows that, in (10a), Morty is the one doing the pleasing and some unspecified person other than he is to be pleased. We might say that *Morty* is the *logical subject* of the verb *please*, and that *someone* is its *logical object*. In (10b), on the other hand, someone besides Morty will be doing the pleasing, and Morty is to be pleased. That is, *Morty* is now the logical object of *please* and *someone* the logical subject. Now despite this difference, which surely must be accounted for in a system of rules representing competence, a phrase structure grammar that attempted to account for competence without outside help would assign (10a) and (10b) identical structural descriptions!

To remedy these and many other shortcomings of the base sub-component considered as a total theory of syntax, we resort to the transformational subcomponent. Transformations are operations defined on P-markers. A given transformation consists of one or more of the following component processes: adjunction, substitution, deletion, and permutation. For instance, both adjunction and permutation of elements take place in the indirect *object transformation*, which takes as input a terminal P-marker such as:

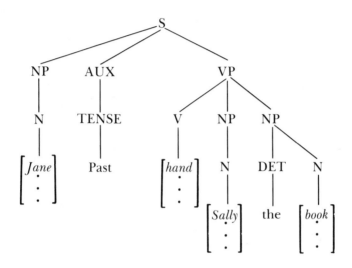

and produce as output a *derived P-marker* (i.e., one which has undergone one or more transformations) such as:

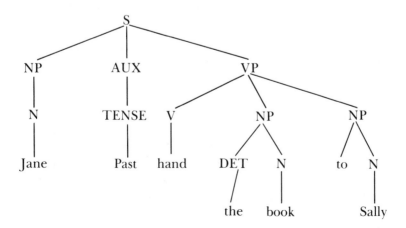

(N.B. This diagram is highly simplified. From now on, feature composition of lexical items will not be indicated unless necessary to the exposition.) Notice that the first and second NP nodes dominated by VP have been permuted, and that the formative *to* has been added to what is now the second NP. As an example of deletion of elements, consider the imperative transformation in English. We assume without discussion that the deep structure of the command *Hurry!* is:

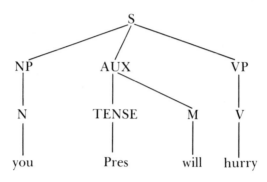

(Note the new rule rewriting AUX as TENSE (M), where M = Modal [*will, can*, etc.].) The imperative transformation deletes the NP and Aux nodes, leaving only:

A transformation featuring substitution of elements is the reflexivization transformation in English, which substitutes the appropriate reflexive element for an NP which is *coreferential* with a preceding one (i.e., refers to the same thing) in a particular context. It takes a structure like:

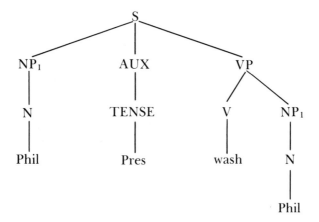

and turns it into one like:

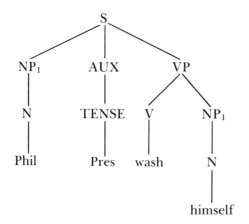

The form of a transformation is

structural description (SD) ⇒ structural change (SC)

The SD is also referred to as the structural index (SI). The symbol ⇒ is to be differentiated from the one →, the first symbolizing transformational, the second phrase-structure, rules. For example, the indirect object transformation illustrated above might be written as:

$$NP^1 + AUX + V + NP^2 + NP^3 \Rightarrow 1 + 2 + 3 + 5 + to + 4$$
$$ 1 2 3 4 5$$

The right-hand expression is the structural change, and the left-side one the structural description or structural index. A comparison with the P-marker represented above for this operation reveals that a transformation rearranges nodes and all that they dominate. That is, the entire structure of one NP was permuted with its preceding NP. At the same time, our transformation reshaped the NP dominating [*Sally*] by adjoining to it a new element (*to*), and in this manner it created a structure never generated by any phrase-structure rule of English grammar. The imperative transformation looks like this:

$$\text{you} + \text{TENSE} + \text{will} + \text{VP} \Rightarrow \emptyset \; \emptyset \; \emptyset \; 4$$
$$\quad 1 \qquad\quad 2 \qquad\quad 3 \qquad\; 4$$

As we saw, this means, delete 1–3 and leave only 4. By a convention known as *tree-pruning*, nodes dominating no terminal symbols are erased from a P-marker which has undergone a transformation. This applies to the NP and AUX nodes of the structure referred to just above. This is why the output tree looks the way it does. The reflexivization transformation might be expressed as:

$$\text{NP} + \text{AUX} + \text{V} + \text{X} + \text{NP} \Rightarrow 1 + 2 + 3 + 4 + \begin{bmatrix} 5 \\ + \text{reflexive} \end{bmatrix}$$
$$\;\; 1 \qquad\; 2 \qquad\; 3 \qquad 4 \qquad 5$$
$$\text{Condition: } 1 = 5$$

Here the syntactic feature [+ reflexive] has been added to the second NP of the SD. Element 4 above is a *variable symbol* which refers to whatever occurs between elements 3 and 5; X can be null, as is the case in the example we show above (*Phil* + Pres + *wash* + *Phil* ⇒ *Phil* + Pres + *wash* + *himself*). Conditions such as the one above can be put on transformations.

The distinction is made between *obligatory* and *optional* transformations. The former are necessary to preserve grammaticality, whereas the latter merely express stylistic options available to the user of the grammar. Not to apply the (obligatory) reflexivization-T to a P-marker meeting its SD would be to generate a nonsentence, for example, *Phil washes Phil. (We assume the identity of the two Phils.) On the other hand, it is just as grammatical to say, *Jane handed Sally the book*, as to say, *Jane handed the book to Sally*. For this reason, the indirect object transformation is optional.

Transformations are subject to two kinds of ordering. There is *intrinsic ordering* and *extrinsic ordering*. The transformational subcomponent of a grammar is actually an ordered set of transformations.

Those rules that apply in a given relative order, because the earlier rule creates the context for the later one, are intrinsically ordered with respect to each other. On the other hand, transformations whose relative sequencing in the grammar must be specified in order to ensure the generation of well-formed strings are said to be extrinsically ordered. For instance, the negative-insertion transformation in English inserts the element *not* into a P-marker of a string such as *Paul was a very young man*, to yield *Paul was not a very young man*. The contraction T in English turns a structure like *Paul was not a very young man* into *Paul wasn't a very young man*. So these two Ts are intrinsically ordered, contraction following negative insertion of necessity. However, given a deep structure underlying a string like *you will wash you*, if we are to be able to generate the correct imperative form, *wash yourself*, then the reflexivization T must first apply to the string to give, you will wash yourself, and only then can the imperative T come along and delete *you* and *will* to produce *wash yourself*. These two Ts, then, are extrinsically ordered in a descriptively adequate grammar of English.

In addition to occurring in a fixed sequence, T-rules are applied *cyclically* to P-markers. To understand this, we must first note that the set of phrase structure rules of the base operates *recursively* in the sense that it contains rules which rewrite some nonterminal A as another nonterminal sequence XBY, and other rules rewriting B as VAW, such that an infinite sequence A...B...A...B...A... is implied. More particularly, the starting symbol S (Sentence) is recursively generated. For this reason, the base produces structures of the form:

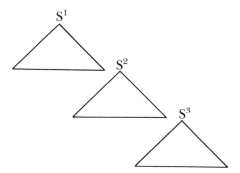

where S nodes are *embedded* in structures dominated by higher S nodes. (The symbol △ is a device for representing all the nodes dominated by a higher node, where the actual content of these is immaterial to a discussion.) Cyclic application of T-rules, then, can be explained with the aid of this diagram:

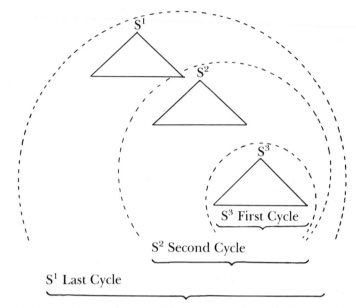

S^3 First Cycle

S^2 Second Cycle

S^1 Last Cycle

The set R of transformational rules—some optional, others obligatory—is applied, in extrinsically specified order, to the innermost S and all it dominates. Then the rules are applied in the same sequence to the next lowest S and all *it* dominates—*including the lower S that it dominates*; finally, the series of rules applies to the highest S and everything *it* dominates, which is to say, the entire structure. Of course, if there are more than three S structures, the process applies as many times as there are S nodes.

At this point we will state explicitly what has been implied throughout our discussion of T-rules: transformations operate not only on deep structures, but also on the derived P-markers that are produced when a T-rule has applied to a P-marker of the base. In other words, a base P-marker is operated on by a T, thereby yielding a first derived P-marker, which is in its turn the input to a second transformation, producing a second derived P-marker, and so on, until the final derived P-marker is produced. This last structure is designated the *surface structure* of the sentence represented.

We have now completed our description of the syntactic component of a TG grammar from the viewpoint of the standard theory. In the briefest terms, we will touch on the role of the *phonological component* of the grammar.

The phonological and semantic components are the two *interpretive components* of a grammar, as we have seen. They process the output of the syntactic, or *generative component* of the grammar. Specifically, the semantic component interprets the deep structures generated by the syntax, whereas the phonological component interprets the surface structures. First a series of rules convert grammatical formatives into sound sequences. At this point, each surface-structure formative has a sound representation in the form of a concatenation of distinctive feature values. A linearly ordered sequence of phonological rules, some context-sensitive rewrite rules, others transformational in format, apply, each to the output of the previous one wherever the rule's SD is met. Some of these rules apply cyclically. What emerges, once all the phonological rules have run their course, is a *phonetic representation* of the sentence generated by the syntactic component. This representation is presumed to form input into the performance device which ultimately results in the production, on the part of the speaker, of an acoustic event associated in his or her mind, and perhaps in the minds of one or more listeners, with the meaning of the sentence corresponding to the abstract sound sequence he or she represents to himself or herself.

We now turn from the phonological to the *semantic component*. The latter, as just mentioned, interprets deep structures and yields *semantic representations* of them. Given a deep structure generated by the syntactic component, the semantic component derives the set of *semantically well-formed* semantic representations, or *readings*, of the particular deep structure. The manner of operation of the semantic component is straightforward. Working "from the bottom up," a set of *projection rules* combine, or amalgamate, the lexical readings (i.e., semantic features only) of the formatives of a deep structure. First the readings of those items which occupy *daughter nodes* of the lowest nonterminals are amalgamated. (Nodes B and C are daughters of node A in the first illustration below; in general, nodes B–N are daughters of node A when A immediately dominates B–N.)

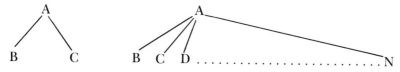

Because a typical lexical entry has multiple readings, it is obvious that even one node with, say, three daughters would yield a considerable number of *derived readings* (amalgamations of individual items' readings) if projection rules proceeded without check. This is not the case, however, because an amalgamation is *blocked* whenever the selection restrictions of a lexical item involved in the proposed amalgamation are not met. For example, the NP *yelping dog* would receive the interpretation, *screaming canine*, but not *screaming meat sausage*. Once the lowest-lying nodes have been assigned derived readings, the new readings of *these* nodes are combined with those of *their* sisters (i.e., B– N in the illustration above), and the process continues ever upward in this manner until the top S node is reached. At this point what has been produced is a set of derived readings for the sentence as a unit. A deep structure associated with a complete semantic interpretation in this way is called a *semantically interpreted underlying phrase marker* (SIUPM). One possible way of construing SIUPMs formally would be as a deep structure each of whose nodes is assigned what we might call a semantic interpretation index, with each index representing a set of simple or derived readings linked with its node.

We can illustrate the process of semantic interpretation by tracing a portion of the operation on this deep structure (cf. Katz and Fodor, 1963):

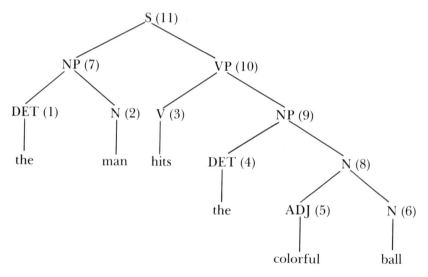

The readings occurring in the lexical entry of the adjective *colorful* are:

Reading 1. (color) [abounding in contrast or variety of bright colors]
Reading 2. (evaluative) [having distinctive character, vividness, or
                        picturesqueness]

Relevant selectional restrictions are:

Reading 1. [—— (physical object) or (social activity)]
Reading 2. [—— (aesthetic object) or (social activity)]

The noun *ball* has the readings:

Reading 1. (social activity) (large) (assembly) [for the purpose of
social dancing]
Reading 2. (physical object) [having globular shape]
Reading 3. (physical object) [solid missile for projection by engine
of war]

Of the six possible *amalgams* (derived readings) of (1) and (2), two
are blocked by the selectional restrictions. The remaining four, which
constitute the derived readings for node 8, are:

Derived Reading 1_8: (social activity) (large) (assembly) (color)
[abounding in contrast or variety of bright
colors] [for the purpose of social dancing]
Derived Reading 2_8: (physical object) (color) [abounding in
contrast or variety of bright colors] [having
globular shape]
Derived Reading 3_8: (physical object) (color) [abounding in
contrast or variety of bright colors] [solid
missile for projection by engine of war]
Derived Reading 4_8: (social activity) (large) (assembly)
(evaluative) [having distinctive character,
vividness or picturesqueness]

Amalgams become progressively large until the highest S node is
reached. Notice that certain semantic phenomena we would expect to
be accounted for in a theory of competence are formally represented
here. For instance, the process of *disambiguation* on the basis of context
has its counterpart in the blocking of combinations not in conformity
with selection restrictions. A collocation whose interpretation is completely blocked by these restrictions would be one which is *semantically
anomalous*, or *meaningless*. Nodes with identical derived readings are
paraphrases of each other, and those with *n* derived readings are, in
isolation, *n-ways ambiguous*.

The full semantic interpretation of a sentence—that is to say, the
output of the semantic component for a given S—consists in the SIUPM
for the particular S, and in a set of statements concerning conclusions
as to semantic anomaly, paraphrases, and such, which can be formally
obtained from the SIUPM. Such an interpretation represents our
semantic competence in its interaction with the given sentence.

Having completed our account of the standard theory, we present
a revised chart to illustrate in some detail the plan of a generative
grammar according to standard theory:

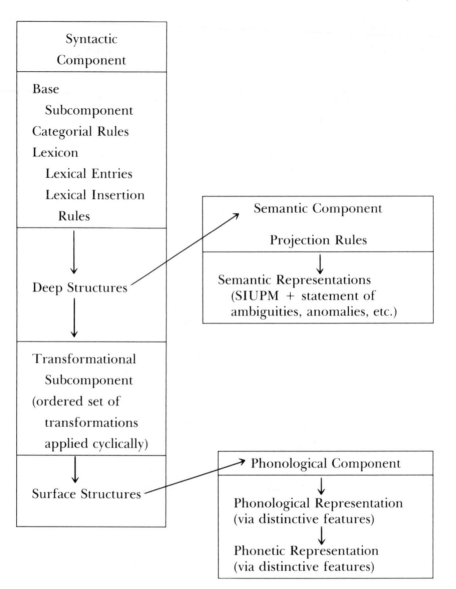

Extension of the Standard Theory

The period from the mid-1960s to the mid-1970s saw challenges issued to the standard theory (ST), in the form of the development of *generative semantics* and of *case grammar*, and witnessed Chomsky's reply and reformulations through a remolded form of the ST which he has called the *extended standard theory* (Chomsky, 1972).

Generative semantics (cf. J. D. Fodor, 1974; Frantz, 1974; Lakoff, 1970a, 1970b; Lakoff and Ross, 1976; McCawley, 1968a,b; Postal, 1972) was in some sense a working-out of ideas implicit in the ST. The latter accepted what has come to be known as the *Katz-Postal hypothesis*, namely the claim that all semantic interpretation in a TG grammar takes place on the basis of the deep structures, or, to put it another way, that no transformations change the meaning of the deep structures to which they apply. Pursuing the logic of KP, generative semanticists argued that every aspect of semantic competence must be based on a specific portion of a deep structure and thus that deep structure must be enriched to permit such interpretation. Deep structures at first became more and more like semantic representations; it was found necessary to merge previously distinct grammatical categories; and the lexical items, originally inserted under terminal nodes of a P-marker to form deep structures, now often corresponded only to groups of nodes dispersed throughout the P-marker. Finally, the level of deep structure disappeared altogether from the generative semanticist account of grammar; arguments were adduced to the effect that certain transformations must precede the insertion of particular lexical items, and that grammatical functions such as subject and object are best defined at the semantic level as are subcategorization features. In this way, the argument continued, the *raison d'être* of the "deep structure" concept was eliminated, and the semantic and syntactic components were but one.

Although the precise locus or loci of lexical insertion was never agreed upon by generative semanticists, they would have supported statements such as the following, which will afford us some feel for this school. Given a base-generated (i.e., semantically generated) structure such as:

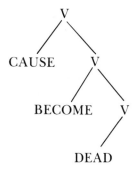

depending upon the particular transformations that are applied, lexical insertion results in "kill," "cause to die," or "cause to become dead."

Another hallmark of the generative semantic approach is the *global rule*, or *global derivational constraint*. Faced with pairs of sentences whose meaning was different and which seemed to be directly related by transformation, GS linguists posited for the two sentences different underlying representations and then had to ensure that the first such underlying structure was not transformed into the second. In order to achieve this result, they had to introduce global derivational constraints, rules with the power to inspect entire derivations and rule out as ungrammatical derivations of which two or more P-marker members stood in some stipulated relation to each other. In contrast, transformations, as interpreted within standard theory, operate only on adjacent P-markers. Global rules were ultimately held to perform the most disparate functions. Yet, powerful as global rules were, constraints even more potent ultimately found their way into the GS arsenal. As GS progressed, its proponents began increasingly to argue that *any speaker judgment* constitutes a grammatical judgment. Social assumptions peculiar to the speaker's culture, the results of deductive reasoning, personal opinions—all these were considered to feed into grammatical competence. Now such factors as these had long been accepted as impinging upon performance, but their admission *en masse* into competence was unprecedented. In order to formalize the contribution of these elements to a grammatical derivation, the so-called transderivational constraints were resorted to. Such rules inspected and related two derivations simultaneously.

Before treating Chomsky's reaction to GS and his and others' development of the extended standard theory, we will briefly consider *case grammar*, the second of the three contemporaneous schools mentioned initially. Associated most closely with C. Fillmore (1966, 1968, 1969) case grammar shares much with GS. Arguing from certain difficulties experienced by the ST in dealing with constructions that simultaneously are exponents of the *grammatical category*, prepositional phrase, and fulfill the *grammatical function* of adverbial of time, location, instrument, and so on, Fillmore considered that such structures occur at the underlying level as NPs to which are attached nodes specifying *grammatical case*. Consistency then requires all NPs to be so specified. The NP and its sister case are dominated, in Fillmore's scheme, by a node whose label (called an *actant* in Fillmore, 1966) specifies the function of the NP/case pair. The base component of a case grammar rewrites sentence, S, as $V + X$, where X is an unordered set of actants. Surface structure is created "from scratch" via T-rules. Fillmore argued for a universal hierarchy of cases such that if two or more were represented in a given sentence, their order of precedence was specified. For instance, given the underlying structure:

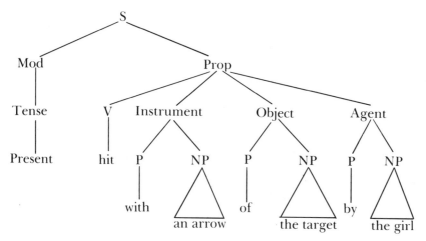

where PS rules are: S → Mod Prop, Mod → Tense, Tense → $\left\{\begin{array}{l}\text{Past}\\\text{Present}\end{array}\right\}$, Prop → V (Location) (Time) (Object) (Instrument) (Source)

(Agent), Location (or—Time, Object, etc.) → P NP, and S = Sentence, Mod = Modal, Prop = Proposition, P = Preposition, NP = Noun Phrase; T-rules yield the string *the + girl + hits + the + target + with + an + arrow*. This is because part of the case hierarchy is the sequence Agent, Object, Instrument.

Work in case grammar continues to the present although Fillmore's own views are relatively far from the schema just presented. During the period under discussion, the school was attacked both by the GS scholars and by proponents of the extended standard theory; the former claiming that case grammar dealt in units which had been shown actually to be complex structures, the latter pointing to the confusion introduced into the syntax by Fillmore's case hierarchy.

The two main differentia of the *extended standard theory* (EST) vis-à-vis the ST are the former's adoption of what has been called the *lexicalist hypothesis* and partial rejection of the Katz-Postal hypothesis. Before we can fruitfully discuss the lexicalist hypothesis, we will have to expand upon certain key characteristics of the GS model. The general tack of GS was progressively to *enrich the syntactic component* of a ST-type grammar as semantic factors were discovered which had no (or inadequate) explanation within the semantic interpretive component of

current ST grammar. For instance, G. Lakoff (1970a) considered the
sentences:

(11a)   John considers Mary's feelings
(11b)   John is considerate of Mary's feelings

In this context, *John considers* seems to mean the same as *John is considerate*,
just as *considers Mary's feelings* and *is considerate of Mary's feelings* do. What
is more, *considers* and *is considerate of* would seem to have the same
selectional restrictions. There was no vehicle within the semantic component of ST grammar to express this aspect of semantic competence.
Therefore, Lakoff proposed to capture the phenomenon syntactically.
Specifically, he argued, *consider* and *considerate* are substantially the
same syntactic object. Further, the deep structures of (11a) and (11b)
are identical except for the presence of a feature [± Adj]. Because the
feature is present in the underlying P-marker of (11b), *is* and *-ate* are
transformationally generated by rules activated by the presence of the
feature [+Adj]. Contrariwise, *of* is deleted (for it is claimed to be
underlyingly present) via a T-rule sensitive to the minus value of the
[Adj] feature. In a similar vein, Bach (1968) conflates verbs, adjectives
and predicate nominals (i.e., lumberjack in *Joe is a lumberjack*). Based
on sentences such as those reflected in (12)–(15):

(12)   The attempt was a failure
(13)   The attempt failed
(14)   Joe is a brute to Jane
(15)   Joe is brutal to Jane

Here again, the similarities between (12) and (13) and between (14) and
(15) are evident; the differences can be handled via the transformational
component—different features trigger different Ts. Bach therefore
subsumed verbs, adjectives, and predicate nominals under the category
verb.

Now it is precisely such collapsing of syntactic categories and
accounting for differences through transformations on the part of GS
that spurred Chomsky to revise ST via the lexicalist hypothesis. Chomsky
first presented his lexicalist hypothesis in "Remarks on Nominalization"
(Jacobs and Rosenbaum, 1970). The thrust of his argumentation here
is that crucial syntactic generalizations are missed by a grammar without
a level of deep structure, and further that the facts point to the
conclusion that deep structure, far from being more abstract than that

in the ST, as the generative semanticists alleged, is actually more shallow than he had assumed in the standard theory.

In particular, Chomsky puts forward three principal arguments, in "Remarks on nominalization," in defense of deep structure. All concern what he calls derived nominals, such as *eagerness* in *John's eagerness to please*, *refusal* in *John's refusal of the offer*, and *criticism* in *John's criticism of the book*. His tripartite argument aims at establishing that such entities are to be considered lexical items, not the end products of transformations applied to closely related verbs and adjectives, such as *eager, refuse*, and *criticize*, respectively.

First is the argument from *productivity*. Although transformations are devices for stating the regular processes of a language, the relation between derived nominals and their putative transformational sources is highly irregular. Large numbers of eligible verbs and adjectives offer no "transformationally related" noun, and similar quantities of nouns point to no possible verbal or adjectival source in the language. Moreover, those verb/noun or adjective/noun pairs which *do* seem related transformationally often differ radically in meaning—and in a great variety of ways—so that the claim that Ts preserve meaning is either abandoned or reduced to triviality via a multitude of *ad hoc* conditions on transformations. So it seems that derived nominals do not come about through transformational processes.

The second argument contrasts derived nominals with *gerundive nominals* (e.g., *eagerness* in *John's eagerness to please* as opposed to *being eager* in *John's being eager to please* or *certainty* in *John's certainty that Bill will win the prize* in contradistinction to *being certain* in *John's being certain that Bill will win the prize*). In a wide variety of ways, derived nominal constructions behave like NPs, whereas the admittedly transformationally derived gerundive nominal constructions do have the character of their transformational sources—sentences. Like NPs, derived nominals pluralize (*John's proof of the theorem/John's proofs of the theorem*), take a Determiner (*the proof of the theorem*), and feature prenominal adjectives (*John's unmotivated criticism of the book*). The same is not true of gerundive nominals (**John's provings* [of] *the theorem*, etc.). Like sentences, gerundive nominals can contain aspect, for example, *John's having criticized the book*. Not so with derived nominals—**John's having criticism of the book*. So derived nominals have the characteristics of lexical nouns, not those of outputs of Ts.

The third and final argument is that derived nominals occur in sentences corresponding to (pretransformational) deep structures, and not in those corresponding to structures which have undergone Ts. On the transformational account of derived nominals, there is no good reason why *lexical transformations* should occur *en bloc* before all the others; but if derived nominals are simply nouns entered in the lexicon, then it is natural that we do not have sets such as *John believed that Mary was ill/John believed Mary to be ill, John's belief that Mary was ill/John's belief of Mary to be ill.* In the first pair, the second expression is derived from the first via the Subject-to-Object-Raising T, but in the second pair the structural index of the T is not met because the second term is a noun and not a verb, so the T is not (grammatically) carried out. There seems to be no explanation for this fact on the transformational account of derived nominals.

In view of arguments of the above types, Chomsky advanced the lexicalist hypothesis, to the effect that lexical items can be entered in the lexicon without specification as to part of speech; in one section of a listing appear all the features common to, say, a verb and its related noun. Then a second portion of the listing is a branching into two or more lexical categories with the specification for each branch featuring the characteristics peculiar to the item *qua* noun or *qua* verb, as the case may be. For instance, a lexical entry we may label *destr* branches into *destroy* and *destruction*, with common features listed under *destr* and idiosyncratic ones under *destroy* and *destruction*, respectively.

Further, in order to deal with similar cooccurrence restrictions on lexical items within Sentence/NP pairs involving derived nominals, (e.g., the matching restrictions on *proved* and *proofs*, respectively, in *John proved the theorem* and *several of John's proofs of the theorem*), Chomsky introduced the *X-bar convention*. This is a new system of representing the rules of the base, supplementing and in some areas supplanting the previous system. This can be presented briefly as:

1. $\bar{X} \to X \ldots$
2. $\bar{\bar{X}} \to [\text{Spec } \bar{X}] \ \bar{X}$
3. $S \to N \ V$

Where X = a lexical category (i.e., N, A, or V); $\bar{X}$ is rewritten as X plus, optionally, any of a system of *complements* of X (constructions related to X and following it); $\bar{X}$ is called the *head* of X... and $\bar{\bar{X}}$ rewrites as the *specifier* of $\bar{X}$, followed by $\bar{X}$. Specifier is defined as Det for $\bar{N}$, Aux for $\bar{V}$, and, for $\bar{A}$, as a system of qualifiers of adjectives including *very* and other formatives. A consideration of these two sets of definitions make it clear that S can be rewritten as $\bar{\bar{N}} \ \bar{\bar{V}}$ (i.e., more or less as NP VP). Note that 1 and 2 above are not rules, but *rule schemata*, because X = N, V or A; $\bar{X}$ = $\bar{N}$, $\bar{V}$, or $\bar{A}$; etc. At one fell swoop Chomsky introduced considerable uniformity, then, into the major sets of base rules via X-bar notation. More important, these schemata are considered descriptive of a *universal base*, that is, they are language universal, so that if the X-bar convention is valid, it expresses a valuable insight into the formal peculiarities of natural languages.

It is by means of the X-bar convention that Chomsky is enabled to explain the common cooccurrence restrictions on related lexical items in S/NP pairs. To return to the example cited above (*John proved the theorem* versus *several of John's proofs of the theorem*), Chomsky provides this pair of tree diagrams:

(A)

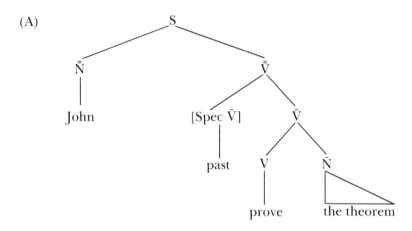

(John proved the theorem)

(B)

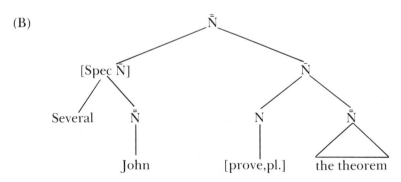

(Several of John's proofs of the theorem)

He observes that, on the one hand, the set of strict subcategorization features for the (prebranching) lexical item *prove* contains one feature for the domain $\bar{N}$ (as in B) and one for the domain $\bar{V}$ (as in A); this explains the differences between A and B. On the other hand, the selectional restrictions refer to the heads of phrases, so that A and B are identical from this point of view. (In both cases, *John* and *theorem* are the head nouns in construction with *prove*.) This move was telling against GS, because one of the major supports for a transformational treatment of derived nominals was the need to explain phenomena such as these. Finally, because lexicalism obviates the need for a host of transformations, EST can be said to posit "shallower" deep structures than GS.

We have mentioned that EST was a modification of ST in two separate domains, both changes being inimical to GS. If the lexicalist hypothesis was the generative semanticists' Scylla, then, their Charybdis was the partial rejection within EST of the Katz-Postal hypothesis. Specifically, EST allowed that grammatical functions (also referred to as *grammatical relations*) be specified at the deep structure (DS) level, but argued that, in a considerable variety of other cases, meaning has to be interpreted on the basis of surface structures (SS).

Of the many arguments advanced for this position we will cite two. Jackendoff (1969) argued that (16) means that *many* arrows are such that they *did not* hit the target, whereas (17) means that *not many* arrows are such that they *did* hit the target. Therefore, the Passive T either changes meaning or else its description must be considerably complicated to prevent the derivation of (17) and (16):

(16) Many arrows did not hit the target
(17) The target was not hit by many arrows
(18) Not much shrapnel hit the soldier
(19) Much shrapnel did not hit the soldier

A similar argument applies to (18) and (19), where the latter corresponds in interpretation to (16), the former to (17). Jackendoff's conclusion is that the surface order of quantifier and negative determines the meaning of sentences such as these, so that semantic intepretation can be completed only after the application of the Passive T.

Chomsky (1972) pointed to "natural answer" relationships between (20) on one hand and (21)–(23) on the other:

(20) Is John certain to *win*?
(21) No, he is certain to *lose*
(22) No, he's likely not to be *nominated*
(23) No, the election won't even *happen*

The italicized words correspond to the *intonation centers* of their respective sentences. They are also the *focused* (questioned) elements, whereas the rest of each sentence constitutes its *presupposition* (i.e., what is assumed already known). Because the phrases in focus are phenomena of surface structure (note that describing natural-answer relationships involves establishing partitions of the sentences that could never correspond to deep-structure), it is concluded that part of the semantic interpretation of such structures as these takes place at the level of SS. Schematically, we can juxtapose ST and EST as follows:

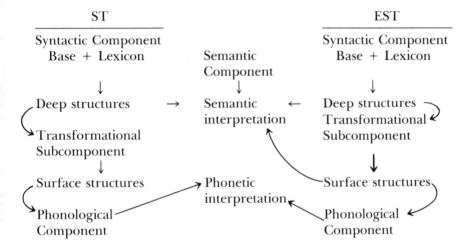

Revised Extended Standard Theory (REST)

Early in the 1970s a major concern of linguists within the TG linguistics camp was the question of the power of transformational grammars. Studies produced by Peters and Ritchie (1969, 1971, 1973) and others claimed that transformational grammars had the power of unrestricted Turing machines and, therefore, postulations within TG were *unfalsifiable*, the grammars were *unlearnable*, and so forth. Many linguists had already started defining constraints on grammars and on transformational rules in order to harness the excessive power of grammars. Ross (1967), in a Ph.D. dissertation, was one of the first to propose formal constraints on rules. Chomsky (1973), among others, produced other sets of constraints. This trend continued through the mid-1970s when a major reorganization of the form of a TG grammar was induced by the introduction of the *trace theory* of movement. The basis of this theory is the claim that when an NP is moved via transformation, it leaves a *trace* (t) at the locus from which it was displaced, and the trace is indexed to the removed NP. In *Reflections on Language*, Chomsky (1975) stated that traces can be considered as logical variables. Thus, an element moved by, for example, the *wh*-movement rule is considered a logical quantifier which must bind its trace/variable. From this assumption it follows that the interpretation of surface structures into the so-called logical forms is possible; that is, given a sentence such as (24), its logical form (25) can be established directly:

> (24)   John thinks that Mary had told him who$_i$ you were going to hire t$_i$
>
> (25)   John thinks that Mary had told him for which person x, you were going to hire x

Fiengo (1974, 1977) treats traces roughly as pronouns and contributes significant additional interpretive roles to the surface structure. In fact in REST practically all semantic interpretation is derived from the surface structure. However, what is here being called surface structure is substantially different from that of the previous theories. In the meantime, the systems of filters and constraints on grammars have reduced transformations to their simplest forms: Move α, where α is some constitutent. Fiengo (1980, p. 21) presents surface structure (SS) as standing at the nexus of four rule systems:

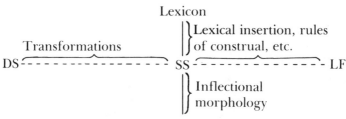

Where DS = deep structure, and LF = logical form.

Government and Binding: GB Linguistics

The diagram representation of EST and its revisions, REST, given at the end of the last section, although consistent with what we have presented so far in this chapter, are not complete in that they do not represent some of the components and subcomponents that we have not discussed; for example, S-structures (an intermediate structure to be distinguished from surface structures), case-marking rules, case filters, deletion rules, conditions on deletions, indexing, binding, and others. Discussions of these will take us far afield from the scope of this study. However, Chomsky (1982) summarizes the "recent versions of EST" as a rule system consisting of three basic parts:

(26a) Lexicon
(26b) Syntax: (i) Base component
           (ii) Transformational component
(26c) Interpretive component: (i) PF component
                            (ii) LF component

The output of the transformational component is S-structure which undergoes case-marking rules and case filters and results in a case-marked S-structure. The PF (phonetic form) component maps S-structure to a surface structure in phonetic representation, and LF component maps S-structure to a representation in logical form which we have already mentioned.

What is now called GB (government and binding) linguistics represents a significant shift in focus from the rule-based system in (26) to the study of a collection of "subsystems of principles." Chomsky proposes (27): "a second perspective from which one can view grammatical processes focuses on principles that hold of rules and representations of various sorts" (Chomsky, 1982, p. 6).

(27a) X-bar theory
(27b) θ-theory
(27c) Case theory
(27d) Binding theory
(27e) Bounding theory
(27f) Control theory
(27g) Government theory

Details aside, these subsystems account for a universal grammar and a set of parameters which can be set in terms of data presented to a person acquiring a particular language. To quote Chomsky (1982, p. 7):

> These systems of principles interact in a variety of ways, and certain relations also obtain between the subtheories of [27] and the subsystems of rules in

[26]. A fully articulated theory of universal grammar (UG) will develop the
properties of these systems and the relations that hold among them.

The grammar of a language is then a set of values for the above
parameters, "while the overall system of rules, principles, and parameters is UG, which we may take to be one element of human biological
endowment, namely, the 'language faculty'."

One of the results of the "significant shift in focus" is the contention
that the base rules such as those in (28) need not be specified because
they can be predicted by the X-bar theory and other subsystems of
principles in (27).

(28a) S → NP INFL VP
(28b) VP → V NP S′

Thus, these rules of the base phrase structure grammar can be derived
from "deeper principles and the specification of parameters that
constitutes the actual grammar."

Concluding Remarks

In this chapter we have followed the development of TG linguistics
from its origins to its most recent embodiment as GB theory. Generative
linguistics is today far from a unified discipline. Among participating
linguists, there are many varieties of transformationalists as well as
practitioners of case, dependency, systemic, functional, and other
theories of grammar. Some of these approaches have been mentioned
or discussed above. Further details concerning these theories will be
presented in subsequent chapters as they become relevant to our study.

Exercises

1. Explain the difference between "pedagogical grammar" and grammar defined in this chapter. Pedagogical grammar is in the sense
   of the following statement: John knows his grammar well; he rarely
   makes errors in his compositions.

2. Give some examples to differentiate between competence and
   performance.

3. Can you think of another analogy other than arithmetic to demonstrate competence versus performance?

4. Discuss the difference between "grammaticality" and "acceptability." Give examples.

5. Describe the role and function of John in the following two sentences:

     (a) John is easy to please
     (b) John is eager to please

   How does the transformational theory help to explain the role of John in these two sentences with identical surface structures?

6. The following sentence is ambiguous:

     Eating apples can be delightful

   Describe the ambiguity in two ways: (a) write a paraphrase for each interpretation; (b) use terms such as "subject," "object," "modifier," "verb," etc. to explain the relationship of *eating* to *apples* for each interpretation.

7. What are the components of a grammar in formal languages, ST, EST, REST, and GB?

8. Early in the chapter we have given a verb phrase rule as VP→V (NP), indicating a verb optionally followed by a noun phrase (e.g. [a] go! [b] leave the room). However this rule does not account for VP in many other sentences, for example:

     You should leave the problem to me
     John left the town to avoid his father
     I would never show a photo of John to Mary
     John saw the man who is the president
     I have bought the book that you mentioned
     I have bought the book you mentioned
     John denied that Mary wrote the paper

   Add additional components to the above VP rule to account for these sentences. (*Hint*: Verb phrases can contain prepositional phrases and sentences with or without complementizers.)

9. As far as you can, draw a phrase marker for any one of the sentences in Exercise 8 together with all lexical features (consult Chomsky [1965] and Radford [1981], if needed).

10. Discuss the following operations and terms: lexical insertion rules, base rules, syntactic features, phonological features, semantic features, deep structure, surface structure, S-structure, and lexical features.

Topics for Research and Term Papers

1. The X-bar theory plays a significant role in EST and is also a major subsystem of GB linguistics. Give some discussion of this theory and as an elementary exercise try to represent the underlying structure of the sentences in Exercise 8 in X-bar convention. (consult Jackendoff [1977] and Radford [1981]).
2. One of the reasons that a simple syntax-directed translation schema cannot be developed for translating English to another natural language is that there is diversity in syntactic rules of different natural languages. Try to develop a partial SDT schema for translating English sentences to another language of your choice. See how far it goes, and explain the problems that arise in this effort.
3. Try to write a context-free grammar in Chomsky Normal Form for as rich a subset of the English language as you can, and then write a Cocke-Younger-Kasami parser for the language generated by this grammar.

References

John Lyons (1968) provides a comprehensive survey of modern linguistics. Other good sources are Robins (1964) and Robins (1967). For foundational expositions on TG linguistics, see Chomsky (1957, 1965) and Bach (1974). Radford (1981) is a comprehensive and readable textbook on EST with a large number of exercises and additional references. Jackendoff (1977) provides a treatise for the X-bar theory. There is much current interest and research in GB linguistics. For details and understanding of implications of the theory, see Chomsky (1981, 1982). For a survey of semantic theories in TG, see J. D. Fodor (1978).

CHAPTER 3

Psycholinguistics

Introduction

Psycholinguistics is that branch of cognitive psychology which seeks answers to the questions: In what form is linguistic knowledge represented in the brain? How is this knowledge used in the processes of speech comprehension and production? What is the normal course of development in the organism of this knowledge and of the ability to use it? Much of the psycholinguistic research of the past fifteen to twenty years—although far from all of it—has been carried on within the framework of assumption and research interests called the *information-processing paradigm*, in which explanations of human mental processes are modelled, with varying degrees of explicitness, on the operations of computers (cf. Lachman *et al.*, 1979). The present discussion will introduce those topics within psycholinguistics and touch on related issues within cognitive psychology, which are most pertinent to the concerns of this study.

A central concern of psycholinguistics has been the comprehension of spoken language. This activity is subdivided into speech perception and comprehension processing. The problem of speech perception is to specify the *code*, or system of representation, into which a listener translates an incoming stream of speech sounds, and to indicate the manner in which this translation is carried out. This code is assumed to constitute the input to the mechanism of comprehension processing, which yields a representation in some form of *semantic code*. The objects of semantic code can interact with other objects stored in memory, to result in inferences and in additions to one's store of knowledge. Whereas, originally, comprehension of spoken language was considered to proceed by discrete steps or stages, more recent theorizing has stressed the overlap and interpenetration of the various constituent processes.

Normally, listeners perceive sound segments of their language *noncontinuously*, or *categorically*, which is to say in an all-or-nothing manner. Although, in fact, there is an infinite range of sounds from, say, the *k* of *kite* to the *g* of *Guy*, English speakers perceive instances of one sound or the other as they process speech, rather than perceiving one of a possible range of intermediate sounds. There has been controversy as to whether categorical perception marks speech perception as fundamentally different from processes involved in perceiving other types of auditory stimuli. Although some scientists (Liberman, 1970) have claimed that, in fact, categorical perception is unique to speech perception, others (Cutting and Rosner, 1974; Pastore *et al.*, 1976) have pointed to phenomena of categorical perception in non-speech auditory perception as well as in visual perception.

The *analysis-by-synthesis* model of speech perception posits an active role for the listener's nonphonological knowledge in the perception of speech. Because input to a speech recognition device is typically highly degraded (e.g., vowels that are not stressed lose many of their distinctive characteristics and approach the neutral vowel *schwa*, the last sound, for instance, of *sofa*; sounds are often run together in rapid speech), mere analysis of input into sound sequences could not suffice as an account of our ability to capture speech. The analysis-by-synthesis model, therefore, claims that a preliminary analysis of the input speech flow is made, then fed into a second component which has access to "higher-level" nonauditory knowledge. The latter component synthesizes a final version of the original input in such a way as to reflect the best hypothesis as to the form of the input, based on a variety of formal considerations. Stevens and House (1972), for instance, mention the contribution to synthesis of constraints on the possible sequences of speech sounds which can form a word (*morpheme structure constraints*). Others have provided experimental evidence for the participation of the full gamut of comprehension processing domains in the synthesis of speech (Martin, 1972; Warren, 1970; Warren and Sherman, 1974).

Because of this integration of speech perception and comprehension processing, it is somewhat artificial to discuss the two processes separately. However, for purposes of study or discussion the two domains are often considered separate. In his review of studies in sentence perception, Levelt (1978) distinguishes two distinct tendencies among recent researchers in the field of sentence perception from the point of view of posited *end points* of comprehension processing. One school of thought, the linguistic camp, views the end-products of comprehension as structural descriptions, more or less in the sense of TG linguistics. A second group of researchers describe the final forms of comprehension processing as nonlinguistic objects referred to as conceptual structures. Levelt is quick to point out that a wide variety of viewpoints have been expressed intermediate to the above two. For example, some have seen structural descriptions as being derived before conceptual structures. Finally, Levelt indicates that those who favor the conceptual structure outlook (the conceptual school in Levelt's terms) tend to include a wider range of mental processes in their accounts of comprehension of spoken language. Linguistic and nonlinguistic context, as well as encyclopedic world-knowledge, is given a prominent place by conceptualists, whereas those of the linguistic school use such concepts to a more limited degree in their theory construction. Another difference among researchers, comments Levelt, is the extent to which conversational context and the concept of meaning as intended by the speaker are integrated into a given theory. Levelt indicates that the task of *communicative* research is one receiving growing attention among psycholinguists.

Having obtained a general idea of the current state of research into comprehension processing, we can backtrack and consider the separate stages of sentence comprehension. Certain stages have been proposed and generally rejected subsequently; other stages are controversial. None is universally acknowledged. Recall that we are discussing *interactive* processes, presented as separate modules for the purposes of exposition.

The key questions, then, in studying the comprehension process, are: What is the code in which the end stage of the comprehension process is represented? How can we characterize the process of using the code to obtain individual objects expressed in the code?

If we consider that the input to the comprehension process is the output of the speech perception process, we can logically begin our discussion of comprehension with the process of lexical access. Note that cognitive operations will be described employing the "computer analogy" to human mental processes which informs modern cognitive psychology. The assumption of some sort of *mental lexicon* is uncontroversial—as a minimum, this must include a large number of sound/meaning pairings corresponding to words, or perhaps to *morphemes* (syntactic elements such as *un-*, *faith*, and *-ful*, in *unfaithful*, or such as *gentle*, *man*, and *-ly* in *gentlemanly*). Many researchers would agree that information equivalent to the syntactic category features, strict subcategorization, and selection restrictional features of theoretical linguistics must also be stored with each lexical entry. Further proposals concerning the contents of lexical entries will be mentioned below.

Lexical Processing

Lexical lookup (access) is the retrieval of information from the mental lexicon during comprehension processing. A considerable body of experimental evidence exists for the claim that a word's* frequency of occurrence in the language to which it belongs is an indicator of its ease of retrieval from the mental lexicon.† For instance, Foss (1969) used the technique of *phoneme monitoring* originally developed by him and by Lynch (Foss and Lynch, 1969) to examine the effects of word frequency in a sentence context. The subjects' task was to comprehend each of sixty sentences presented orally and, in addition, to indicate immediately the occurrence of the first /b/ sound of the sentence. Responses were registered by means of a button that subjects were to depress as soon as they heard /b/. A timer measured the interval between the initial occurrence of /b/ (in the so-called target word) and the pushing of the button. Pairs of sentences (borrowed from Foss and Hakes, 1978, p. 105) such as

(1) The travelling *b*assoon player found himself without funds in a strange town
(2) The itinerant *b*assoon player found himself without funds in a strange town

constituted the test sentences. Note that the difference between (1) and (2) is that the relatively low-frequency *itinerant* has been substituted for the more high-frequency *travelling*; moreover, the two terms are roughly synonymous, and the other words of the sentence have been held constant. In interpreting the *reaction times* (RT) displayed by his subjects, Foss posited a fixed-capacity decision mechanism able to devote a greater proportion of time to the task of phoneme-monitoring when it

* In the immediately following discussion, we use the term *word* as if it were clear that items are stored in the mental lexicon as words. This is, however, a highly controversial assumption; many researchers would point to experimental evidence supporting the claim that we store lexical items as morphemes others opt for larger units.
† Such evidence is to be found early in Solomon and Howes (1951), and in Howes and Solomon (1951). Critiques of this work are Eriksen (1963) and Erdelyi (1974). Additional support for the frequency/ease-of-retrieval contention is in Brown and Rubenstein (1961).

is less taxed by a concurrent sentence-processing operation. Specifically, he predicted that the processing difficulty introduced by the presence in sentences such as (2) of relatively low-frequency words would lead to longer RTs than those of their counterparts such as (1). In fact, Foss's results were significant and of the predicted character (see, however, Levelt [1978] for a critical discussion). Such evidence is consistent with theories of lexical access which portray the mental lexicon as grouping high-frequency items in such a manner that they are tested before other items for conformity to phonological input (Forster and Bednall, 1976).

However, theories of this stamp have been criticized for failing to indicate how phonological information gathered during speech perception interacts with lexical search. In line with this position are theories which have the lexicon organized on phonological principles; a virtue of these explanations is that they account for phenomena such as the mistaken retrieval of a word for a similar sounding one. On the other hand, those advocating the importance of frequency measures would claim that a phonologically organized lexicon fails to capture the effects of frequency.

Still another approach to lexical access explains the frequency effect as a consequence of a more general principle. The more recently we have heard (or, in some versions, spoken, read, or thought) a word, the more readily available it is to us for reuse, according to this theoretical orientation. The fact that frequent words are easy of access would be due to our tendency as individuals to employ frequently the most common terms of our language; hence such items are among those which, at a given moment, we have used most recently.*

The *logogen* theory of Morton (1979) has been claimed to encompass all of the phenomena described above. In Morton's words,

> a logogen . . . is the device which makes a word available as a response and it does so by collecting evidence that the word is present as a stimulus, appropriate as a response, or both. . . . The evidence for the presence of a particular word can come from the outside world, by vision or hearing, or from other processes in the brain such as those concerned with context, which we can globally term the Cognitive System. . . . When a logogen has collected sufficient evidence, the appropriate response is made available. The amount of evidence necessary for this is called the *threshold* of the logogen. Those who like metaphors based on energy can imagine inputs to the logogen system increasing the levels of activation in particular logogens; when the level exceeds a certain amount in any logogen the response is triggered. (pp. 112–113)

---

* Early exponents of this view were Howes (1954) and Daston (1957). More recently, see Scarborough *et al.* (1977).

Both frequency and recency effects can be accounted for within this framework by hypothesizing that a logogen's threshold is permanently reduced by a small amount each time its corresponding word is spoken, seen, heard, written, or thought (cf. Morton, 1979, pp. 136ff).

Sentence Processing

Input synthesis is the name given to the ensemble of processes involved in the comprehension of sentences. These include speech recognition and dictionary lookup, and most scholars would agree that some form of syntactic analysis is involved as well. Some have questioned, however, the assumption that detailed syntactic analysis is a necessary part of the processing of all sentences. Slobin (1966) had 6- and 7-year-old children listen to a sentence and indicate by depressing a button whether or not a picture presented immediately after the sentence correctly represented the situation described by the sentence. He measured the verification times, that is, the amount of time it took each subject to accomplish each of the tasks presented, and hypothesized that length of verification time was directly proportional to difficulty of comprehension processing for a given sentence. His feelings were that sentences in the active voice were more rapidly comprehended than passive voice analogues, with one quite significant exception—the so-called *nonreversible passives*. Sentences which fail to "make sense" when cast in the active voice (e.g., *The boy was stung by the hornet*, but *The boy stung the hornet*), showed no difference in verification time from their corresponding actives (here, *The hornet stung the boy*). Walker *et al.* (1968) conducted a probe latency study in which adult subjects were required to determine whether a certain word (the "probe word") had occurred in a given sentence. Again, latencies were longer for passive than active sentences in the reversible cases (e.g., *John liked Bill, Bill liked John*) but no significant difference in probe latency was found for nonreversible cases (see also Herriot, 1969). The foregoing results have suggested to some that syntactic structure is computed in the course of sentence comprehension only where more than one noun of the sentence can potentially serve as the logical subject. Otherwise the comprehension process is telescoped by the omission of any detailed syntactic analysis. That is, recognition of the presence of the set {*stung, boy, hornet*} would be sufficient for understanding.

Many researchers have seriously questioned the validity of picture verification tasks on the grounds that there is no way to determine how much "verification time" is devoted to the activity of evaluating the picture rather than the sentence. More generally, a certain amount of doubt has been cast on the reliability of results of sentence-comprehension experiments based on successive measurement, the measurement of responses after the stimulus has been presented and processed.* (Contrast simultaneous measurement procedures where measurement coincides with stimulus presentation.) Moreover, the results of Forster and Olbrei (1973), who had subjects judge as quickly as possible after stimulus presentation whether a given sequence of words constituted a sentence, were that both reversible and nonreversible passives had longer latencies than their respective actives. The authors of the study concluded that their findings failed to confirm claims that syntactic processing can be short-circuited under certain conditions. The details of this controversy aside, there is general agreement with Levelt's (1978) contention that syntax plays an important role in sentence understanding. But in what form does syntax influence sentence processing, and which nonsyntactic factors play roles?

The basic question of input synthesis, then, is, What are the structures that the listener constructs mentally during sentence comprehension? Modern psycholinguistics began with a research program aimed at investigating the "psychological reality" of the key structures of TG grammar—phrase markers, surface structure, deep structure—and of its notion of the transformation. Early results were consistent with a comprehension model closely based on grammatical operations of the kind spelled out by TG linguists.

Three early experiments took *degree of grammaticality* as defined within theoretical linguistics as an independent variable, and established its significant effect on sentence perception against a background of noise (Miller and Isard, 1963), the ability of subjects to repeat sentences presented to them (Epstein, 1961), and subjects' performance on sentence-memorization tasks (Marks and Miller, 1964).

The research program being addressed by such experiments was presented to psychologists by Miller (1962a). In his paper, traditionally considered to mark the inauguration of modern psycholinguistics, Miller first gave a summary presentation of Chomsky's early approach to language and then asked

> Are these systems of rules nothing more than a convenient way to summarize
> linguistic data or do they have some relevance for the psychological processes
> involved? If human speech is a skilled act whose component parts are related
> to one another in the general manner that the linguists have been describing,

---

* Contra-verification time as a measure of comprehension, see Tanenhaus *et al.* (1976), Gough (1966), Forster and Olbrei (1973). A critique of successive measurement procedures is to be found in Levelt (1978).

what measurable consequences can we expect to find? What measurable effects would such skills have on our psychological processes? (p. 750)

The remainder of Miller's discussion is largely devoted to providing an extended answer to these questions. He first considers whether one can identify solid empirical evidence for the psychological reality of syntactic categories. Citing Ervin (1961), Miller notes the observation of psychologists working in the field of word association that the syntactic category of a response in such experiments had tended to match that of its stimulus word. Moving to slightly larger syntactic constituents, he points to experimental evidence that triplets consisting of an English grammatical formative (words such as *of, with, to, by, but*) flanked by nonsense syllables (e.g., *tah-of-zum) are more easily memorized than pairs composed of a nonsense syllable followed by a grammatical formative (e.g., tah-of*). His explanation is that the triplets simulate concrete instances of "natural syntactic constituents" and hence are more easily remembered.

Continuing his effort to persuade his readers that the psychological problems posed by grammatical concepts are well defined, Miller advances to sentence level and argues for the essential role of syntactic analysis in sentence comprehension, based on the syntactically ambiguous sentence, *They are eating apples.* In order to identify the appropriate meaning of the sentence in a given situation, we must use contextual clues as to the intended syntactic structure. He cites as evidence "for the existence of perceptual units larger than a single word" the results of an experiment of his own (Miller, 1962b). These results indicate that under time pressure subjects are better able to recall words presented in sentential context than in nonsentential context. As a final piece of evidence concerning the crucial role of syntactic structure in sentence perception, Miller cites another of his experiments which demonstrated the relative ease of recall and of repetition of the so-called right-branching structures as opposed to self-embedded structures.*

Miller notes that typically a subject gradually apprehended the structure of a center-embedded sentence over two or three repetitions of it. Summarizing the role of syntactic form in sentence comprehension, Miller states that

---

* A sentence featuring an extensive *right-branching* structure, and cited by Miller, is *This is the cow with the crumpled horn that tossed the dog that worried the cat that killed the rat that ate the malt that lay in the house that Jack built.* Recasting the same sentence as a *center-embedded* structure, we get *The rat that the cat that the dog that the cow tossed worried killed ate the malt that lay in the house that Jack built.* For further discussion of these terms, see Chomsky's earliest writings as listed in the discussion of TG linguistics in Chapter 2. Miller uses *right-recursive* instead of *right-branching*; the two terms are equivalent here.

just as we induce a three-dimensional space underlying the two-dimensional pattern on the retina, so we must induce a syntactic structure underlying the linear string of sounds in a sentence. And just as the student of space perception must have a good understanding of projective geometry, so a student of psycholinguistics must have a good understanding of grammar. (p. 756)

In support of the psychological reality of transformations, Miller refers to preliminary work by K. McKean, D. Slobin, and himself which indicated the existence of a relationship between the number of transformations TG theory posits as separating two sentences and the amount of time necessary for subjects to match a pair of transformationally related sentences. In addition, he offers an interpretation of experimental work of Jacques Mehler's concerning recall of transformationally related sentences. Miller proposes a quantitative accounting of Mehler's data, based on the hypothesis that listeners recode sentences they hear in such a way as to represent both the original phrase-structure form of the sentence and each of the transformations it has undergone. The fit of Miller's predictions with Mehler's data is quite close.

Concluding his paper, Miller looks forward to the development of a research program exploring "verbal systems" as a means of gaining insight into the structure and workings of the human mind.

In light of Miller's programmatic statements, we can briefly discuss the early psycholinguistic studies referred to above. Epstein's (1961) experiment actually predated Miller's paper. His work provided early evidence of the psychological reality of syntax. His findings were that subjects more efficiently recalled sequences of nonsense words which have been provided with English "grammatical endings" (*inflectional and derivational* indices) and grouped into "sentences" than sequences without such additional features. For instance, a sequence like, *A haky deebs reciled the dison tofently um flutest pay* were better remembered than one such as *haky deeb um flut recile pav tofent dison* even though the latter series of elements is shorter than the first. Factors which had been used as independent variables in such inquiries in the past were controlled. All the "words" had zero frequency in English, meant nothing, and could not be claimed to be associated with other "words" in the sequence as a result of subjects' previous experience with them.

Marks and Miller (1964) presented subjects with four different types of sentential or sentencelike sequences of English words. The first type consisted of actual sentences, such as *Fatal accidents deter careful drivers*, *Noisy parties wake sleeping neighbors*, and *Rapid flashes augur violent storms*. The second type was limited to intercalated versions of several of the sentences, such as *Noisy accidents deter sleeping storms*, in which semantic constraints on sentences were violated. A third group of structures were simply scrambled versions of the original sentences— *Wake parties sleeping neighbors noisy*, that violated syntactic conditions on sentences. Finally, both kinds of violation were combined to yield sequences such as *Rapid noisy deter wake careful*. Subjects' recall of word sequences of Types 1–4 became decreasingly rapid with each step downward in type. The Miller and Isard (1963) experiment was similar, except that only Types 1–3 were used, and that they tested for recognition of utterances against a noisy background instead of for recall. The two studies taken together were interpreted as evidence that the notion of "degree of grammaticality" as defined by TG linguistics could successfully serve as an independent variable in the psycholinguistic laboratory.

Once the relevance of linguistic concepts to psychological processes had been demonstrated to the satisfaction of a significant number of scientists, more specific hypotheses began to be tested. Fodor and Bever (1965) used a technique which had been developed by Ladefoged and Broadbent (1960), called *click localization*, to test the theory that listeners code the sentences they hear, in the form of *surface structure* representations, at least at some point in comprehension processing. Because high-level constituents of an S, such as NP and VP, would presumably constitute the units of perception in this case, Fodor and Bever predicted that the possibility of attending to external stimuli would be greatest when a subject had just finished computing one major syntactic unit and had not yet begun to compute the next one; conversely, the least attention would be available for such stimuli when the subject was in the process of computing a major unit. In other words, they expected attention to be at its height between the processing of NP and VP or between the processing of two clauses. Accordingly, they presented to the subject a short burst of noise, a "click," at selected points in time while the subject was simultaneously hearing a sentence. In click studies, the sentence is often presented in one ear while the click is sounded in the other ear. The instructions supplied each subject were to listen to the sentence and to the click, then to write down, the very moment the listening experience had concluded, both the sentence and the exact point within the sentence at which the click had occurred. What Fodor and Bever (1965) found in their experiment was that clicks were not only perceived most accurately when they occurred at clause and other major syntactic boundaries, but clicks actually having been sounded at points other than boundary points were perceived to have occurred nearer to these boundaries than was the case. In other words, major syntactic boundaries "attracted" clicks. It is to be noted that the significance of these findings was somewhat diluted by the tendency of subjects to report clicks earlier than they had actually occurred under all conditions and not only under the influence of syntactic boundaries. Other click studies will be discussed later.

Additional studies that pursued the tack of demonstrating the psychological reality of TG constructs were that of Clifton and Odom (1966), which showed that subjects perceive sentences to be dissimilar in proportion to their transformational remoteness from each other, and that of Gough (1965), which revealed that sentences are less rapidly evaluated for truth and falsity (in an experimental setting such as Gough's) the greater the number of transformations linguists would analyze them as having undergone.

The theory of sentence comprehension implicitly or explicitly assumed by the authors of the foregoing studies was labelled (and vigorously attacked) by J. A. Fodor, T. Bever, and M. Garrett, in a series of articles (Bever, 1970; Fodor and Garrett, 1966, 1967; Fodor, Garrett, and Bever, 1968). The theory which they called the *derivational theory of complexity* (DTC) is as follows: A mental representation of the surface structure of an incoming utterance is constructed by the listener. On this basis one or more "reverse transformations" (operations that simply have the inverse effect, in the direction surface-to-deep structure, of transformations proceeding in the opposite direction) are triggered and applied, resulting in a deep structure representation. (Particular word sequences were held to trigger reverse transformations, e.g., NP^1 was V by NP^2—*the boy was hit by the car*—sets off an operation which yields the prior sequence NP^2 V NP^1—*the car hit the boy*.) Each reverse transformation is assumed to take time to apply. When all appropriate reverse transformations have run their course, the input utterance is represented in an underlying form analogous to the deep structure of TG linguistics. It follows from this account of sentence processing that the more transformations required to process a sentence, the longer it takes to comprehend it. This was the prediction that was most widely tested, and the one which Fodor, Bever, and Garrett attacked most sharply.

Among other criticisms, the above authors pointed out that sentence length had been confounded with transformational relationships in a large number of studies claiming to demonstrate the role of transformational history in sentence-processing tasks, that is, the corresponding passive sentence of, say, *Joe hit the ball*, is not only, *ex hypothesi*, more complex, but also simply longer than the original sentence. Counter-evidence to a number of key experimentally based claims for DTC was adduced by Fodor, Bever, and Garrett. Capitalizing on Miller's (1956) generally accepted claim that we possess a short-term memory store which can hold seven items (give or take two) at a time, Savin and Perchonock (1965) had performed an experiment which indicated that the residual space in short-term memory decreased as the transformational complexity of a sentence being kept there increased, other things being equal. Their subjects were able to memorize fewer words while storing a sentence that is transformationally relatively complex than while storing a transformationally simpler sentence. In Fodor and Garrett (1966), however, an experiment by Fodor, Bever, Garrett, and Mehler is reported in which the particle-movement transformation is applied to sentences such as *Mary picked up the handkerchief*, resulting in transforms such as *Mary picked the handkerchief up*, but failing to lead to a decrease in the number of words which subjects could concurrently store while remembering the new sentence. Because the particle-movement transformation did not seem to behave according to the predictions of the DTC, this experiment constituted ammunition to be used against the theory. A similar, later experiment by Fodor and Garrett (1967) had similar results. Further, it was argued, the DTC seems implausible *prima facie*—transformations often delete elements of underlying structure and otherwise reduce complexity. (For instance, *the dirty cup* was considered to be transformationally derived from *the cup which is dirty*, *it amazed Bill that John left the party angrily* from *that John left the party angrily amazed Bill*.) It therefore stands to reason that the first and third constructions cited just above, and others like them, are not harder to comprehend than their underlying counterparts. Finally, some had taken results of certain click studies to indicate that, contrary to the assumption, necessary to the DTC, that surface structure is computed as an early step in sentence processing, the only boundaries to which clicks are attracted are sentence boundaries, that is, those at underlying, not surface, structure. This would mean that some other experimental evidence would have to be adduced in support of the idea that surface structure is recovered during sentence comprehension. Many of the arguments put forth by Fodor, Bever, and Garrett against the DTC have since been questioned, as have the experimental results which partially underlie their contentions. We will briefly return to some of these objections below. However, the fact is that psycholinguists in general have turned away from the DTC in light of the arguments of Fodor, Bever and Garrett or of others.

For their part, Fodor, Bever, and Garrett, in rejecting the DTC, reasoned that a weaker but seemingly more plausible connection between sentence processing and grammatical theory would be that both assign to a given string similar underlying structural descriptions (more or less similar deep structures). That is, even if the comprehension process involves recovering deep structure, it does not necessarily involve doing so by means of reverse transformations. Instead of transformations, a range of *comprehension strategies* was posited by the authors cited above to account for the construction of deep structures from input utterances. One or more strategies interact with a given input to yield a hypothesis as to its deep structure, which is then tested against the input for fit. If the first guess is rejected, other strategies are called upon to produce new hypotheses as to the input's deep structure. Finally a match is decided upon, or, in some cases, the input is simply not comprehended.

A basic principle of the *strategy model*, as it has been called, is the *verb complexity hypothesis* (VCH). It predicts that comprehension is slower for sentences featuring verbs with richer possibilities of subcategorization, than for sentences with less versatile verbs. For instance, contrast *Betty met Sam* with *Betty knew Sam* and with *Betty knew that Sam was a New Yorker*. The verb *know* can appear in more kinds of sentential structure than can the verb *meet*. (Note the impossibility of *Betty met that . . .* ; *meet* clearly cannot take, as a direct object, a sentence preceded by *that*.) When a verb is accessed from the lexicon, a specification of the syntactic configurations into which it can enter is accessed along with it. The greater the number of possible structures into which a given verb can be introduced, the greater the number of possible interpretations it implies, the longer it will take to process a sentence featuring the verb. Fodor, Garrett, and Bever (1968) conducted an experiment one of whose aims was to test the VCH. Given a set of sentences, each of which featured "complex" main verbs (i.e., verbs with relatively many subcategorization possibilities), and a parallel set differing from the first only in that, for each complex verb, a "simple" verb (one with relatively few subcategorization possibilities) has been substituted, subjects were able to paraphrase the sentences of the second set more accurately (to a statistically significant degree) and more quickly (to a statistically nonsignificant degree). Moreover, it was discovered that unscrambling "scrambled sentences" from the second set was easier for subjects than performing the same task on sentences of the first set. All results were interpreted as supporting the VCH. Holmes and Forster (1972) later found that verb complexity makes a slight difference in the average number of words recalled from a sentence briefly presented to subjects when the latter are questioned immediately after such presentation. Again, the authors' conclusion was that the VCH was supported by their results.

Alongside the above experiments which are claimed to support the VCH, we find five studies conducted by Hakes (1971) which cast doubt on the theory. Hakes obtained mixed results on a paraphrase task; sometimes sentences with complex main verbs were more difficult to paraphrase than corresponding simple-verb sentences, but sometimes not. More damning were the outcomes of a group of phoneme-monitoring tasks included in Hakes's studies—here latencies showed no effect at all from verb complexity. The unequivocal testimony of the phoneme-monitoring results has been considered especially revealing by some, because, although paraphrase is a successive measurement technique (cf. Levelt, 1978) and hence perhaps unreliable as an index of what is going on in the listener's mind during the perception process, phoneme-monitoring is a simultaneous measurement task which properly controlled would seem more likely to provide insights into comprehension processes (however, see Levelt's [1978] critique).

Although experimental evidence for the VCH is less than compelling, other aspects of the strategy model have fared better. The essence of the strategy model is the acceptance of the proposal that sentence comprehension involves the recovery of deep structure but the rejection of any role for (reverse) transformations in the comprehension process. The deep structure representation is achieved through a process of perception which operates directly on the incoming sequence of words (or morphemes) and takes the form of a set of perceptual mapping rules (Bever, 1970) according to which hypotheses are set up for the deep structures of an input string, tested against the string, and accepted or rejected based on such tests. In an early study whose aim was to test one proposed rule, Fodor and Garrett (1967) pointed out that the relative pronoun *that* indicates important aspects of the deep structural organization of a sentence of the form *The man that the woman saw is my brother*. Namely, it warns the listener of the presence of an embedded sentence whose subject will be the next NP and whose object will be the preceding NP (i.e., *the woman saw the man*). A version of the same sentence, but without the relative pronoun—"The man the woman saw is my brother"—does not permit the use of a perceptual strategy which helps construct the deep structure based on the clue offered by the presence of *that*; hence it should be more difficult to comprehend. This is what their results would seem to indicate: sentences with the relative pronoun were easier for subjects to paraphrase than those without them. Fodor and Garrett achieved the additional end of casting further doubt on claims for the psychological reality of transformations, while making their own results more convincing by controlling for the presence of transformations as such. (It could have been objected to

Fodor and Garrett that their results were due simply to the intervention of an additional transformation in the pronounless sentences, one which deletes pronouns in preclausal position.) What they did was to add adjectives to the pronounless sentences (e.g., *The tall young man the nice old woman saw was my brother*) and to obtain paraphrases of these new sentences. The introduction of the adjectives involved transformations so that if the mere presence of transformations were the relevant variable such paraphrases would be more difficult than paraphrases of the original pronounless sentences. In fact, these new sentences were not more difficult. The overall results were taken as confirmation of the existence of a perceptual rule that uses relative pronouns as clues to deep structural organization. Hakes and Foss (1970) confirmed Fodor and Garrett's results using the phoneme-monitoring technique; later, Hakes (1972) reconfirmed them using both phoneme-monitoring and paraphrasing (see also Hakes and Cairns, 1970).

As the strategy model has developed, many individual strategies have been presented and argued for most notably by Bever and his associates. Versions of the strategy model have been formalized (Kaplan, 1972, 1975); critics of the model consider such formalization necessary before the approach can be evaluated in any meaningful way.

More recently, Bever and others have given the strategy model a functional orientation. The notion of *functional clause* was developed by Tanenhaus and Carroll (1975) based on evidence that the clause as a perceptual unit could not be identified with the deep-structure clause of linguistic analysis. Units of perceptual segmentation, called functional clauses, are defined with reference to factors such as clause length, clause type (main, subordinate, etc.), certain surface syntactical features, and functional completeness. (A clause is functionally complete to the extent that it overtly expresses a full set of grammatical relations; contrast the functionally incomplete *After reading the book, I discussed it with Joe* with the functionally complete *After I had read the book, I discussed it with Joe*.) Commenting on the work of Tanenhaus and Carroll, Carroll and Bever (1976) state that

> their research with the click-location technique suggests that functional completeness and not deep and surface grammatical structure predicts the suitability of a linguistic sequence as a segmentation unit. And this, in turn, suggests that purely structural theories of segmentation make the correct predictions only when they are in accord with functional theories. (p. 338)

In summing up the state of the art of sentence comprehension research in their article, Carroll and Bever (1976) firmly place the study of perceptual processes at the heart of the subject matter with the analysis of rule systems which imply linguistic structures corresponding to those processable by listeners playing a decidedly ancillary role.

> Treating the sentence as a perceptual object is like starting the study of three-dimensional visual perception with the perception of simple geometric forms. The same distinction arises between the formalization of the mapping between two dimensions and three and the actual visual processes that perceivers use to recognize three-dimensional objects. To take the grammatical rules as perceptual processes would be like taking the laws of perspective as direct models of psychological processes; to claim that the different levels of linguistic structure are perceived objects would be like claiming that a two-dimensional structure is analyzed independently of the three-dimensional object it represents. In both cases, the manifest stimulus is analyzed only insofar as the analysis supports a usable interpretation. Miller and his colleagues demonstrated that the sentence is a relevant unit in speech perception. Since there is an arbitrarily large number of sentences this required a theory that would describe the set of *possible* sentences: A generative linguistic grammar provided such a theory, and thereby became a hypothesis about the structure of linguistic knowledge. The grammar provides a structural description of each sentence, independent of perception. Such an independent description of what people *know* about a sentence sets the goal of what they must *discover* about a sentence when they comprehend it. This raised the question of how the grammar that describes linguistic knowledge is employed during sentence comprehension. The answer to this question offered by research has evolved in three phases. The first phase was to take the *grammatical transformations* literally as psychological processes. When this proposal broke down, the next phase was to argue that the linguistically defined *structures* are psychologically "real." Further research shows that structures are pertinent behaviorally only as a result of perceptual operations. Accordingly, our current focus must be on structures only insofar as they reveal properties of the psychological processes themselves. (p. 339)

Consistent with the orientation expressed in the preceding quotation is an article by Carroll, Bever, and Pollack (1981) which stresses the manipulable nature of linguistic intuitions—a subset of which constitutes the touchstone of theoretical linguistic constructs—and calls for a theory of the behavioral processes underlying linguistic intuitions.

The burden of the work of Marslen-Wilson (1973, 1975, 1976), who has approached the problem of language comprehension from a perspective quite different from that of Fodor, Bever, Garrett, Carroll, and their associates, has been the contention that phonetic, syntactic, and semantic processing take place concurrently from the beginning of utterance perception and the concomitant denial of the notion, widely entertained by other workers, that syntactic processing takes place only when the elements of a perceptual unit (phrase, clause, etc.) have been gathered in short-term memory. Actually, Marslen-Wilson's subsequent claims were more radical than this. An idea of the evolution of Marslen-Wilson's theorizing over the past five to ten years is provided in the

"Conclusion and Overview" section of an article by Marslen-Wilson, Tyler, and Seidenberg (1978).

Although Marslen-Wilson's work runs in direct contradiction to the bulk of the work of the authors cited above, he notes that the direction taken by Carroll and his associates is consistent with his own. (The comment is made in Marslen-Wilson, Tyler, and Seidenberg [1978] and refers specifically to the article by Carroll, Tanenhaus, and Bever [1978]). It is interesting to observe the similarities between the ideas expressed in the two articles which appear in the same volume. Marslen-Wilson *et al.* label their approach to sentence comprehension the "on-line interactive hypothesis" and state (p. 225) that "the basic assumption of the on-line interactive hypothesis" is "that the listener attempts to fully interpret utterances word-by-word as he hears them." A major independent variable for them is "informational completeness"—the availability at a given point in the processing of a sentence (word-by-word) of the information necessary to set up a semantic interpretation. Carroll *et al.*, for their part, find that two of the experiments they present "suggest, contrary to our initial hypothesis, that listeners use context on-line as they are processing and organizing sentences" (p. 200). The chief independent variable for Carroll *et al.* is "functional completeness: the property of having a complete, coherent, and explicit set of grammatical relations. Functional completeness guarantees the listener a good propositional unit" (p. 211).

Marslen-Wilson *et al.* first point to a series of experiments using the research technique of shadowing, which consists simply in requiring the subject to repeat verbal material presented to him or her, as immediately as he or she can. The results of these experiments, taken as a group, are claimed to indicate that listeners process syntactically and semantically the input utterances in an on-line (immediate) fashion. Among other reasons for this contention is the fact that errors in repetition in these experiments could be linked to semantically and/or syntactically problematic words or expressions immediately preceding the locus of the mistake.

At this point, Marslen-Wilson's hypothesis was that on-line processing is the mode of sentence comprehension, that is, that utterances are understood word-by-word as they are perceived. The listener was conceived of as actively constructing both syntactic and semantic representations of input utterances, using the totality of information derived from the utterance, at any one point, to facilitate construal of subsequent utterance elements.

Because this model of sentence processing was in obvious conflict with the then popular assumption that processing takes place a clause at a time (Marslen-Wilson calls this assumption the *clausal processing hypothesis*) the next wave of experiments sought to establish that in fact semantic interpretation is continuous and is uniformly available at any point in a sentence—not just at clause boundaries. This is what was found. This work also indicated that contextual information conveyed by previous sentences is immediately incorporated into the interpretation of a given sentence.

The next experimental focus was the claim of the clausal processing hypothesis that "syntax is autonomous," that is, that while a clause is being syntactically processed it is immune to the effects of any semantic variables. What Tyler and Marslen-Wilson (1977) did was to present to subjects a syntactically ambiguous phrase, such as *landing planes* or *folding chairs*, which, if construed as the subject of a clause, would require a singular verb for the sake of proper agreement on one syntactic construal and a plural verb on the other construal. The task was to utter as immediately as possible a "probe word" presented visually just after the subject has heard the syntactically ambiguous phrase. The probe word is an appropriate verb, either in singular or plural form. Now the interest of the task is that, immediately prior to hearing the phrase, the subject hears a sentence which offers contextual information that disambiguates the phrase (i.e., makes clear which of the possible construals is intended). Marslen-Wilson and Tyler found that latencies of probe utterance were significantly reduced when the disambiguating contextual information had been presented along with the phrase. For them, this result had far-reaching import.

First, it undermined the thesis of syntactic autonomy because clearly semantic variables had influenced clause-internal syntactic processing. Second, it occasioned a radical revision of their own processing model. Because syntactic and semantic factors seemed to interact on-line, the rationale for the positing of a separate syntactic representation based upon which a semantic account of an utterance is derived, disappears. In a series of articles, Marslen-Wilson and his associates, then, argued against the necessity of including a separate syntactic component in a sentence-processing model.

This, then, is the course of affairs that has led Marslen-Wilson to the following description of an on-line hypothesis:

> The listener is seen as attempting a direct word-by-word mapping of the input onto a single internal representation. In its most developed form, this representation could correspond to what we call comprehending a sentence, and would not be describable in strictly linguistic terms, since it would have to contain the products of inferences drawn from the listener's non-linguistic knowledge. So it would not, in itself, be either a syntactic or a semantic

representation, although for its construction it would draw upon both syntactic and semantic aspects of the input. (Marslen-Wilson, Tyler, and Seidenberg, 1978, p. 242–243)

These conclusions of Marslen-Wilson *et al.*, referring as they do to representations not describable in strictly linguistic terms and to inferences drawn from the listener's nonlinguistic knowledge, lead us naturally to turn from the linguistic school of sentence-processing theorists to the conceptual school of Levelt. However, in order to appreciate the viewpoints of the researchers of the conceptual school, we must briefly consider some key notions of cognitive psychology in general and of the study of human memory in particular.

Cognitive Psychology and Memory

Two groups of authors are considered to have sounded keynotes in the development of modern cognitive psychology: G. Miller, E. Galanter, and K. Pibram on the one hand, and A. Newell and H. Simon on the other. In a ground-breaking book, Miller, Galanter, and Pibram (1960) focused on human intelligence, not as something a given individual has more of than another has, which was how intelligence had been approached previously, but as a quality universally possessed by humans, which involves the setting and solving of problems in an open-ended way and without preset limitations on the character of problems or solutions.

Miller, Galanter, and Pibram introduced the seminal concept of the *plan*, or procedure (often hierarchically organized) for achieving some goal. Human goal-directed behavior is considered to reflect such plans, which can be spelled out in the following flowchart-like manner (adapted from Bower, 1975). Of course, one can choose how fine-

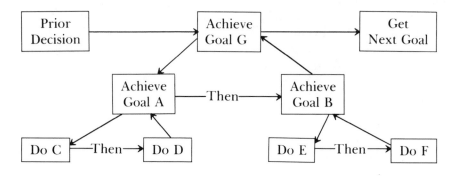

grained a representation one will work with, because clearly a behavior included at the bottom level in one version of a plan (e.g., the behavior of brushing one's teeth) can be analyzed further, in which case subordinate processes would be given a place in a new, more detailed plan.

Newell and Simon first enunciated and disseminated the idea that the Turing machine can be thought of as a *general-purpose symbol manipulator* rather than simply as a number-manipulator and, furthermore, that the human cognitive system and the man-made computing system can be considered to be *systems of symbol-manipulating processes.* Thus arose what has been designated the "computer analogy" to human cognitive operations. A broad spectrum exists among cognitive psychologists as to how close-fitting this analogy is considered to be. Some view the symbol manipulations of computer systems as "psychologically real" models of human cognitive subsystems. Others never explicitly compare the operations of a computer to those of the mind. Most theorists occupy conceptual way stations somewhere between these two extreme points. It is important to keep in mind, in this regard, that no comparison is being drawn at the level of hardware but only at the highly abstract level of symbol manipulation.

What Newell and Simon (1972) have called an *information-processing system* represents events in the external world via symbols and symbol structures; possesses a memory which stores *types* of symbol structures and permits the system's receptors to recognize *tokens* corresponding to these types (i.e., objects standing in a relation of representation to the types); and processes input symbol configurations by means of a set of elementary operations which can be combined in an indefinite number of ways and which can themselves be represented symbolically within the system. A representation of the human perceptual and memory systems from an information-processing point of view, as conceived by Bower (1975, p. 37), is given as an example of an application of Newell and Simon's ideas.

As defined by Newell and Simon (1972, p. 20–21), an *information-processing system* comprises a sensory system, a response generator, a memory, and a central processor. Their "definitions and postulates of an information processing system" are:

1. There is a set of elements, called *symbols.*
2. A *symbol structure* consists of *tokens* (equivalently, *instances* of or *occurrences*) of symbols connected by a set of *relations.*
3. A *memory* is a component of an IPS capable of storing and retaining symbol structures.

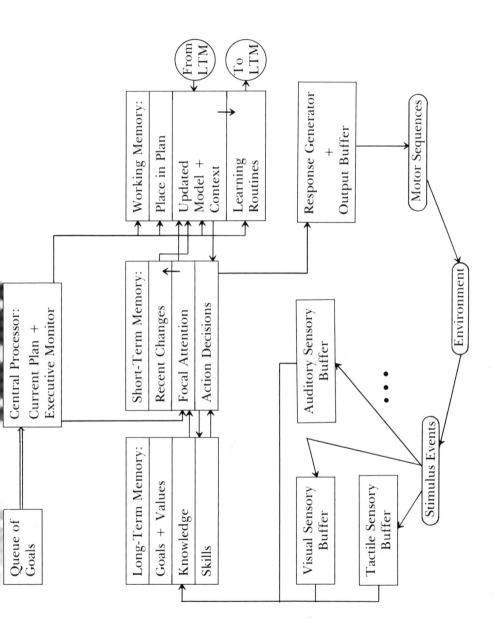

4. An *information process* is a process that has symbol structures for (some of) its inputs or outputs.

5. A *processor* is a component of an IPS consisting of: (a) a (fixed) set of *elementary information processes* (eip's); (b) a *short-term memory* (STM) that holds the input and output structures of the eip's; (c) an *interpreter* that determines the sequence of eip's to be executed by the IPS as a function of the symbol structures in STM.

6. A symbol structure *designates* (equivalently, *references* or *points to*) an object if there exist information processes that admit the structure as input and either (a) affect the object or (b) produce, as output, symbol structures that depend on the object.

7. A symbol structure is a *program* if (a) the object it designates is an information process and (b) the interpreter, if given the program, can execute the designated process (literally this should read, "if given an input that designates the program").

8. A symbol is primitive if its designation (or its creation) is fixed by the elementary information processes or by the external environment of the IPS.

The indefinite term *object* is used in the definitions above to encompass at least three sorts of things:

1. symbol structures stored in one or another of the IPS's memories, which are often usefully classified into (a) data structures and (b) programs (see Item 7 in the preceding list;

2. processes that the IPS is capable of executing; and

3. an external environment of sensible (readable) stimuli. *Reading* consists in creating in memory internal symbol structures that designate external stimuli; *writing* is the inverse operation of creating responses in the external environment that are designated by internal symbol structures.

Newell and Simon use as an illustration an IPS for translating Morse code. Given the stimulus (−,−,−), internal symbol structures are created in short-term memory corresponding to the stimulus. The structure would consist in tokens of three dash symbols connected by the "next" relation. A match of the symbol structure is found in long-term memory associated with the letter *S*. The original structure is now supplanted in STM by *S*; as more structures are read and associated with letters, new symbol structures are formed which will ultimately be recoded as words.

Corresponding to each of our sensory systems (tactile, auditory, visual, etc.) is a preliminary analyzer which interacts with the environment, imposing a considerable degree of structure on incoming stimuli, and transfers an image or *icon* of a given stimulus onto a sensory buffer where it is retained for further analysis for a very brief period. This takes place automatically with some degree of consciousness. Moreover, input preanalysis for the various senses proceeds in parallel. Icons placed on a sensory buffer are subject to feature extraction and, on this basis, to stimulus identification (pattern recognition). The latter process includes recognition of both individual objects and of whole complexes of objects (i.e., in the visual mode *scenes*).

Symbol structures corresponding to those stimuli which are in conscious focus, when activated in memory, are said to be held in *short-term memory*. In contradistinction to long-term memory, which is discussed later, short-term memory may be thought of as an extremely small (about seven-member) group of currently active symbol configurations ordered by recency of activation and in intimate and rapid contact with input-processing operatons. In connection with the limited storage capacity of short-term memory, Miller (1956) showed that this capacity is both quite small and quite constant, but that, within these limits, much variation is possible because symbol structures can be "chunked" in memory. Given a sequence of, for instance, eighteen letters to remember, one can retain them without too much trouble if they happen to spell out six consecutive three-letter words of English, but probably not unless some such "chunking," or organization into familiar larger patterns, can be effected. (Of course, what will be remembered is a sequence of six words, not one of eighteen letters). Items can also be recoded for the purpose of facilitating their retention in short-term memory—stimuli perceived in the visual mode can be recoded into auditory form, words from a foreign language can be translated into one's own language for better retention.

Working memory, or *intermediate-term memory*, (ITM) can be defined as follows: "We may think of the working memory as containing a description of the *setting*, framework, or context within which the more dynamic alterations of the world before us are taking place." (Bower, 1975, p. 162). Working memory has as one of its "primary functions . . . to build up and maintain an internal model of the immediate environment and what has been happening in our world over the past minute or two." Our ephemeral impressions of the passing scene registered in short-term memory (STM), then, are integrated into ITM; "new information 'updates' rather than completely casting aside the current model." ITM shapes our perceptions by relating our fund of general knowledge concerning typical outcomes of processes and events we witness, to our moment-to-moment sensations. The latter are registered in STM, the former repose in LTM (long-term memory).

Long-term memory is a truly impressive store, containing, as it does, our entire fund of background knowledge concerning ourselves, our world, and everything we feel we know or actually do know about the fabric of reality.

In order to continue to prepare ourselves to understand the conceptualist psycholinguistic theories of comprehension, we now look somewhat more closely at some of the principal results of the study of human memory.

Many cognitive psychologists accept the distinction drawn by Tulving (1972) between *episodic* and *semantic memory*. What do episodic and semantic memory have in common? Both are

> information processing systems that (a) selectively receive information from perceptual systems . . . or other cognitive systems, (b) retain various aspects of this information, and (c) upon instructions transmit specific retained information to other systems, including those responsible for translating it into behavior and conscious awareness. (p. 385)

What are the differences between episodic and semantic memory?

> The two systems differ from one another in terms of (a) the nature of stored information, (b) autobiographical versus cognitive reference, (c) conditions and consequences of retrieval, and probably also in terms of (d) their vulnerability to interference resulting in transformation and erasure of stored information, and (e) their dependence upon each other . . . . Episodic memory receives and stores information about temporally dated episodes or events, and temporal-spatial relationships among these events. A perceptual event can be stored in the episodic system solely in terms of its perceptible properties or attributes, and it is always stored in terms of its autobiographical reference to the already existing contents of the episodic memory store. The act of retrieval of information from the episodic memory store, in addition to making the retrieval contents accessible to inspection, also serves as a special type of input memory store. The system is probably quite susceptible to transformation and loss of information. While the specific form in which perceptual information is registered into the episodic memory can at times be strongly influenced by information in semantic memory—we refer to the phenomenon as encoding—it is also possible for the episodic system to operate relatively independently of the semantic system. Semantic memory is the memory necessary for the use of language. It is a mental thesaurus, organized knowledge a person possesses about words and other verbal symbols, their meaning and referents, about relations among them, and about rules, formulas, and algorithms for the manipulation of these symbols, concepts, and relations. Semantic memory does not register perceptible properties of inputs, but rather cognitive referents of input signals. The semantic system permits the retrieval of information that was not directly stored in it, and retrieval of information from the system leaves its contents unchanged, although any act of retrieval constitutes an input into episodic

memory. The semantic system is probably much less susceptible to involuntary transformation and loss of information than the episodic system. Finally, the semantic system may be quite independent of the episodic system in recording and maintaining information since identical storage consequences may be brought about by a great variety of input signals .... A short list of examples may help the reader. ... The following memory claims are based on mnemonic information stored in episodic memory: (a) I remember seeing a flash of light a short while ago, followed by a loud sound a few seconds later; (b) Last year, while on my summer vacation, I met a retired sea captain who knew more jokes than any other person I have ever met; (c) I remember that I have an appointment with a student at 9:30 tomorrow morning. ... Now, consider some illustrations of the nature of information handled by the semantic memory system: (a) I remember that the chemical formula for common table salt is NaCl; (b) I know that summers are usually quite hot in Katmandu; (c) I know that the name of the month that follows June is July. (Tulving, 1972, pp. 384–386)

Over the past decade-and-a-half, a major theory of episodic memory has been put forward and developed, has fallen off seriously in acceptability among psychologists, and has been supplanted not by a generally agreed upon theory but by a variety of attractive approaches to particular aspects of episodic memory. The theory in question is that of Atkinson and Shiffrin (1968). Fundamental to this theory is the distinction between *structural features* and *control processes* of episodic memory, where the former term refers to those involuntary operations of episodic memory which seem to be omnipresent, to underlie the system known as episodic memory, and to be invoked regardless of the content or context of informational input; whereas the latter term denotes task-specific operations of the memory system used consciously by the subject.

Experimental findings referred to by Atkinson and Shiffrin indicated to them that the structural features of episodic memory consist of three memory stores, each different from the others in degree of permanence of stored information, in format of stored information, in apparent purpose for which it is being stored, and in manner of information loss. The stores are the *long-term store*, the *short-term store*, and the *sensory register*. The sensory register simply records the data of sense verbatim and holds it for less than a second after which it disappears either through spontaneous decay or through erasure by incoming data. Actually, corresponding to each sense modality there is a sensory register so that it would be most correct to speak of sense registers in the plural. No control processes are involved in the acquisition of information by the sensory registers—the process is automatic and unconscious. However, material in a sensory register can be selected for processing in STM; as they are processed they are subject to control processes. In fact, control processes work exclusively on the contents of the short-term store, and that is what gives the store its particular character. Input can come to it both from sensory registers and from long-term memory and can be kept active indefinitely through repetition, or rehearsal. Unless it is rehearsed, however, material in the short-term store disappears within 15 to 30 seconds. It is possible that information changes basic format in entering STM from a sensory register; for instance, certain visual information may enter STM in auditory format. The short-term store functions to hold a small amount of information, then, drawn from either LTM, a sensory register, or both, and to process it consciously to permit the organism to adjust its behavior and thoughts in accordance with current environmental characteristics. Finally, the long-term store is a cast repository of information and rules for cognitive routines, to whose storage capacity science has been able to assign no limit. Data enters it both consciously and unconsciously from STM, and, as has been noted, data flow from it to STM as cognitive processes are being carried out. Memories can be lost, of course, from the long-term store; this can happen because of interference during searches of long-term store, interference from new material being entered, simple decay, or for other reasons. Of course, at the time, all of the above claims made by Atkinson and Schiffrin were supported by experimental results; in the present context, however, it is impractical to provide anything more than a summary of their theory, leaving aside its underpinnings.

As already stated, the model of episodic memory presented (i.e., the Atkinson and Schiffrin model) has waned in influence. Among other reasons for the demise of the theory were experimental findings that seemed to falsify three claims considered basic to a so-called multistore model of episodic memory (i.e., one which posits separate memory stores). The first claim was that each store accepts a different type of code; the second was that each store should have its particular rate for forgetting items; the third was that some estimate must be able to be made of the capacity of the stores. Although no single theory has replaced that of Atkinson and Schiffrin, approaches to episodic memory which bear on psycholinguistic research and theory can be cited.

In a bellwether article, Craik and Lockhart (1972) offered a critique of multistore models and proposed a new theoretical orientation, the *depth of processing* or *level of processing* approach. This orientation places more emphasis on processes of memory than on structure.

Craik and Lockhart begin their article with a statement of the case for multistore models, as they see it. They present a useful chart summarizing what they call "commonly accepted differences between

Feature	Sensory Registers	Short-Term Store	Long-Term Store
Entry of information	Preattentive	Requires attention	Rehearsal
Maintenance of information	Not possible	Continued attention Rehearsal	Repetition Organization
Format on information	Literal copy of output	Phonemic Probably visual Possibly semantic	Largely semantic Some auditory and visual
Capacity	Large	Small	No known limit
Information loss	Decay	Displacement Possible decay	Possibly on loss Loss of accessibility or discriminibility by interference
Trace duration	¼-2 seconds	Up to 30 seconds	Minutes to years
Retrieval	Readout	Probably automatic Items in consciousness Temporal/phonemic cues	Retrieval cues Possible search process

the three stages" (p. 672) of episodic memory as represented within multistore memory models. (The chart should be comprehensible based on the foregoing account of memory and serves to summarize much of that discussion.) Craik and Lockhart (1972, pp. 671–684) note the intuitively satisfying nature of the model summarized above.

> Such multistore models are apparently specific and concrete; information flows in well-regulated paths between stores whose characteristics have intuitive appeal; their properties may be elicited by experiment and described either behaviorally or mathematically. All that remains, it seems, is to specify the properties of each component more precisely and to work out the transfer functions more accurately. (p. 673)

Craik and Lockhart (p. 170) claim that the multistore formulation is unsatisfactory in terms of its capacity, coding, and forgetting characteristics, and they prepare to present their own model with the following remark:

> there are some basic findings which any model must accommodate. It seems certain that stimuli are encoded in different ways within the memory system. A word may be encoded at various times in terms of its visual, phonemic, or semantic features, its verbal associates, or an image. Differently encoded representations apparently persist for different lengths of time. The phenomenon of limited capacity at some points in the system seems real enough and, thus, should also be taken into consideration. Finally, the roles of perceptual, attentional, and rehearsal processes should also be noted. (p. 675)

Craik and Lockhart then present their own model. The heart of this presentation is:

> Many theorists now agree that perception involves the rapid analysis of a number of levels or stages. . . . Preliminary stages are concerned with the analysis of such physical or sensory features as lines, angles, brightness, pitch, and loudness, while later stages are more concerned with matching input against stored abstractions from past learning; that is, later stages are concerned with pattern recognition and the extraction of meaning. The conception of a series or hierarchy of processing stages is often referred to as "depth of processing," where greater "depth" implies a greater degree of semantic or cognitive analysis. After the stimulus has been recognized, it may undergo further processing by enrichment or elaboration. For example, after a word is recognized, it may trigger associations, images, or stories on the basis of the subject's past experience with the word. Such "elaboration coding". . . is not restricted to verbal material. We would argue that similar levels of processing exist in the perceptual analysis of sounds, sights, smells, and so on. Analysis proceeds through a series of sensory stages to levels associated with matching or pattern recognition and finally to semantic-associative stages of stimulus enrichment. One of the results of this perceptual analysis is the memory trace. Such features of the trace as its coding characteristics and its persistence thus arise essentially as byproducts of perceptual processing . . . Specifically, we suggest that trace persistence is a function of depth of analysis, with deeper levels of analysis associated with more elaborate, longer lasting, and stronger traces. Since the organism is normally concerned only with the extraction of meaning from the stimuli, it is advantageous to store the products of such deep analyses, but there is usually no need to store the products of preliminary analyses. It is perfectly possible to draw a box around early analyses and call it sensory memory and a box around intermediate analyses called short-term memory, but that procedure both oversimplifies matters and evades the more significant issues. (p. 675)

As a rough analogue to what had been called short-term memory, Craik and Lockhart refer to *primary memory*; this is a holding state, in which memories can be kept active by *recycling*, or rehearsing them, at a given level of processing. However, the more fully a memory is integrated with the individual's existing store of information, the more permanent it becomes. That is, processing at greater depth takes place when relationships are found between the new data and what is already known.

Since the publication of the cited article, the theory has undergone development in two separate directions. First, Craik and his co-workers (Cermak and Craik, 1978; Craik and Jacoby, 1975) have elaborated the Craik and Lockhart theory so as to take account of recent episodic memory studies. Second, a number of researchers have provided evidence for the existence of a range of depths of semantic processing from shallow to deep. (Previously, it had been considered that semantic processing automatically meant deep processing.) For instance, in an often cited study, Mistler-Lachman (1972) found that the same ambiguous test sentences, when presented to subjects as part of tasks each of which involved deeper semantic processing than the previous one, led to task latencies which varied proportionally with the depth of processing hypothesized for the task.

There has been no shortage of critiques of the depth of processing approach. Some workers have pointed out that what Craik and Lockhart call depth of processing is simply the type of processing which is optimal for the accomplishment of certain tasks; "shallower" processing is, however, most efficient for the realizing of other tasks. With this in mind, Bransford and others (1978) have proposed conceiving alternate modes of processing nonhierarchically as *transfer-appropriate processing* rather than as levels of processing. In this connection, Tulving (1978) argues that phenomena previously explained within the depths-of-processing format can be accounted for more convincingly by his *encoding specificity* principle (cf. Thomson and Tulving, 1970; Tulving and Osler, 1968). This principle states that ability to recall items is intimately linked to the reevocation of the *context* in which the item was first encoded. The applicability of the encoding specificity principle to a wide area of recall tasks has been demonstrated (Dewer *et al.*, 1977; Winograd and Rivers-Bulkeley, 1977). The major research task facing proponents of encoding specificity is that of discovering the qualitative and quantitative nature of "memory context"—what is encoded along with a memory, and what is necessary to trigger its retrieval?

A focal question of episodic memory research has been that of the form in which memories for events and experiences are stored. The classic hypothesis in this domain is that of Paivio (1969, 1971). His *dual coding hypothesis* posits two separate internal codes, an analogue code (i.e., one which represents perceptual items) and a linguistic code (one that represents linguistic items). Controversy has surrounded Paivio's assumption that the analogue code is virtually photographic, that is, consists of a copy, more or less, of external events that it takes the form of *imagery*. A final tenet of Paivio's theory is that image code can be translated into linguistic code, and vice versa. As has been the case of other theories of memory we have noted, it would be impractical to cite the specific evidence on which Paivio based each of the foregoing claims.

Rebutting Paivio's claim that analogue memory takes the form of mental imagery, Pylyshyn (1973) has pointed out that the very manner in which we forget portions of memories of events is evidence that such memories have been processed and classified and not simply recorded verbatim. He notes that, rather than indiscriminately forgetting all partial memories of a scene which cooccur within some region of our image—say, everything in the left foreground of the image—we forget instead categories of items; perhaps we retain memories of the principals of some event but lose those of nonessential participants in the scene.

Posner (1973; Posner and Rogers, 1978) as well as others has suggested that there are at least three distinct memory codes, namely, verbal, visual, and motor, which retain their integrity in storage, which can be isolated experimentally, and which are accessible in different degrees to different individuals (i.e., are sensitive to individual differences).

We began our discussion of memory with a presentation of Tulving's distinction between episodic and semantic memory. Having examined some theories of episodic memory, we now turn to a brief account of semantic memory.

The study of semantic memory is directed toward discovering answers to the questions: What is the structural organization of concepts in memory, and, What is the mechanism by which such concepts are combined and interrelated in the process of sentence comprehension? There are a number of different broad approaches to these issues, and we will briefly consider two of these. The first approach seeks to answer the first of the two questions just referred to as well as a subpart of the second question. The second orientation, a discussion of which follows that of the first approach, attempts to respond to both queries integrally.*

The first group of theories to be dealt with squarely face the problem of characterizing conceptual organization in memory. In addition, because the experiments discussed and conducted in support of such theories typically involve subjects' evaluations of simple classificatory sentences such as *Dogs are mammals*, and *Dogs are plants*, at least a restricted part of the general issue of sentence comprehension is confronted.

The classic theory in this tradition is that of Quillian (1968, 1969). Working within the artificial intelligence framework, Quillian developed a computer model of human text comprehension for a very restricted body of English texts. Quillian conceived semantic memory structure by means of a representational device directly related to implementation

---

* A useful tripartite analysis of recent approaches to the study of semantic memory as well as a highly interesting presentation of at least one such approach along with a discussion of the relation of work in psychological semantics to that in linguistic semantics are to be found in Smith (1978). In particular, Smith has a useful discussion of competence versus performance approaches to semantics.

by computer but only metaphorically linked with human memory processes. In Quillian's scheme, conceptual organization is represented as a *network* in which labelled *nodes* correspond to *concepts* and a system of *links* connecting the nodes stand in one-to-one correspondence with postulated relations between the concepts. The following figure (adapted from Collins and Quillian, 1969) gives an idea of the network format.

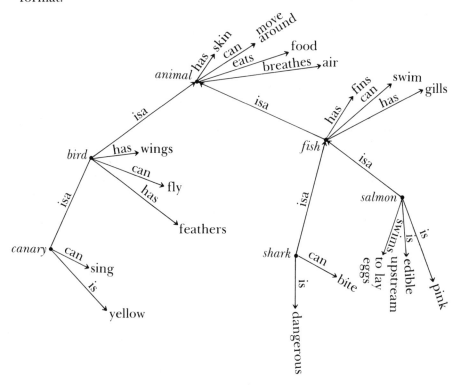

Note that the *isa* links express category relations, whereas the remaining links specify attribute relations. Key characteristics of this model are that concepts are hierarchically ordered (*shark* is subordinate to *fish*, which is subordinate to *animal*; the same is true of the series *canary*, *bird*, *animal*), and that the assignment of links is governed by what has been called the principle of *cognitive economy* (cf. Conrad, 1972). This is the notion that a given attribute is assigned to the highest node to which it properly applies. For instance, although birds, fish (and, perforce, canaries, sharks, and salmon) all have skin, the attribute *has skin* is stored with the node *animal*, because this is the highest node to which the property *has skin* applies. In using his or her conceptual network to verify a test sentence such as *A salmon is a fish* or *A canary is an animal*, a subject locates a path (a series of connected links, or a single link) connecting the concepts corresponding to the two nouns and verifies (or fails to verify) that the path is labelled in a way consistent with the assertion being made by the sentence under consideration. An important point is that, in verifying a sentence such as *A canary breathes air*, the subject is, *ex hypothesi*, making use of information not directly stored with the concept "canary" itself; in other words, he or she is making an inference. Recall that in defining semantic memory, Tulving stipulated that, unlike episodic memory, semantic memory permitted the retrieval of information that was not directly stored in it. Quillian's conception of the problem of semantic memory includes features consciously designed to capture this aspect of semantic memory.

Collins and Quillian (1969) experimentally tested Quillian's system, the Teachable Language Comprehender (TLC). It is clear that, given TLC, latencies on sentence-verification tasks centered on questions such as *Is a canary an animal? Do sharks have skin?* and *Are salmon fish?* should increase in proportion to the number of links to be traversed by TLC (and, metaphorically, by the subject), other factors being held constant. This is exactly what Collins and Quillian found for true sentences, however, results for false sentences proved intractable.

A variety of subsequent experimental results seriously undermined TLC and have led to the revision of Quillian's theory as well as to the appearance of competing theories. Schaeffer and Wallace (1969) showed that the facilitating effect of node proximity, within TLC, on sentence-verification judgments was actually reversed for a particular group of "false " judgments. Wilkins (1971) demonstrated that latencies were reduced by concept *typicality*; that is, given a sentence such as *Robins are birds* and one such as *Ravens are birds*, subjects are sooner able to evaluate the former than the latter. This correlates with experimentally gathered data indicating which instances of a given category (e.g., *birds*) subjects supply first when asked to think of an item from the category. Of course, this runs counter to TLC, according to which all category instances are equal. Conrad (1972) provided evidence to the effect that attribute links are associated not only with the highest possible node, as in TLC with its principle of cognitive economy, but with lower nodes in addition. Rips, Shoben, and Smith (1973) conducted experiments which imply that the categories in terms of which humans organize their conceptual systems are, at least in the semantic areas tested, not organized according to strict class inclusion, as in TLC. Rosch (1973, 1975; Rosch *et al.*, 1976) has further undermined certain aspects of

TLC through her experimentally supported concepts of *typicality* and *base level.*

One of the developments in semantic memory research since the eclipse of TLC has been the emergence of a revised version of Quillian's theory, put forth by Collins and Loftus (1975) and called the *theory of spreading activation.* What Collins and Loftus have done is to remove the hierarchical nature of Quillian's network and to build in semantic distance features, so that, for example, the aforementioned results on item typicality within a given semantic domain are accounted for. Technically, the longer the links, in the new theory, the more remote the associations in memory between the concept nodes connected by these links. Parallel to the concept network (the schema just described links concepts and not words) is a word network, or lexicon, organized according to sound similarity. Other research findings which tended to throw doubt upon TLC have been accommodated within the new framework. The mode of operation of a spreading activation network is as follows: When a subject thinks about, reads about, sees an exemplar of, or hears about a concept, that concept's node is activated, or energized. This energy spreads outward from the activated node in all directions along whatever links enter and leave it. As the energy spreads, it loses strength and finally dissipates completely. In a sentence-verification task, then, such as that of a responding true or false to *Sharks have skin*, the nodes corresponding to both *shark* and *skin* are activated in the subject's semantic memory network when he or she hears the sentence. Energy spreads outward from each of the loci of activation, and both excitations converge on and strengthen the links most immediately connected with each of the two nodes. A separate decision mechanism intervenes to judge whether one of the strengthened paths between nodes corresponds to the proposition under consideration. Because many nodes have typically been traversed in the course of the subject's reacting to such stimulus sentence, it is clear that the decision device must keep track of the nodes of origin of each wave of energy if it is to evaluate a proposition.

A second semantic memory theory which has been proposed in the wake of the foundering of TLC is the feature comparison model of Smith, Shoben, and Rips (1974). According to these researchers, concepts are stored as sets of *semantic features*; each concept's feature collection comprises both a set of *defining features* and one of *accidental features*. Sentence-verification is viewed as a two-stage process in which, first, a global feature comparison is made between the concepts corresponding to the nouns in a sentence for evaluation, for example, *Canaries eat food*, or *Salmon breathe air*. That is, a similarity index is computed on the spot for the two concepts (feature sets). A high positive or negative score results in an immediate decision that the proposition is, respectively, true or false. However, a mediocre value for the similarity index means that the decision process continues to the second stage. At this point a detailed feature-by-feature comparison of the two concepts is made, but in this case only of the *defining features*. A careful comparison is effected at this point and a verification judgment is rendered on this basis. This theoretical approach captures the fact that subjects seem to make "snap judgments" on exceedingly clear-cut sentence-verification tasks, whereas latencies increase where decisions of an apparently more "technical" nature are involved. By weighting the values of the features within each concept set, Smith, Shoben, and Rips have been able to account for "typicality" effects such as those pointed up by the research of Rosch *et al.* (1976).

A final type of semantic memory model is associated with the psychologists Bartlett (1932) and Rosch (1975), among others, and with the philosophers Cassirer (1923) and Putnam (1975), again *inter alia*. Called the prototype, or stereotype, or schema model, this approach considers that concepts are stored in memory in the form of exemplars (*prototypes*) along with a set of rules or procedures for accessing less-than-perfect instances of a concept in response to stimuli evoking the concept in question. For instance, in an experiment conducted by Rosch, subjects in effect had to classify words according to whether or not they belonged to certain semantic classes; given the class *bird*, they were to indicate, as part of their task, if a robin is a bird, and if a turkey is a bird. In cases where the subjects' task was basically confined to making the judgment just specified, they judged the prototypes (e.g., *robin*, for the class *birds*) to be members of their appropriate class more rapidly than they so judged nonprototypes. This would be due to the intervention of whatever rules or procedures are necessary to determine class membership of less-than-typical items of the set, according to prototype theory.

Conceptual Comprehension

We have completed our excursus on cognitive psychology and are ready to discuss properly psycholinguistic matters once more. In the initial discussion above of broad theoretical orientations to the problem of sentence comprehension, we mentioned Levelt's distinction between researchers of the linguistic school (those who believe that the end products of sentence comprehension appear in memory in the form of a code which is linguistic in nature) and those of the conceptual school (for whom the end product of sentence processing is a nonlinguistic, conceptual object.) To illustrate the conceptual viewpoint, Levelt cites Roger Schank, a researcher in the field of artificial intelligence. Schank (1975a) had said that a comprehension device should associate a linguistic input with what we will call a *conceptual structure*. A conceptual structure consists of concepts and the relationship between them. Levelt (1978) comments, "The end-term, therefore, is a nonlinguistic object. The aim of the theory is to explain how a linguistic object (a text, sentence, etc.) is mapped onto a non-linguistic object (a conceptual structure)" (p. 2).

Conceptualists have claimed support from research on memory which suggests that sentence comprehension gives rise to memory structures in nonlinguistic code. Bransford and Franks (1970) and Franks and Bransford (1972), presented subjects with a number of sentences, a proportion of which concerned the same situation; later, subjects were questioned concerning the sentences. Specifically, they were asked which of a list of sentences appeared in the original group. Bransford and Franks found that there was no difference in recognition scores between sentences that had been on the original list and those that had not so long as the sentence in question was on the same topic as, and was consistent with, those of the original group. Moreover, whether or not a given sentence had actually been in the original group, subjects were likely to claim that it had been in proportion to the number of propositions it contained that were consistent with the situation depicted in the original sentence group. These results meant to Bransford and Franks that, "individual sentences lost their unique status in memory in favor of a more wholistic representation of semantic events."

A major claim of the conceptualists has been that encyclopedic, that is, general knowledge, plays a major role in forging the interpretation of a given sentence (Barclay, 1973; Fillenbaum, 1966, 1968; Johnson, Bransford, and Solomon, 1973). A series of research findings of the 1960s and 1970s indicated to a large number of psycholinguists that nonlinguistic factors played a major role in sentence interpretation. First were a group of studies showing that response latencies, in tasks involving certain types of responses to sentences, could not be predicted on the basis of syntactic variables alone, but essentially involved variables drawn from the realm of extralinguistic knowledge. For instance, in the sentence-verification tasks of Slobin, referred to above, reversible passives were verified more slowly than reversible actives, but there was no significant difference in latencies for nonreversible sentences. Yet reversibility has to do with world knowledge. It is part of our nonlinguistic information store that dogs bite men but men do not bite dogs (Olson and Filby, 1972; Wason, 1965). A second research result (Clark and Chase, 1972) indicated that words considered to be *marked* (less likely to occur) by grammatical theory could be made to support shorter latencies than unmarked items of a corresponding nature, in an otherwise similar context, simply by manipulating the subject's center of attention (or at least requesting the subject to focus on one thing and not on another). A third experimental result was interpreted as bearing on the form of representation of the end-products of sentence comprehension. Danks and Sorce (1973) interpreted the results of a recall experiment they conducted to indicate that subjects decide on the basis of the exigencies of the task they are presented with, whether to encode a test sentence in a linguistic or imagal manner. Finally, Danks and Schwenk (1974) produced experimental data which seemed to indicate that ordering rules obtaining among adjectives would predictably be violated by subjects responding to certain extralinguistic contexts.

Two additional phenomena that theorists of the conceptual school consider essential to any account of sentence comprehension are *presuppositions* and *intended meaning*. Presuppositions are implied premises of a statement or question. For instance, the question, *Have you stopped regularly beating your wife?* presupposes that the man being addressed has been regularly beating his wife. Note that, in order to comprehend such a question, it is necessary, if only for an instant, to assume the veracity of the background assumption, or presupposition. Experimental results indicate that subjects normally incorporate presuppositions, in whole or in part, into their store of general knowledge (Hornby, 1974). *Intended meaning* is inferred both from indirect statements, such as *I'd like to add some salt to these beans*, which, in context, can be intepreted to mean, *Please pass the salt*, and from distorted or highly unconventional utterances. One experiment which illustrates subjects' strong tendency to extract what they consider normal "intended" meanings from utterances is that of Fillenbaum (1974). Fillenbaum demonstrated not only that subjects "corrected," in the course of interpreting them, "non-normal" meanings of utterances, but also that, when confronted with their corrections and asked if their interpretations coincided with the original sentence (which was also displayed to them for a second time), a large percentage of subjects insisted that their interpretation must have corresponded to the speaker's actual intended meaning.

It will be recalled that Marslen-Wilson and his associates have put forth and defended conceptualist theories of sentence comprehension, theories which stress "on-line" processing of linguistic input and involve the interaction of (at least) syntactic, semantic, and world-knowledge effects in this process. Another, perhaps more inclusive, theory of comprehension is that of Clark (Clark, 1976, 1978; Clark and Haviland, 1977; Haviland and Clark, 1974). Clark has accepted the challenge of accounting for the gamut of phenomena mentioned above as being considered essential to comprehension theories by researchers of the conceptualist school. He sets the process of sentence comprehension in the broadest possible framework.

> Comprehension, in short, calls on people's general capacity to think—to use information and solve problems. Although people develop specialized strategies for comprehension, these are still built on their general ability to solve problems—to set up goals, search in the memory for pertinent information, and decide when the goals have been reached. Indeed, in inferring what is meant, people consider non-linguistic factors that are far removed from the utterance itself, and their skill at solving this problem is sometimes taxed to the limit. Comprehension is a form of thinking that should not be set off from the rest. (Clark, 1978, p. 320)

Clark's views on natural language comprehension are not fully conceptualist, and are certainly not exclusively linguistic. In his article (Clark, 1978), he explicitly situates his approach between these two extremes, labelling it *intentional*. Although allowing for the computation of the literal meaning of a sentence, Clark sees this process as far from sufficient for its understanding. According to the intentional view of comprehension,

> comprehension is conceived to be the process by which people arrive at the interpretation the speaker *intended* them to grasp for the utterance in that context . . . Unlike the [linguistic] view, this view requires listeners to draw inferences that go well beyond the literal or direct meaning of a sentence . . . . But unlike the [conceptualist] view, it limits the inferences to those that listeners judge the speaker intended them to draw. . . . In this view the speaker's intentions are critical, but they can ever only be inferred. (p. 275)

Clark's understanding of the comprehension process is as follows: In interpreting a given sentence, the listener sets for himself or herself the *goal* of determining the interpretation which the speaker intends for him or her to make. In order to meet this goal, the listener draws on a *data base* consisting of "the sentence uttered; the time, place and circumstances of the utterance; the speaker's beliefs about the listener [as perceived by the listener]; [and] general knowledge" (p. 298). In utilizing this data base toward these ends, the listener is constrained by a set of *boundary conditions*, that is, "various tacit agreements between speaker and listener about how language is to be used" (p. 298). The process, then, of comprehending a sentence, or, in Clark's words, "solving for the intended interpretation," can be thought of, on the intentional view, in these terms:

1. Build a candidate interpretation of the sentence.
2. Test the candidate interpretation against the boundary conditions.
3. If it passes all the tests, accept it as the intended meaning. Otherwise begin at 1 again.

One starts by constructing the simplest possible interpretation of the given sentence, in certain cases, the "literal" or "direct" interpretation. (In other cases, according to Clark, it may not be possible, in principle, to speak of a literal interpretation at all, i.e., in those cases, the acceptability of an interpretation will be completely context-dependent.) The putative interpretation is examined in the light of the appropriate boundary conditions for the type of utterance being interpreted. These conditions take the form of tacit agreements between listener and speaker of the kind described by Grice (1975) (on whose work Clark explicitly bases the general framework of his "boundary conditions") as falling under the *cooperative principle*. Grice has claimed that, in communication, speaker and listener normally share an unspoken accord to the effect that the speaker will obey four "maxims": (1) be informative, (2) be relevant, (3) be truthful, (4) be clear. That is, communication is a sort of cooperative partnership in which each participant normally begins by assuming that the other is behaving in a maximally cooperative manner, as defined by the four maxims above. Clark has elaborated Grice's principle, applying it in some detail to each of several communication tasks, for example, interpreting indirect requests, interpreting "linguistic shorthand" expressions, and so forth. For instance, drawing on the work of Austin (1962) and Searle (1969), Clark specifies the tacit agreement that speakers and listeners engage in when, respectively, making and interpreting a request, whether direct or indirect.

Requests are made so as to satisfy these four "felicity conditions":

1. *Preparatory condition.* The speaker believes the listener *is able* to carry out the requested act.
2. *Sincerity condition.* The speaker *wants* the listener to carry out the requested act.
3. *Propositional content.* The speaker predicates a *future* act (the one being requested) of the listener.
4. *Essential condition.* The speaker counts his utterance as an attempt to get the listener to carry out the requested act. (Clark, 1978, p. 300)

Such conditions function as correctives in the interpretation of direct or indirect requests. Clark (1978) gives the example of speakers who "*imply* they are requesting something merely by suggesting that one of the felicity conditions for that request is fulfilled" (p. 301). For instance, in the proper context, a speaker can say, "This soup needs salt," or, "Why don't you pass the salt?" and be correctly interpreted by the listener(s) to have in mind as a *final interpretation* (net, or basic, interpretation), the message conveyed directly by the utterance "Please pass the salt." In arriving at such results, the listener shuttles back and forth, as it were, between Steps 2 and 1 of the three-step algorithm for sentence comprehension set. Finally, having determined the interpretation (actually, in many cases of which the indirect request is one, the listener derives a set of interpretations) that is consonant with the sentence uttered; with the context in which it was uttered; with the listener's beliefs concerning the speaker and his or her knowledge of and attitude toward the listener; and with the listener's encyclopedic knowledge store, the listener halts the comprehension process, considering that he or she has inferred what the speaker meant by the utterance.*

According to Clark and Haviland (1977), the cooperative principle supports what they call the *given/new contract* between speaker and listener. In framing his or her communication, the speaker draws on a judgment as to the extent of the listener's current knowledge about the situation or topic to be discussed. Whatever the listener can be assumed to know already is considered by the speaker *given* information. To this nexus of information the speaker will seek to adjoin an additional element, the *new* information. Specific linguistic devices, claim Clark and Haviland, set off the new from the given portion of the message. The speaker encodes his or her message so as to effect this demarcation linguistically, and the listener, in processing a sentence, proceeds according to the given/new contract, that is, assumes that those linguistic vehicles, which normally serve to express new information, are being used by the speaker to do exactly that. Clark and his associates have provided experimental evidence for the claim that comprehension is impeded when the given/new contract is violated. This facilitating effect upon communication of adherence to the given/new contract is explained via the supposition that the listener uses the given information

---

* Of course, the terms *halts* and *considering* are used in a metaphorical sense, as this sort of inference is conducted at an unconscious level, as Clark is careful to point out. Moreover, the whole computer analogy is metaphorical; something is going on, subconsciously during the inference process which can be considered analogous to the algorithm outlined, but the brain is not a computer, and we are quite far from being able to guess how it does in fact carry out the processes which we represent at a high level of abstraction.

to locate in memory the data complex being referred to, and then adds to it a representation of the new information.

An integral part of Clark's theory has been his *propositional theory* of the encoding of linguistic input. As put forth by Chase and Clark (1972), this theory posits a polyvalent propositional code into which both sentential and pictorial information is converted via input synthesis. For example, the sentence, *The plus is above the star,* would be encoded, (true *(plus above star)*), where the innermost parentheses contain the most embedded proposition, with each successive pair of parentheses containing a predication about the immediately subordinate proposition. However, the "picture"

```
+
*
```

could be assigned the same internal representation because Chase and Clark showed that the representation assigned to a picture is task-dependent, that is, given certain tasks, subjects seem to encode the above "picture" as follows: (true *(star below plus)*).*

Returning from the intentionalist to the conceptualist school, we note that in reviewing the development of the field of sentence comprehension from the 1960s through the mid-1970s, Levelt (1978) cites two causes for a shift in the direction of investigation in this field, at least among some researchers:

> Firstly, the development of simultaneous-measurement techniques† has made it possible to demonstrate the highly interactive nature of sentence-under-standing processes: semantic decisions which can take place quite early in the sentence can affect syntactic and phonetic decisions, and it is not quite clear any more what occurs prior to what. Secondly, developments in artificial intelligence and related fields became increasingly influential in psycholinguistic theory construction. (p. 61)

As already noted, conceptualists view the endpoint of linguistic processing as a nonlinguistic object. They deny what Levelt calls the

---

* For further details of the propositional code, see Chase and Clark (1972). The propositional theory of Clark was further elaborated in Clark (1974) and in Carpenter and Just (1975).

† Levelt is referring here to the techniques of phoneme-monitoring, click-monitoring, and shadowing, as well as to what he calls "psychophysiological measurement," that is, techniques whose aim is to study physiological correlates of mental functioning during sentence perception, for example, studies of pupil size or of galvanic skin response under such circumstances.

immediate linguistic awareness hypothesis (ILA), according to which it is the case in all linguistic processing that at some stage a representation of the underlying linguistic structure of the sentence being processed is derived. Often they view processing as task-dependent, that is, varying radically in characteristics and procedures according to the immediate task the comprehender is attempting to accomplish. Incorporating the concept of task-dependent processing into the definition of "understanding" held by some conceptualists, then, we can say with Levelt that some conceptualists use the term "to denote the process mediating between syntactic input and the completion of a linguistic or non-linguistic task (memory task, paraphrase task, verification task, etc.)." Marslen-Wilson, Bransford *et al.* (1972), Barclay (1973), and Franks (1974) are among the researchers who reject ILA and who propose a *constructive* view of sentence processing in which semantic and world-knowledge contributions to sentence processing are given preeminence. As Levelt notes, such an approach dovetails nicely with the Craik and Lockhart depth-of-processing memory model. On the one hand, this theory of memory is difficult to reconcile with a model of comprehension that divides the process into series of analyses (syntactic, semantic, etc.) each feeding on the previous one. On the other hand, the Craik and Lockhart model predicts the task dependency claimed by some conceptualists, as well as the prominent role of semantic processing in comprehension.

A formal device which has been used widely by conceptualists who formalize their theories is the *augmented transition network* (ATN). (It should be noted that ATNs have also been used to implement linguistically oriented theories. In fact, the first use of this formalism within psycholinguistics was by Kaplan [1972] who employed ATNs to represent a number of the comprehension strategies developed by Bever.) For those who reject the proposition that sentence processing entails the casting of input into deep-structurelike form, it is natural to embrace formal processing devices not bearing features derived from or inspired by transformational grammar. ATNs are such devices.

ATNs are formal devices based upon finite-state automata but given much more power by the addition of the ability to impose conditions of any desired complexity on the transition from one state to another and by recursive call of subnets. These augmentations endow ATNs with Turing machine power. An important condition which can be imposed on a given transition is the calling and successful execution of "subroutines," that is, shift of control to subordinate networks which continue to process the input and, when they have been able to accomplish the tasks set out for them, return control to the original, superordinate network. As processing of linguistic input proceeds, in a left-to-right manner, more abstract representations of input can be constructed, and/or actions can be carried out such as the semantically triggered lookup of some concept in long-term memory.*

Before concluding, we take brief note of a development which has taken place over the last decade, namely, the emergence of *global theories of comprehension*. These are attempts to encompass the range of comprehension activities of humans, not simply those attached to particular tasks. The ideal global comprehension system seems to have, grossly, four components. These are: an interface parser, which converts physical stimuli to symbol structures; a representational network encompassing both encyclopedic and linguistic knowledge; control systems that operate the overall device; and an output-synthesis mechanism. No single system proposed to date has all these features, but each has some combination of them. Some of these systems span the gap between artificial intelligence and psycholinguistics, whereas others are more firmly in the AI camp.

Schank, one of the global theory modelers, who was cited above as a conceptualist, has developed computer models of language comprehension which he intends as psychological theories of human language processing. Schank's approach to processing sentences is *predictive* and top-down (as opposed to the bottom-up strategies of the more linguistically oriented psycholinguists). For instance, in presenting his conceptual dependency theory of language processing, Schank (1975a) writes:

> In this schema, analysis can proceed by making predictions about what the meaning of the sentence is going to be as it is being input. A person reading or listening to a language he understands does not wait to see or hear every word of a sentence in order to understand what that sentence is going to say. He extracts meanings, makes inferences, and puts them into a world model. He makes predictions before the sentence is completed. Mostly he tries to make sense of what he expects to hear. The important point is that the procedures he uses to do this are the same ones that he uses in listening to his own language. It is, after all, the same world, and his world model is not dependent on his language. Thus, language analysis . . . is much easier than language generation because all the rules you need for generation are not employed in analysis. In fact, only a few of them are. As long as all the words are known, the semantic model can complete the task. . . . The rules of French grammar are not crucial in understanding French, but the rules of the world are. (p. 13)

---

* Brief but clear introductions to ATNs can be found in Anderson and Bower (1973), pp. 121–124, and Levelt (1978), pp. 55–59. See also Woods (1970), Wanner (1980), and Wanner and Maratsos (1978).

The 1970s saw at least seven major global comprehension systems advanced.* Few of their claims have been tested experimentally, and many consider the further formalization and testing of predictions of such models to be a major task for the 1980s.

Exercises

1. *Perception* is defined by the dictionary as the process of obtaining information about the world through the senses. Elaborate on this definition for language perception.

2. Describe the role of frequency in lexical processing.

3. What are the structures that listeners construct mentally during sentence comprehension?

4. How successful were Miller's attempts to demonstrate the psychological reality of transformational rules?

5. Clark and Clark (1977, p. 12) state, "Loosely speaking, the surface structure says how the sentence is to be spoken, and underlying representation says how it is to be understood." Discuss.

6. Describe the "verb complexity hypothesis."

7. Discuss "shadowing" techniques and their consequences as described by Marslen-Wilson.

8. Describe and contrast the structure and functions of short-term and long-term memories.

9. What is the difference between episodic memory and semantic memory?

10. Construct a detailed semantic memory network for the world of edible fruits produced by shrubs, plants, and trees.

* Anderson and Bower (1973), Kintsch (1974), Norman, Rumelhart, and the LNR Research Group (1975), Schank (1975a), Frederiksen (1975), Anderson (1976), and Miller and Johnson-Laird (1976).

11. What is a sensory register?

12. Describe the Gricean hypothesis for cooperative or communicative "contract" between speaker and hearer.

13. Contrast propositional versus conceptual hypotheses for language and knowledge representations.

Topics for Research and Term Papers

1. Write a critical survey of various approaches in lexical processing.

2. Write a critical survey of various approaches in sentence processing.

3. Discuss at least two positions and approaches to cognitive psychology and memory.

4. Write an essay describing various approaches to conceptual comprehension. (Consult the references given in the text.)

References

Much of the material discussed in this chapter can be found in greater details in Foss and Hakes (1978), Lachman *et al.* (1979), and Anderson (1976). Much of the data has been drawn from the first two sources. For surveys of studies in sentence perception, see Levelt (1974, 1978). Other specific references are given in the text of this chapter.

PART II

Views and Reviews

CHAPTER 4

Language Comprehension

Introduction

This chapter is concerned with human information processing in the specific context of language comprehension. The latter, in turn, is one aspect of the general process of human perception which includes visual, aural, somesthetic, and even emotive processes. We will provide a survey of studies in language perception and draw some conclusions concerning the activity of building computer models for language comprehension, with obvious implications for information science (e.g., communication with databases) and for artificial intelligence.

Although animals and even plants share some aspects of perception, it is believed by many linguists and cognitive psychologists that the faculty of language (to be distinguished from communication) is a characteristic of *Homo sapiens* alone. In any case, it is evident that all normal human beings have the capacity for language comprehension and use. How much of the language knowledge is innate as a part of human genetic inheritance and how much is acquired? Do other mammals have a language capacity to any extent? These and many other similar questions are of great interest and importance for linguists, psychologists, and philosophers among others. We are, however, not concerned here with the nature of language and its acquisition inasmuch as we are concerned with how the process of language comprehension and language use (linguistic "performance") works.

There is no paucity of apparently diverse computer models for natural language processing, "understanding," and communication. These models can, if suitably designed, be useful tools for testing hypotheses about language comprehension and may eventually provide for better computer systems for information processing. These models must, however, be regarded as instantiations of computer analogy in the sense of cognitive psychology. Unfortunately, they are often taken for or presented as representations of the real world, and this is the source of the follies and dangers which are so aptly enunciated by Weizenbaum (1976), himself one of the early builders of such models, and is what led to the over pessimistic views of Dreyfus (1979) about the whole enterprise of artificial intelligence.

The specifics of human language comprehension which are intimately tied with the working of the brain are, by and large, unknown and remain among the most challenging and baffling questions of science and of human behavior. On the other hand, functioning computer models must necessarily contain well-formed and complete working components. To the extent that we know about language and working of the brain, our knowledge (which is indeed meager) can be implemented in the model in one way or another. But for all that is unknown or uncertain, usually, *ad hoc* and engineering gimmicks are used to make the model work. Again, the danger lies in confusing the metaphor for the real. There are also more basic theoretical issues concerning the learning theory and the theories of computability, complexity, and decidability. Outstanding questions on these topics demonstrate that a realistic model of any human complex mental behavior is not attainable unless and until such questions have been satisfactorily resolved. We will have some further comments about these issues later. While we are on the topic of model building, there is another general issue: It is a well-known fact that our perception or internal representation of a real object is never identical to that of the external object (Held and Richards, 1972). In some cases it is possible to demonstrate this difference through scientific analysis and computations. Now, do we build a model according to what we perceive of the real world or according to what it really is? The latter undertaking implies being able to describe a real object beyond and above what our senses tell us.

A further issue has to do with yet another aspect of perception. Do we see objects (trees, pictures, etc.) as whole units and instantaneously, as is advocated by the Gestaltists, who believe that perception is a phenomenon of physiologically inborn or innate aspects of brain, or is the process a synthesizing activity, as advocated, for example, by the structuralists, who believe in a process of learning and combining of simpler and atomic elements into more complex objects? In other words, in visual perception when we look at a square, do we immediately see a complete shape (square) or do we perceive dots and lines and construct a square? There are arguments and experimental data in support of both positions. Some introspectionists even tried to "prove" the phenomena of structuralism by devising experiments to analyze or to take apart perceptions of objects according to their constituent elements. These are by no means the only questions concerning perception which remain controversial and inconclusive.

However, modern psychology tends to leave such questions which are essentially epistemological in nature to philosophy and to deal with tangible mechanisms of perception. The roots of this tendency go back at least to Berkeley and Kant who took the *representation* of perception as the object of observation rather than the *thing-in-itself*. Berkeley observed that the material world in general exists simply and solely in our representation, and that it is false and indeed absurd to attribute to it an existence outside all representation and independent of the knowing subject (cf. Schopenhauer, 1958, vol. 2, p. 4). The Kantian principle is also based on the proposition that the objective world exists only as representation. At an earlier date, in the seventeenth century, Kepler had made more direct observations concerning perception. Crombie (1964) provides the relevant details of these observations in terms of the modern theory of perception. In his study of the optics of the eye, Kepler took the eye as a *camera obscura*, as suggested by Leonardo da Vinci, and treated visual perception as a purely mechanical process, an optical instrument like "clockwork" rather than a "divine living being." Thus, he arrived at a correct analysis of the mechanics of visual perception through physical and mathematical (geometrical) modeling, and he left the speculative questions concerning sensations and their relations to mental processes to "natural philosophers." A lesson we can draw today from Kepler's study is that "our principal efforts should be directed toward an objective anaysis of the properties of perceptual processes rather than toward attempting to find an internal portrait of the external world " (Held and Richards, 1972, p. 3).

Language Perception

A student of language perception will inevitably face a broad range of phenomena and apparently diverse approaches and goals. Researchers in speech recognition have been concerned mainly with the identification of the critical units of acoustic signals such as phonemes or even bundles of features which make up a phoneme and syllable boundaries. Linguists have generally, with a few exceptions, been concerned with sentence recognition. Artificial intelligence (AI) researchers and psycholinguists have been interested in larger units of discourse: paragraphs, texts, stories, and such. The ultimate goal for everyone is to comprehend connected discourse. The traditional linguistic approach has been to assume "levels of linguistic analysis" in which "atomic" units are recognized and are then constructed into more complex constituents (cf. the discussion of visual perception above). In modern linguistic theory, this position has its roots in what is now called the Standard Transformational Theory (Chomsky, 1965; Katz and Postal, 1964), where the initial stages essentially derive an underlying or deep structure for the input signal which is then subjected to semantic interpretation in the subsequent levels of analysis. This position became highly controversial, and, in the case of language perception, the "deep structure hypothesis" and the notion of strict sequential levels of analysis have few supporters at present. Moyne (1980) proposed a model of language use in which various components and subcomponents of an input signal are subjected to an on-line simultaneous analysis. The procedures, however, have certain delay factors to allow for the cases where the processing of one component may be dependent on a total or partial processing of another. This view was already expressed by, among others, Levelt and Flores d'Arcais (1978) in their introduction to the *Studies in the Perception of Language*:

> This broad range of processes covered by the term "language perception" implies several levels of perceptual decision-making, corresponding to several different levels of perceptual organization. The fact that these different levels can be considered separately does not, however, necessarily mean that perceptual processing at one level depends on the completion of the processing at some lower level. In fact, much of the available evidence points to a different conclusion: that language perception is a highly interactive process in which structure can be derived simultaneously at many different levels. (p. xiii)

It is in the sense of the above position that at the beginning of the introduction to this chapter we state that language comprehension is one aspect of the general process of human perception. Those who disagree with this position treat language perception and language comprehension as distinct phenomena. For them perception is the initial stage, mainly syntactic analysis (parsing, etc.); comprehension is the subsequent activity where inferences, use of knowledge, context considerations, and in some cases semantic analysis are carried out. There are other variations in which there is no explicit distinction between perception and comprehension but two-stage or multistage understanding is assumed with various degrees of autonomy for the stages. The *coding hypothesis* of the 1960s assumed that sentences are processed in short-term memory and are stored, in the form of their deep structures, in the long-term memory (see Johnson-Laird, 1974; Lachman *et al.,* 1979; Levelt, 1966, for discussion). Although Levelt (1978) points out that this hypothesis was proved to be false (Bransford *et al.,* 1972; Fodor *et al.,* 1974), the notion of multilevel analysis is still supported (Clark and Clark, 1977; Clark and Lucy, 1975; Cutler, 1976). This is also the approach which was widely used in many of the models for natural language processing by computers (Culicover *et al.,* 1969; Moyne, 1962; Petrick, 1976; Sager, 1978; Woods, 1967). On the other hand, the "global" approach assumes, as stated earlier, that there is a parallel and interactive processing of all aspects of language comprehension—phonological, syntactic, semantic, pragmatic, and so on (Marslen-Wilson, 1973, 1975; Moyne, 1980; Schank, 1972, 1975a; Wilks, 1973b; Winograd, 1972). The issues and the controversies are by no means settled.

Miller and Johnson-Laird (1976) in a major work on language perception set themselves the task of developing a theory of perception through a series of proposals and revisions. Their basic approach for a possible solution to the problem of language perception is an analytic approach. The problem should be divided into subproblems which interact strongly with each other yet which can be studied independently. Linguists have done this by identifying components or fields such as phonology, morphology, lexicon, and such. Miller and Johnson-Laird state, however, that this kind of subdivision is more difficult in psychology because in psychology everything is supposed to be related to everything else. They reject the associative and verbal behaviorist theories of stimulus–response–reinforcement as inadequate and claim that there does not exist a coherent theory of learning and, furthermore, that we cannot have such a theory for language acquisition until we know what is acquired.

The first "revision" proposed by Miller and Johnson-Laird is to change the learning theory about association of words (labels) with objects to learning rules relating perceptual judgments to assertible utterances. Sentence verification theories, however, are psychological theories, and they do not have absolute truth verification as in logic and mathematics. There are certain differences between these two assertions:

(1) Two is a prime number
(2) Cat is on the mat

(1) is always true, but (2) is true only in special situations. Thus, in sentence verification, we must have at least a two-stage process: (a) discovering the proposition that the sentence expresses in a particular situation and (b) verifying the truth of the proposition in that situation. The first step is *pragmatics*, the second is *semantics*. There are, however, other problems, for example, how to verify the truth of

(3)   John believes that the cat is on the mat

The solution proposed for this particular problem is this: For all cases "A believes that P," divide all possible worlds into two: (a) those compatible with what A believes and (b) those incompatible. Now, "A believes that P" is true for all worlds which are compatible with what A believes.

Pragmatic theory is naturally involved in language perception. The listener must be able, at the time of the utterance, to identify a particular *cat* and a particular *mat* to which the speaker refers. He can then formulate a proposition and supply it for semantic verification. Such formulation can be something like this:

(4)   P is true iff $F(x) = 1$

$F(x)$ describes a mental computation to be performed, such as attending to x and judging whether it is F. If the result of computation is that $F(x) = 1$, then P is true. For example, in verifying the utterance "snow is white," a person must look at snow and judge its color or he must remember what it looks like. Thus, various perceptual predicates appear to bias language comprehension. The problem for psychology is to characterize the mental computations that a person performs when he learns and applies the aforementioned rules and formulations.

Thus, we need a theory of mental computation, and a further "revision" proposed by Miller and Johnson-Laird is to develop a theory like that of computation for computers. $F(x)$ is, therefore, assumed to be a program of instructions to be executed. A clause is first "compiled" into a procedural language suitable for execution by the computation system. Obviously, there must be a syntactic apparatus present in this system; the infinitely many sentences of a language cannot be individually associated with a specific program. The language use model proposed has a system of lexical, syntactic, and semantic components which is said to be much more integrated and interactively related than those proposed by linguists.

The final "revision" deals with the notion of *lexical concept*: every word in a language is a label for a lexical concept which expresses anything capable of being the meaning of that word.

Clark's theory of sentence comprehension (Clark, 1978) is based on the Gricean view that the meaning of a sentence is the meaning intended by the speaker of that sentence in a given context. On this view, it is the task of the listener to determine what the speaker-intended interpretation of an utterance is, using the sentence's iiteral meaning, knowledge of the real world, contextual information, and knowledge of certain tacit conventions of speech. Because the listener must go beyond literal meaning and draw inferences to determine a sentence's interpretation, sentence comprehension is seen by Clark to be a form of problem solving. The goal of Clark's theory, therefore, is to determine what inferences people make beyond literal meaning and how they make them.

Clark characterizes language comprehension in the following way: The listener has a goal, namely, interpretation of an utterance. His data base is "the sentence uttered; the time, place and circumstances of the utterance; the speaker's beliefs about the listener; (and) general knowledge." (p. 298). Tacit conventions of speech, called boundary conditions by Clark, limit the range of possible interpretations which can be assigned to a sentence. As we noted in Chapter 3, Grice (1975) sketches four conventions which govern ordinary discourse: (a) be informative, (b) be relevant, (c) be truthful, and (d) be clear. Violation of any of these conventions leads to what Grice calls *implicature*, that is, an inference that goes beyond the literal meaning of a sentence.

An example of the violation of a convention leading to an implicature is the case where A says to B: "What a beautiful girl!" in referring to an obviously ugly one. B realizes that the girl referred to is ugly, that A knows this to be the case, and that A knows B realizes that the girl is ugly. Because A has violated the truthfulness convention but is obviously trying to communicate something, B must infer that A does not mean to be taken literally and instead is implying that the girl referred to in the utterance is quite ugly.

Given the listener's goal, data base, and the boundary conditions, the listener employs three mental operations to derive the correct interpretation of the sentence. First, the listener constructs a possible interpretation of the given utterance. Next, the listener evaluates the interpretation in relation to the data base and the boundary conditions. At this stage, if the interpretation appears to be correct, the listener accepts it. If the interpretation does not match up with the information in the data base and the boundary conditions, the listener must go back to the first step and construct another possible interpretation of the sentence.

Clark illustrates the ways in which a model such as this might be used to explicate sentence comprehension in the processing of indirect requests, definite reference, and shorthand expressions. These three types of expressions are used because each of them requires that the listener make inferences not found directly in literal meaning in order to be understood.

Although English has a construction for expressing requests directly, namely, the imperative, it is possible to express requests indirectly through a variety of constructions such as (5), (6), and (7), depending upon the contexts within which they are uttered.

(5) This soup needs salt
(6) Can you reach the salt?
(7) If you don't pass the salt, I'll scream

In order for an imperative to successfully convey a direct request, it must satisfy the four "felicity" conditions discussed by Austin (1962) and Searle (1969, 1975), which we will repeat here for ready reference:

1. *Preparatory condition.* The speaker believes the listener *is able to* carry out the requested act.
2. *Sincerity condition.* The speaker *wants* the listener to carry out the requested act.
3. *Propositional content.* The speaker predicates a *future* act (the one being requested) of the listener.
4. *Essential condition.* The speaker counts his utterance as an attempt to get the listener to carry out the requested act. (Clark, 1978, p. 300)

In order for constructions such as (5), (6), or (7) to be interpreted as requests, however, they must express, in statement or question form, at least one of these four tacit agreements about what a request is. Thus, for example, (6) is understood as a request, given the correct context, because it asks whether or not the listener can fulfill the preparatory condition. Sentences (5) and (7) express, with varying degrees of intensity, the sincerity condition, because they indicate a desire on the part of the speaker for the listener to fulfill a request. Therefore, uttered in the correct setting, (5) and (7) are understood as requests. As Clark points out, "speakers can *imply* they are requesting something merely by suggesting that one of the felicity conditions for that request is fulfilled" (p. 300).

It is important to note, however, that making requests in this way requires speakers to employ one interpretation to imply others. Clark suggests that speakers are, in fact, constructing "chains of interpretation." An example of one such chain is the utterance of sentence (5) in a mealtime situation. When A utters sentence (5) to B, A is making the assertion that the soup referred to needs salt. What is being communicated by sentence (5) does not, however, end with the information conveyed by this assertion. The assertion is used by A to make an indirect request. Thus (5) is used to make both an assertion and a request. Making the assertion, however, is not A's goal. The assertion is just a link in the chain of interpretation. Sentence (5) itself cannot be used to make a request. Rather, the assertion made by sentence (5) is used to make the request. The chain of interpretation for this utterance is this: statement (5) → assertion → request.

Clark's essential claim is that every utterance involves a chain of interpretation of varying degrees of length and/or complexity. Each utterance has an initial interpretation and a final one; intervening between these may be zero or more intermediate interpretations, depending upon context and other factors. The interpretations of an utterance are ordered with respect to one another.

Clark proposes the following model of the process by which listeners arrive at the speaker-intended interpretation of indirect requests:

Step 1.   Compute the direct interpretation of the utterance.

Step 2.   Decide if the interpretations computed so far are what were intended. Are there sufficient and plausible reasons for the speaker to have intended these interpretations alone in this context?

Step 3a.   If yes, proceed to Step 4.

Step 3b.   If no, use the immediately preceding interpretation to compute an additional interpretation by way of the tacit agreements on speech acts. Then return to Step 2.

Step 4.   Utilize the utterance on the basis of its collection of interpretations and assume that the final interpretation is the ultimate reason for the utterance.

Clark claims that there is psychological evidence for the plausibility of this model. First, because the model assumes that direct and indirect requests are given the same interpretation by listeners, subjects should show similar response time patterns for both types of requests in psycholinguistic experiments.

In a 1975 study by Clark and Lucy, this claim was tested. Subjects were shown a picture and a request and were asked to indicate whether or not the picture showed the request being carried out. Both direct and indirect requests were used so that their verification times could be compared. Clark and Lucy also used requests of both a positive and negative nature so that the response time patterns for these could be compared to response time patterns for affirmative and negative assertions in earlier studies such as Clark and Chase (1972). In these earlier studies, in which subjects were given a picture-sentence comparison task, true affirmative sentences were responded to more quickly than false affirmatives. Both true and false affirmative sentences elicited quicker response times than true negative assertions, with false negatives eliciting the longest response times.

Clark and Lucy found that both direct and indirect requests showed the same response time patterns although response times were slower overall for indirect requests than for direct requests. (This is to be expected, given Clark's model, because the chain of interpretation for indirect requests would be longer than for direct requests and would thus, presumably, require more processing time.) Furthermore, Clark and Lucy found that requests of either type differing in positive or negative force showed the same response time patterns as affirmative and negative assertions, a finding that is to be expected from earlier studies. Thus, requests of a positive nature (e.g., "Can you open the door?" "Please open the door") were responded to faster than requests of a negative nature (e.g., "Must you open the door?" "Please do not open the door"), with "true" requests (i.e., those verified by their matching pictures) eliciting faster responses than "false" ones (i.e., those not verified by their matching pictures) in each category.

Other findings of the Clark and Lucy study are possible evidence for another notion implicit in Clark's model, namely, that the final interpretation of an utterance depends upon processing of previous interpretations. If this is, in fact, the case, subjects should take longer to respond to a sentence whose initial interpretation is difficult and requires more processing than to a sentence with a simpler initial interpretation.

Clark and Lucy examined response times to pairs of sentences like (8) and (9).

     (8)   I'll be very happy if you open the door
            I'll be very sad if you open the door
     (9)   I'll be very sad unless you open the door
            I'll be very unhappy unless you open the door

The first sentence in each of these pairs is positive in character; the second is negative. There is a crucial difference, however, between the way in which the final interpretations of pair (8) and pair (9) are arrived at because the direct interpretation of (8) depends on the word *if* and the direct interpretation of (9) depends on the word *unless*. *If* is easier to process than *unless*, which is semantically negative and has the approximate meaning "only if not." According to Clark's model, it should take longer to arrive at the final interpretation of (9) than of (8) because the initial interpretation of (9) is more difficult. This was, in fact, the study's finding. The sentences in pair (9) were responded to about 1 second slower than the sentences in pair (8).

Although this finding seems to confirm the chain-of-interpretation hypothesis, Clark points out that another interpretation of the result is possible. He suggests that it is the idiomatic character of certain indirect requests which may be responsible for the results here. It is possible that indirect requests like "Can you open the door?" are processed in the same way as idioms like *kick the bucket*, which is to say that the interpretation of "Can you open the door?" as a request is not mediated through its literal meaning but is, in fact, one of its direct interpretations. It is unclear which indirect requests, if any, are treated idiomatically and whether idiomaticity is an either/or property of sentences and phrases, or operates in a continuum. Until these issues are definitively settled, the results for sentence pairs such as (8) and (9) are indeterminate.

Evidence that listeners not only construct chains of interpretation but also remember them may be found in a study by Jarvella and Collas (1974).

More evidence that listeners construct chains of interpretation is found in a study by Keenan, Macwhinney, and Mayhew (1977). Keenan *et al.* constructed sentences like (10) and (11), both of which, in their direct interpretations, are assertions. Sentence (11), uttered in a certain context has, in addition to its literal interpretation, an indirect interpretation in which it is also a reprimand.

(10)   I think there are two fundamental tasks in this study
(11)   I think you've made a fundamental error in this study

Clark claims that sentence (10) has only one memory code because it has only one interpretation, whereas (11) has two, corresponding to its two interpretations. If this is the case, subjects should remember (11) more easily than (10) and should more easily be able to differentiate between (11) and paraphrases of its literal meaning because these paraphrases may be structured so that they do not lead to the reprimand interpretation of (11).

Keenan *et al.* claimed that sentences like (11) had "high interactional content" which, in Clark's terminology means that they were characterized by a chain of interpretation. Sentences like (10) were labelled "low interactional content" meaning, in Clark's theory, that they had only their direct, literal interpretations.

Keenan *et al.*'s subjects attended a talk and a few days later were asked to say if certain sentences had been uttered during the conversation they had attended. Subjects were given sentences that had been uttered during the talk, paraphrases of real sentences from the talk, and sentences that would have been expected to occur at the talk but that did not.

Keenan *et al.* found that high interactional content sentences had a higher incidence of recognition than low interactional content sentences. Subjects also showed greater accuracy in differentiating between high interactional content sentences and their paraphrases than they did for low interactional content sentences.

The difference between high and low interactional content sentences disappeared in Keenan *et al.*'s control study where subjects not present at the discussion session were told to study a list of these sentences and given the same recognition task. In this condition, subjects showed about equal recognition levels for both types of sentences. Clark suggests the reason for this is that when not presented in context, all sentences were probably interpreted directly so that all met Keenan *et al.*'s definition for low interactional content.

Interpretation of definite reference, like interpretation of indirect requests, is an example of the problem-solving character of natural language comprehension, since it requires the listener to draw inferences about what is being referred to, given an utterance, its context, and the tacit agreement governing the use of definite reference in speech.

Clark formulates a model of this process as follows (p. 302):

Step 1. Compute the description of the intended referent.

Step 2. Search memory for an entity that fits this description and satisfies the criterion that the speaker could expect you to select it uniquely on the basis of this description. If successful, go to Step 4.

Step 3. Add the simplest assumption to memory that posits the existence of an entity that fits this description and satisfies the criterion that the speaker could expect you to select it uniquely on the basis of this description. If successful, go to Step 4.

Step 4. Identify this entity as the intended referent.

The tacit agreement governing this process is Clark and Haviland's given/new contract, which has been formulated as follows (p. 307):

> The speaker agrees to use a definite noun phrase only when he has a specific referent in mind and is confident that the listener can identify it uniquely from its description in the noun phrase.

The model proposed by Clark would predict that identifying the referent in a sequence of sentences like (12) would not require the construction of a bridging assumption because at Step 2, when required to search memory for an entity satisfying the description *the woman*, a listener would find just such an entity in the following sentence (12).

(12)   I met a man and a woman yesterday—the woman was a doctor
(13)   I met two people yesterday—the woman was a doctor

Given a sequence of sentences like (13), however, the listener would be required to construct a bridging assumption because the first sentence in the sequence contains no such entity. The bridging assumption which the listener would have to construct is: One of the two people mentioned is a woman and the other is not.

Clark concedes that this model of definite reference interpretation leaves a great deal unspecified. For example, nothing is said here about how memory actually operates or when memory search stops and bridge building begins. Nor is the process of bridge building itself examined or defined.

Clark's theory is significant in suggesting what inferences listeners must make beyond literal meaning in order to understand language in a discourse situation. Many of his suggestions are intuitively plausible; however, he offers no formalism for representing the proposition in sentences or inferences and he has no system of rules which constrain the class of possible propositions or inferences. This renders his theory empirically unverifiable. As Anderson (1976) points out, although psycholinguistic experiments by Clark and others seem to bear out his theory, it has not been shown that other possible explanations cannot equally well account for the phenomena Clark has observed.

Clark's theory skips what must certainly be the first step in comprehension, namely the encoding of the incoming string. Clark's model starts after the encoding has taken place and shows how the encoded material is used. The assumption Clark makes is that the incoming string ends up in some propositional form. However, he offers no independent evidence that this is so and does not give us the means to test this contention empirically because he offers no formalism for rendering incoming strings into propositions.

If these omissions in Clark's theory can be filled in, it offers an excellent process model to work with for AI researchers.

Sentence-Processing Mechanism

There are currently several views and claims concerning the nature of the human sentence-processing mechanism (HSPM). For example, Cairns (1980) and Forster (1979) argue that the processor is composed of several autonomous dedicated subprocessors, whereas Marslen-Wilson (1976) and Marslen-Wilson and Tyler (1980) present data which they claim show that the view of serial subprocessors is mistaken and that HSPM is instead a set of interactive devices. In this section we will consider these and other conflicting assertions about the form and function of the HSPM in the hope of determining some of the minimal requirements for a model of human language comprehension.

One of the most impressive feats of the HSPM is that it takes the continuous speech stream and segments it into separate words (or perhaps some other tokens). Very little has been said about just how this segmentation occurs, but there is a large literature on lexical processing which is fed by the segmentation process. Marslen-Wilson and his colleagues have done a lot of work on lexical processing using a technique called "shadowing" (see Marslen-Wilson and Tyler, 1980, and the references therein). In a shadowing experiment the subject is presented with two simultaneous utterances, one in each ear through headphones, and is asked to ignore one and attend to ("shadow") the other. The subject is asked at the same time to repeat the shadowed utterance as quickly as possible (cf. Cherry, 1953). Most subjects have a delay of 400 to 750 ms, that is, two to three syllables (Marslen-Wilson, 1976). Some subjects, however, have a delay of only 250 ms or one syllable. A serial model of processing would predict that these close shadowers were just doing a very "shallow" analysis. If this were true, one would expect that their retention would be less good when compared with more distant shadowers, and also one would predict different patterns of errors between the two types of shadowers. However, Marslen-Wilson has found that close shadowers remembered as much as distant shadowers and that there was no detectable difference in their errors. In experiments reported by Marslen-Wilson (1976) it was found that close shadowers were just as affected by syntactic and semantic anomalies in the sentences that they shadowed as the distant shadowers were. It was also discovered that phonologically mutated words were restored significantly more often if the first syllable or two were the same as a word that was semantically and syntactically correct. Marslen-Wilson and his colleagues take these findings as evidence that lexical access does not just rely on phonetic input but also uses syntactic and semantic information available from the context (both inter- and intrasentential context).

To verify the predictions resulting from this "on-line interactive" model of language comprehension, Marslen-Wilson and Tyler (1980) used what is called three-word monitoring tasks: subjects were told to look for the same word (*identical*), a rhyming word (*rhyme*), or a word belonging to some taxonomic category (*category*). They also varied the prose contexts. The three different types of prose were: *normal prose* (normal English), *syntactic prose* (content words changed for words of the same form class), and *random word order* (scrambled sentences). The idea is that the different tasks require different types of information (phonological versus semantic) and that the different prose types allow the processor access to different combinations of phonological, syntactic, and semantic information. An autonomous serial-processing model would predict that prose type would not affect performance on the identical task and that rhyme would be done faster than category. However, Marslen-Wilson and Tyler found that syntactic and semantic contexts did facilitate monitoring for the identical lexical items. Reaction times were significantly slower in syntactic prose and random word order. They failed to find any evidence of the temporal priority of the availability of phonological information over semantic information (i.e., there was no difference in performance on the rhyme and category tasks). Thus, they claim, they have supported the predictions of their on-line interactive model and have shown contradictory evidence for the predictions of an autonomous model.

On the other hand, Cairns (1980) has postulated that there are at least three separate operations that make up "lexical processing." These are: *retrieval* (the item is looked up in the lexicon); *post-access decisions* (the retrieved items are checked against the input stream to insure the correctness of retrieval); *integration* (the newly retrieved lexical items are incorporated with other processed lexical items). Marslen-Wilson and his associates claim that their work covers at least the first stage of these processes, but Cairns claims that it does not. Cairns cites the work of Cairns and Kamerman (1975), Swinney (1979), and Tanenhaus, Leiman, and Seidenberg (1979) as showing that the retrieval and post-access decision processes are completed within approximately 250 ms. Recall that Marslen-Wilson showed in his experiments a delay factor of 250 ms for close shadowers. However, this result may be misleading since he does not say what proportion of the population these close shadowers are. Our survey of literature, including the report of Marslen-Wilson and Welsh (1978), produces an average shadowing latency of about 800 ms. This may appear to support the views of the autonomists because it would lead one to believe that most of the shadowing phenomena occur well after lexical access. There are also problems with the conclusions of Marslen-Wilson and Tyler (1980). First, there was no check made to see if normal language comprehension was going on. Although it seems reasonable to assume that it was in normal prose, there seems to be no reason to assume that it was in the syntactic prose or random word order conditions. In fact, it is not even clear that "normal language processing" is a well-defined notion over random strings of English words. Furthermore, in the one example that they give of syntactic prose, it appears that the strict subcategorization requirements were violated. If this is so for a large portion of their syntactic prose sentences, then normal parsing could not have taken place, as they claim it did. Another major problem is that when the subjects were told to monitor for a particular word (say, *lead*), the test sentences do not appear to have contained words that were identical in the initial syllables (say, *letter*). All the subject had to do was to hear the initial phoneme or two and respond. This would in fact explain how the subjects were able to respond well before the word was finished. No lexical access was necessary, only phoneme matching. Thus, on the basis of these arguments, one could presumably maintain the view that the process of lexical access is autonomous (see Chapter 9, however).

Another issue in lexical accessing over which there has been some controversy concerns what is retrieved when an ambiguous lexical item is processed. It has been argued that in a semantically neutral context, all the meanings of a lexically ambiguous word or token are retrieved based only on phonological information (see Cairns, 1973, and Foss, 1970). The decision process in which one meaning is chosen to be incorporated into the rest of the sentence was shown by Cairns and Kamerman (1975) to be completed within two syllables (approximately 500 ms) after the ambiguous item had been accessed. The question then was this: Could semantic context bias retrieval process such that only one meaning is retrieved? Some studies have shown that biasing context could reduce on-line processing difficulties found with ambiguous lexical items (e.g., Cairns and Hsu, 1980; Swinney and Hakes, 1976), whereas other studies have shown that it could not (e.g., Foss and Jenkins, 1973; Holmes, Arwas, and Garrett, 1977). The controversy is claimed to have been settled recently by the studies reported in

Tanenhaus, Leiman, and Seidenberg (1979) and Swinney (1979). These studies claim that a lexically ambiguous item will prime words related to all of its different meanings for a very short time and then only the contextually correct meaning is primed. Tanenhaus *et al.* identified this time period as 200 ms, whereas Swinney found that it occurs within 750 ms. Thus, it appears to be the case that context does not effect lexical access but only post-access decisions.

Moyne (1981, 1984) has proposed that the question of autonomy has not been resolved and, furthermore, it is undecidable. He proposes a dual processing model in which, depending on various circumstances, both autonomous and nonautonomous processing is possible.

After a word has been retrieved from the lexicon, it must be syntactically associated with the other words processed (i.e., placed into a phrase marker of some sort). It is usually assumed in the linguistic literature that this phrase marker is a tree diagram (see, for example, Fodor and Frazier, 1980; Ford, Bresnan, and Kaplan, 1982; Kimball, 1973). Langendoen and Moyne (1981) have argued, however, that the parser produces a labelled bracketing construct which is not just a variation of the tree structure in the usual sense, but a "reduced" postfix notation where category markers pile up at the end of the string and several opening brackets may be closed by a single closing bracket. We will return to this proposal later; at present, its relevance is to point out that although it may be conceptually easier to consider the HSPM as building and manipulating trees or the equivalent bracketing representations of them, there are other approaches which may facilitate the parsing mechanism in terms of "reduced" or contracted notations. The postfix representation has the added attraction of supporting the evidence obtained by various studies that suggest that the HSPM is more active during the processing of the ends of major constituents than it is during the processing of the beginnings of such constituents (Carroll and Tanenhaus, 1975; Fodor, Bever, and Garrett, 1974).

The internal structure of the syntactic processor has recently been the focal point of some disagreement. Frazier and Fodor (1978) and Fodor and Frazier (1980) have presented arguments that there must be two stages to syntactic parsing. The first stage (their PPP, preliminary phrase packager) can only view a few words at a time (they postulate that six is about the right number) and assigns all the structure that it can before shunting that material to the second-stage parser (their SSS, sentence structure supervisor). The SSS takes the data passed from the PPP and finishes the phrase marker for the whole string. As a motivation for a two-stage parser, Frazier and Fodor (1978) point out that without some sort of chunking, the parser should be overloading toward the end of a long sentence, given that the capacity of working memory is severely limited. However, sentence length is not a good measure of processing complexity. The demand on working memory can be reduced to a reasonable level if one recalls that in addition to clearing partial analyses from the first-stage parser and shunting them to the second-stage parser, there are other possibilities; for example, arriving at the comprehension of a sentence before all the levels of analyses have been "completed." Furthermore, there is the phenomenon that material with more structure places less demands on the storage space. However, as motivation for their own particular model of two-stage parsing, called the *sausage machine* (SM), Fodor and Frazier present sentences like 14–18 below:

(14) Joe bought the book for Susan
(15) Joe bought the book that I had been trying to obtain for Susan
(16) John read the note to Mary
(17) John read the memo, the letter and the note to Mary
(18) Though Martha claimed that she will be the first woman president, yesterday she announced she'd rather be an astronaut

In (15) the prepositional phrase (PP) *for Susan* cannot be associated with the verb *bought*, as it is in (14), because when the PP is encountered by the PPP, *bought* is no longer within its viewing window. Sentences (16) and (17) exhibit the same kind of closure differences with fewer structural and lexical differences. Sentence (18) shows that this cannot be entailed by a principle such as *right association* (Kimball, 1973), but is in fact *local association* as the SM requires. Thus, it would follow that the parser in some way will be sensitive to the size of the constituents it is dealing with.

Wanner (1980) and Ford, Bresnan, and Kaplan (1982) point out that size is not the only consideration affecting the parser. Compare (14) with (19) and (20).

(19) Joe included the book for Susan
(20) Joe said that Bill died yesterday

The only difference between (14) and (19) is the verb, yet the closure preferences are different. In (14) the PP is associated with the verb; in (19) it is associated with the noun phrase. The PPP should be able to "see" the whole in (20), so there should be no preference in analysis. However, low right attachment is clearly preferred. Ford, Bresnan, and Kaplan take these examples as evidence that what is important to the parser is not the length of the constituent to be attached and the length of the constituent being attached into, but rather the lexical properties of the verb. By assuming a relationship of *strength* that holds between the various lexical forms of verbs (i.e., what grammatical functions they require for the string to be grammatical) and postulating four parsing principles, Ford *et al.* can predict not only closure biases, but also the lexical expectations noted by Fodor (1978), various garden-path effects, and gap-finding strategies. They argue that not only does the SM fail to account for the effect engendered by changing verbs, it cannot do this in principle because the SM is based on the transformational theory of grammar. Ford, Bresnan, and Kaplan argue that transformational grammar cannot capture with strict subcategorization what their lexical-functional grammar does with its functional structures. Consider the sentences:

(21)   A library can be found in the barracks for the recruits
(22)   In the barracks can be found a library for the recruits

Using the evidence from locative-inversion sentences like (21) and (22), Ford *et al.* show that the functional subject has the same closure properties as any other argument of the verb in final position, as opposed to verbal complements which are not arguments. We should point out, however, that although it is true that strict subcategorization cannot account for this given that these are strictly local statements and, therefore, cannot mention the subject, it is not clear that a transformational grammar cannot capture the notion of the "argument" of a verb. Indeed, there has recently been much discussion about the fact that a grammar must define this notion (cf. Chomsky, 1981). However, the difference between sentences (16) and (17) does not seem to be lexical. Thomas Maxfield (personal communication, July, 1981) suggests that an adequate model of syntactic parsing will have to be sensitive to both size and lexical properties of the verb. In light of the above discussion of sentences (14) and (19), consider these two sentences (due to Edwin Battistella):

(23)   John read a book for Mary
(24)   John read a book for linguists

As in (14) and (19), the only difference between (23) and (24) is in one lexical item. However, we cannot find a solution for this pair of sentences analogous to the one that Ford *et al.* propose for (14) and (19), because even if we assume that nouns have an argument structure (and it is not clear that they do), in this construction what makes the difference would be the argument, not the predicate. One could claim (as Ford *et al.* do, in places, as well as Frazier and Fodor, 1978) that the difference is based on semantic/pragmatic information. This would require giving up the autonomy in the sense of Forster (1979) and Cairns (1980).

We should point out one other aspect of a viable parser which does not appear to be controversial but has implications for the autonomy view. The parser must be able to predict what it is about to encounter. This is built into the augmented transition network models such as Wanner (1980) and Wanner and Maratsos (1978). It is also a consequence of the parsing mechanism postulated by Ford *et al.* already discussed. Fodor and Frazier (1980) also provide for this in their SM, arguing that it is necessary for the speed and efficiency of the parser.

Text-Processing Mechanisms

Linguists have been concerned with sentences and their structures as discrete objects of grammatical analysis. Much of the relatively small effort for language use and understanding in the linguistic sense of *performance*, or in the cognitive and psychological view of comprehension and perception, has been concerned with sentence processing. The current research in artificial intelligence with its emphasis on text and discourse processing is bound to have impact on our understanding of human language comprehension.

In modern linguistics, among the first to discuss grammatical analysis beyond sentences was Zellig Harris (1952; reprinted with commentaries in Fodor and Katz, 1964). Harris conceived of discourse analysis as a study dealing with two sets of distinct but related problems: first, the problem of continuing descriptive linguistics beyond the limits of a single sentence at a time; second, the problem of correlating what we now call "knowledge of the world" or one's cultural heritage with language (i.e., "nonlinguistic and linguistic behavior"). Recall that this contribution by Harris was made in the era of descriptive linguistics; hence it was not so much concerned with the comprehension of a connected discourse as it was with the classification and semiformal analysis of a text. Harris developed a methodology for partitioning a text into a group of equivalent classes. Each class consisted of elements or sequences of elements which had identical or equivalent environments within sentences. If some element or elements did not have equivalent environments, they were classified with elements with which they had close grammatical ties. Next, similarity and other relations between equivalent classes were investigated, and attempt was made to establish the distribution of classes in the text and the patterning of class occurrences. These operations made

> no use of any knowledge concerning the meaning of the morphemes or the intent or conditions of the author. They requir[ed] only a knowledge of morpheme boundaries, including sentence junctures and other morphemic intonations (or punctuation). (Fodor and Katz, 1964, p. 383)

Thus, this approach to discourse analysis was intended to provide information about the structure of a text or its type classification, and about the role that each element played in such a structure. Descriptive linguistics at this time was concerned with the role that each element played in the structure of a sentence.

Katz and Fodor (1963), proposing a semantic theory as a component of the generative transformational grammar (cf. Moyne, 1974), concluded that the upper bound for the scope of a semantic theory of linguistic *competence* was a sentence. They further assumed that for the purposes of their theory, a discourse could be regarded as a single sentence which consists of the sequence of sentences in the text connected by some appropriate sentential connectives (or conjunctions). Kiefer (1975) proposed that there is a coordination operation that takes sentences as input and yields texts. However, Lang (1973) and Kiefer (1977) argued that such coordination operations and sentence conjunctions were constrained by certain conditions which divided a discourse into "bound" and "free" texts. Katz and Fodor assumed that any text that did not meet the criterion of sentence coordination was beyond the scope of semantic theory.

Whatever the implications of these observations, it is clear that humans comprehend and, particularly, disambiguate utterances not only in the context of a discourse (both "bound" and "free") but also in the context of their encyclopedic knowledge of the real world. For the purposes of this study, concerned with mechanisms for human comprehension, the approaches taken in cognitive science and psychology are relevant (see Chapter 12 of Lachman *et al.*, 1979, and the references therein for a survey). As an example of a theoretical approach relevant to comprehension, we will give an outline of the proposal by Kintsch and his associates.

The goal of Kintsch's theory is to account for the processing of texts and to explain how it is possible for readers to understand texts, remember them, and paraphrase or summarize them. Kintsch's essential claim is that the processing of a text is cyclic and results in the representation of its meaning as a hierarchically ordered set of propositions which he calls a text base.

A text base has both a macrostructure and a microstructure. The macrostructure is a structural representation of textual coherence at a global level. It may be likened to an outline of the text. The microstructure of a text, its detailed representation, is arrived at in the following way:

Sentences, appearing in the same order as they do in the text, are broken down into one or more propositions. Each proposition consists of a relational term and one or more arguments. In Kintsch's notation, verbs, modifiers, and sentence connectors function as relational terms, whereas nouns or whole propositions are used as arguments. (Kintsch notes that it is the ability of the system to embed propositions inside other propositions that gives language its descriptive richness.) It is assumed that relational terms and arguments are concepts. To distinguish between these and lexical items, they are written in capital letters.

Propositions in the text base are listed, numbered, and hierarchically ordered according to reference, that is, the repetition of arguments. One or more propositions are placed at the superordinate level of the hierarchy; all propositions sharing an argument with the superordinate propositions are ordered at the next (second) level of the hierarchy. Propositions sharing an argument with the propositions at the second level are placed at the third level, and this process continues until all propositions have been ordered in a like manner. The resulting connected graph is the microstructure of the text base and, if it has been successfully executed, indicates that the text is coherent at the local level. The example that follows of a text base hierarchy is reproduced from Kintsch (1978).

It is clear from the work of Clark, Kintsch, and others (see discussion of Clark) that the drawing of inferences is necessary to language processing. For that reason, in Kintsch's model, a text can be processed only if it is both referentially and inferentially coherent. Referential coherence can be fairly easily accounted for in Kintsch's theory because it involves only overlap of arguments. Accounting for the inferential coherence of a text base is far less simple, and, in fact, Kintsch does not attempt to do so, merely stating that, where gaps in the text base do occur, missing inferences in the form of bridging propositions are supplied as needed for processing.

Kintsch suggests, however, that a complete language comprehension system would look to the semantic components, the speaker's store of real-world knowledge, and schema theory, to describe and formalize how a reader/speaker supplies missing inferences.

Fragment of an Episode from a Short Story and the Corresponding Text Base

Text	Text Base
This Landolfo, then, having made the sort of preliminary calculations merchants normally make, purchased a very large ship, loaded it with a mixed cargo of goods paid for out of his own pocket, and sailed with them to Cyprus. (The episode continues with a description of how this endeavor finally resulted in Landolfo's ruin.)	1(PURCHASE,agent:L, object:SHIP) 2(LARGE,SHIP) 3(VERY,2) 4(AFTER,1,5) 5(CALCULATE, agent:L) 6(PRELIMINARY,5) 7(LIKE,5,8) 8(CALCULATE, agent:MERCHANT) 9(NORMAL,8) 10(LOAD,agent:L, goal:SHIP, object:CARGO) 11(MIXED,CARGO) 12(CONSIST OF, object:CARGO, source:GOODS) 13(PAY,agent:L, object:GOODS, instrument:MONEY) 14(OWN,agent:L, object:MONEY) 15(SAIL,agent:L, object:GOODS, goal:CYPRUS)

The role which the semantic component plays in the drawing of inference is dealt with in the discussion of propositional notation. However, because schema theory is important not only to the question of inferential coherence, but also to the formulation of macrorules and macrostructures, it will be examined here.

Kintsch defines a schema as a "a representation of a situation or of an event; it is a prototype or norm and specifies the usual sequence

Level 1   Level 2   Level 3   Level 4

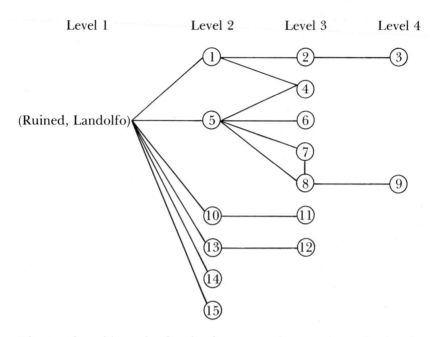

The text base hierarchy for the fragment of a text shown in the above table. Propositions are indicated only by their numbers; shared arguments among them are shown as connecting lines.

of events that is to be expected" (Kintsch, 1978, p. 78), given a particular situation. Schemata are not exact constructs, but represent a loosely structural framework within which a great deal of real-world information may be represented.

Kintsch cites evidence which indicates that schemata are important in both comprehension and recall. In a 1975 study by Kintsch and Van Dijk, subjects were asked to read two stories of comparable length and syntactic complexity and write summaries of each of them. One, a selection from Boccaccio's *Decameron*, conforms quite closely to the familiar scheme representing the notion of a story in the tradition of the European literary style. The second, an Apache Indian folktale, does not conform to this scheme. The subjects, American college students, were expected to find comprehension and summarization of the first story simpler than the second because of the first story's familiar structure.

This expectation was borne out by the findings of the study. The subjects' summaries of the *Decameron* story resembled each other more closely than did their summaries of the Apache story, indicating that the subjects had a clearer notion, for the European story, of what a summary should contain. In fact, when subjects were given a scrambled version of the *Decameron* story, they were able to write coherent summaries comparable to those written for the unscrambled version, even though the scrambled stories took longer for them to read. It should be noted that the Apache story was by no means incomprehensible to subjects. However, there was significantly less agreement among subjects as to what would be relevant or appropriate in the summary of such a story.

Kintsch and Van Dijk suggest that subjects' having a schema for one story and not the other is responsible for these results. The claim is not that comprehension without a schema is impossible, but rather that a schema facilitates comprehension and recall and/or summarization. They hypothesize that a schema "functions like an outline with empty slots. When one reads a story, this outline is filled in and becomes the macrostructure of that story" (Kintsch, 1978, p. 82).

Kintsch and Van Dijk (1978) point out that different schemata for the same story may be activated by different readers or by the same reader in different situations. Thus, for example, the typical story schema may be evoked for a subject who reads a *Decameron* story in order to summarize it, whereas a completely different schema (overriding the typical story schema) may be evoked for a subject who reads the story as part of his research on literary style or for the purposes of reviewing it.

The derivation of the macrostructure of a text is, as has been noted, schema-based and largely inferential. In Kintsch's theory, macrorules are used to derive macrostructures from microstructures. Macrorules must distill the essence of the microstructure and organize it globally so that the gist of the text is preserved and detail unnecessary to the macrostructure is eliminated. Because the process of comprehension involves inference, and the interpretation of a given proposition is frequently dependent upon prior propositions, one constraint which must be placed on the application of macrorules is that they may not destroy a proposition which is needed for the interpretation of another proposition. Otherwise, macrorules may apply freely as needed in order to accomplish any of the following: (1) deletion of unnecessary detail, (2) generalization, that is, substitution of several specific propositions in the microstructure with one general proposition which contains them, and (3) construction of propositions which are logical conditions or consequences of propositions in the microstructure.

As propositional notation is clearly essential to the construction of both microstructure and macrostructure in Kintsch's text-processing model, he offers several motivations for using propositional notation rather than some other form of representation for sentences. First, he notes that a propositional notation renders experimental procedures simpler because it provides a means of drawing equivalences between paraphrases and simplifies scoring.

A second motivation for the use of a propositional notation is that it is necessary to a description of how inferences are drawn as part of the language comprehension process. The concepts which function as arguments and relational terms in this notation are a shorthand for knowledge complexes which will be fully specified in the lexicon. This information in the lexicon is necessary to explain how, given a sentence such as (25), the reader infers (26), or any of a number of possible inferences based upon knowledge of the structure of the verb *buy* and knowledge of what buying involves in the real world.

(25) John bought a ring for Mary
(26) John used some form of money in this transaction

Finally, as Kintsch points out, a notation such as this makes an empirical claim, namely, that semantic information is central to processing and that surface syntactic characteristics of the signal are not. Although never explicitly stated, the claim seems to be that the proposition is a unit of processing and/or recall.

Kintsch cites a number of studies which appear to bear out this contention. A study by Wanner (1974), later replicated by others, showed that it is possible to predict whether or not a lexical item is a good recall cue for a particular sentence based on its propositional analysis. Given pairs of sentences such as (27) and (28), Wanner hypothesized that the word *detective* would be a better recall cue

(27) The governor asked the detective to cease drinking
(28) The governor asked the detective to prevent drinking

for (27) than for (28). An examination of the propositional analysis of the sentences reveals why this should be the case. The propositional analysis of (27) is (29), whereas (30) corresponds to (28). The word concept *detective* appears as the argument

(29) (ask, governor, detective, (cease, detective, (drink, detective)))
(30) (ask, governor, detective, (prevent, detective, (drink, someone)))

in three propositions in (29), but as the argument in only two of the propositions in (30). By contrast, the lexical item *governor*, which occurs with the same frequency in the propositional representations of (27) and (28), is an equally effective recall cue for both of the sentences.

A plausible explanation for these findings seems to be that the more frequently a word concept appears in the propositional representation of a sentence, the better a recall cue its corresponding lexical item will be, an indication that propositional structure has perceptual reality.

A 1973 study by Kintsch and Keenan demonstrated that the length of time subjects need to read a text relates predictably to the quantity of the text which they are able to remember. They noted, however, that what determines reading time for a particular text is the number of propositions in sentences and not their lexical length.

In the study, subjects were given sentences of equal length which differed in the number of propositions they contained. They were asked to read the sentences and press a response button when finished. Subjects were then expected to write the sentences, although their written responses were not required to be a word-for-word replication of the test sentences.

Kintsch and Keenan found that "for each additional proposition recalled, an additional 1.5 sec of reading time was required. Therefore, the number of propositions, not the number of words, determined reading time in this study" (Kintsch, 1978, p. 63).

Kintsch and Keenan were also able to predict which propositions were more likely to be recalled by subjects on the basis of the structure of the text base. They hypothesized that more superordinate propositions would be recalled by subjects than subordinate propositions. This was, in fact, shown to be the case. Kintsch and Keenan found that approximately 80% of the first-level propositions of short paragraphs were recalled, but subjects recalled only 30% of subordinate propositions.

Kintsch cites a study by Meyer (1975), who obtained similar results by placing a paragraph in different parts of a text and observing the effect of its position in the text on recall. Subjects demonstrated greater recall for the same paragraph when it was located higher in the text base hierarchy than when it was placed lower.

Kintsch suggests that the reason superordinate propositions are easier for subjects to recall is that throughout the cyclic processing of the text, they are subject to implicit repetition. An examination of Kintsch's process model shows how such implicit repetition takes place.

Kintsch's processor operates cyclically on propositions, not sentences. The size of the chunks fed to the processor depends upon the difficulty and/or familiarity of the material, and the skill of the reader. This flexibility which Kintsch builds into his model insures that it will be able to accurately reflect real differences in comprehension and recall for different types of materials and readers of varying abilities.

Given that the number of propositions has been determined for a particular type of text, a chunk of that size is inserted into the processor where it is scanned for referential coherence. Where inferences are needed, they are supplied. The processor is attempting to construct connected graphs. A number of propositions (their selection dependent upon the complexity of the material and reader strategies) are held in a short-term buffer so that incoming propositions can be checked against them for argument overlap.

Thus, in each cycle, a number of new propositions are processed whereas a number of propositions retained in the buffer are reprocessed as part of referential checking. Old propositions in the buffer are replaced with newer propositions and then shunted to long-term storage to be reproduced or recalled with a production probability which is a function of textual organization and reader strategies. Kintsch suggests that good readers look to the recency of propositions in the text and their position in the hierarchy in deciding whether or not to retain them (and thereby reprocess them) in the short-term buffer. If readers use a strategy of retaining propositions high in the text base hierarchy, Kintsch's hypothesis about their implicit reprocessing is explained.

Kintsch's theory of text processing explains a wide range of phenomena and is eminently testable. For these reasons it bears further examination and elaboration.

Exercises

1. Language development among humans seems to represent a high degree of abstraction, which is a cognitive process. How would you differentiate this capacity from various and sometimes complex means of communication which exist among dolphins, elephants, prairie dogs, birds, bees, and so on? It has been claimed, however, that apes can develop a certain amount of syntactic abstraction ability. See Premack (1976) for a nontechnical presentation of this view. See also the debate between Chomsky and Premack in *The*

Sciences, Journal of the New York Academy of Sciences, November 1979.

2. In Chapter 2, we discussed language as an object of linguistic study. In this chapter we have considered language from the view point of cognitive psychology. What are some of the common and differing parameters in these two approaches?

3. There are other human capacities for abstraction: music, mathematics, dance movements, and so on. Is it language faculty which is unique in humans or is it abstraction faculty? The capacity to think? *Cogito ergo sum!*

4. Compare Miller and Johnson-Laird's position on language perception with that of Clark. Consult the additional references provided, if necessary.

5. Discuss the "the chains of interpretation" hypothesis with the help of the following two sentences:

   (a) I remember there were several men in this room
   (b) I remember you saw several men in this room

6. Discuss the opposing views concerning the autonomy of subcomponents in sentence processing.

7. Discuss the controversy concerning lexical accessing of ambiguous lexical items.

8. What are the arguments for the two-stage parsing of the "sausage machine" proposal?

9. Give an outline of Kintsch's theory of text processing (discourse), and compare this with the views of Zellig Harris.

References

Levelt and Flores d'Arcais (1978) contains surveys and research articles on the state of the art in language perception. Clark and Clark (1977) is an excellent text on psycholinguistics and cognitive processes in language comprehension. Lachman, Lachman, and Butterfield (1979) is an informative book on human information processing, and the volume edited by Cole (1980) contains articles by different authors on current research on speech perception. Winograd (1983) is an excellent source for current models and practical algorithms for computer processing. The following journals, among others, often contain articles on the current research on the topics discussed in this chapter: *Cognitive Psychology, Cognition, Journal of Verbal Learning and Verbal Behavior, Memory and Cognition.*

CHAPTER 5

Grammars and Parsing Strategies

Introduction

There is no doubt that the development of the generative transformational grammar and the transformational theory in linguistics have had the widest and the most profound impact on all aspects of studies on human language processing in the last several decades. Many of the divergent and competing theories extant today started as offshoots or variants of the TG theory although it is sometimes hard to see the connection now. It is for these reasons that we have devoted a chapter in the first part of this study to the transformational theory. In this chapter we will discuss various other grammatical studies, parsing strategies, and theoretical proposals that have relevance for the plan and goals of this study.

Lexical Functional Grammar

The original work which has evolved into lexical-functional grammar (LFG) was seen as an extension of transformational grammar (Bresnan, 1978). However, at this stage in the development of the two theories, it is difficult to imagine that they are closely related. Much of the work done in the LFG approach to linguistics explicitly assumes the *competence hypothesis* (Ford, Bresnan and Kaplan, 1982; Kaplan and Bresnan, 1982). The competence hypothesis states that a model of performance must incorporate a theoretically sound grammar which is separate from and independent of the computational mechanisms that operate on it (i.e., the processor) and from the nongrammatical factors that effect the processor. Although these may be studied separately, it is assumed that the requirement that they fit together into a coherent system places more severe restrictions on the form and function of the separate components than may be imposed on any one individually. It is generally accepted in transformational theory that a theory of performance will include a linguistically sound theory of competence (see Chomsky, 1965, for example), but considerations of performance do not usually enter into the arguments about competence.

LFG is like transformational grammar in that LFG postulates two distinct levels of syntactic description, one which feeds the phonological component and one which feeds the semantic component. In transformational grammar, the two levels are both taken to be phrase markers whereas in LFG, one level is a phrase marker and the other level is a different type of structure. The level in LFG which is a phrase marker is called the *constituent structure* (c-structure) of a sentence. C-structures define the hierarchical relationships in sentences. These are the "surface" relationships and are given by a "slightly modified" context-free phrase structure grammar (Kaplan and Bresnan, 1982). There are no movement or deletion rules in the syntax of an LFG. The c-structure feeds directly into the phonological component. The other level of syntactic description is called the *functional structure* (f-structure). F-structures explicitly represent grammatical relations and are the input to the semantic component. F-structures give enough information so that the semantic component can determine the proper predicate–argument relationships. An f-structure consists of a set of ordered pairs, an attribute and its value. The relationships between predicate–argument structures and surface configurations is defined in two stages. Lexical entries give a direct mapping between semantic arguments and grammatical functions. Syntactic rules identify grammatical functions with particular morphological and structural configurations. The lexical form of a verb (which is the value of the PRED[icate] attribute in a sentence's f-structure) defines a mapping from grammatical functions to thematic (or semantic) relations. Consider the lexical form (1) of the verb *hand* as it occurs in (2):

(1) 'HAND ⟨(SUBJ) (OBJ2) (OBJ)⟩'
(2) John handed me the book

This lexical form will cause the semantic component to interpret the subject of *hand* as the agent, the second object (OBJ2) as theme, and the object as goal.

The context-free grammar that produces the c-structures is augmented so that it annotates various nodes as to the grammatical function which it serves and also so that it produces the *functional description* (f-description) of a string. An f-description is a finite set of statements which specify properties of the string's f-structure. It is an intermediary between a string's f-structure and c-structure. An f-description may be used to determine if an f-structure for the string has all the properties required by the grammar, or it may be used to construct the f-structure of the string associated with the f-description. The statements of an f-description are equations describing the values of attributes in an f-structure. These equations are possible because an attribute may have only one value in any given f-structure. We can, therefore, specify an f-structure by specifying the values of its grammatical functions. That is, by the equation

(3) f_1 (SUBJ) NUM =

   SG (where f_1 is a variable ranging over f-structures)

we can determine that the subject of f_1 is singular in number.

An f-description is built up using the grammar and the lexicon. The f-description is derived from schemata associated with certain elements on the right-hand side of the phrase structure rules and with syntactic features and semantic forms of lexical items. The schemata have the form of the statements to be derived, except that where the statement has an f-structure variable (f_1, f_2, etc.) the schemata have a *metavariable* ($\uparrow$ and $\downarrow$ for immediate domination and $\Uparrow$ and $\Downarrow$ for bounded or long-distance domination). The phrase structure rules then have the form:

(4) S → NP   VP
       ($\uparrow$ SUBJ) = $\downarrow$ $\uparrow$ = $\downarrow$

and lexical entries have the form:

(5) *girl*:   N ($\uparrow$ NUM)   = SG
                    ($\uparrow$ PRED)   = 'GIRL'
     handed: V ($\uparrow$ TENSE) = PAST
                    ($\uparrow$ PRED)   = 'HAND ($\uparrow$ SUBJ) ($\uparrow$ OBJ2) ($\uparrow$ OBJ)'

Lexical entries contain a specification of the preterminal category under which the item may be inserted and a set of schemata. Although the schemata have two different sources, they are treated uniformly by a tripartite instantiation procedure.

The first step in the instantiation procedure is to attach the schemata to the appropriate nodes in the c-structure. Then f-structures variables (called the node's " $\downarrow$ -variable") are added to the root node and to each node where a scheme contains a metavariable (this will be all the subtrees which correspond to a subcomponent of the f-structure). The third step in instantiation is substitution. In the substitution step, first all the $\downarrow$ metavariables at a node are replaced by that node's $\downarrow$ -variable and then that $\downarrow$ -variable replaces all of the node's daughter's $\uparrow$ metavariables. In this instantiation process, the schemata become the f-description of the sentence.

Long-distance metavariables are treated basically the same as immediate domination metavariables. They are introduced by certain phrase structure rules, for example, NP → e and by certain lexical
$$\uparrow = \Uparrow$$
forms, for example, who: N ($\uparrow$ PRED) = 'WHO' (certain details have
$$\uparrow = \Uparrow$$
been omitted; see Kaplan and Bresnan, 1982, for a full discussion). The difference between the two kinds of metavariables is that corresponding $\Uparrow$ and $\Downarrow$ may be infinitely far apart whereas corresponding $\uparrow$ and $\downarrow$ must be attached to mother and daughter nodes. The distance between $\Uparrow$ and $\Downarrow$ is, theoretically, limited only by the occurrences of *bounding nodes*. Kaplan and Bresnan argue against the Chomskian notion of subjacency with NP and S (possibly $\bar{\text{S}}$) as universal bounding nodes (see Chomsky 1977, and references therein). They support their argument with evidence from NP's that do not appear to be bounding nodes (cf. *Who$_i$ did John see* [$_{NP}$ *a picture* [$_{PP}$ *of* e$_i$]]). They propose instead that there is a universal theory of bounding nodes which specify in which syntactic configurations they may appear (Kaplan and Bresnan leave this theory to future research). Bounding nodes are then defined by the PS rules of a language taken from the finite stock offered by the universal theory. Bounding nodes are marked by enclosing a category on the right-hand side of a PS rule in a box, for example,

$$(6) \quad \bar{\text{S}} \rightarrow (\text{that}) \ \boxed{\text{S}}$$
$$\text{NP} \rightarrow \boxed{\text{NP}} \ \text{N}$$

However, this notion of defining bounding nodes runs into trouble with the genitive NP construction in English (e.g., *I reported Joe's dying to Bill*). Genitive NP's create *wh*-islands (i.e., they are bounding nodes), yet there is no simple way to state this. A plausible rule for this construction (from Ford, Bresnan and Kaplan, 1982) is

(7)   NP → NP's VP

In order to account for the fact that a ⇓ variable cannot control the constituents on the right-hand side of the rule, both elements must be enclosed in boxes even though VP's are not usually thought of as possible bounding nodes. Also, this is just a convoluted way of saying that the element on the left (i.e., the whole NP) constitutes a barrier to control dependencies.

Given that an f-description can be produced for every string that is generated by the PS rules, an algorithm may be devised which will attempt to construct an f-structure out of the f-description. This is done basically by "solving" the equations of the f-descriptions. That is, equals are substituted for equals until all the variables can be eliminated (except for the ↓-variable of the VP node which equals the f-structure of the whole sentence, given the schema attached to the VP node by [4]). However, just because a string has an f-description does not imply that it also has a valid f-structure. If the f-description gives more than one value to any given attribute, then the algorithm halts and marks the string as ungrammatical.

Strings that would be marked as ungrammatical because of violations of strict subcategorization in transformational theory are handled differently in LFG. Subcategorization plays no role in LFG (Ford, Bresnan and Kaplan, 1982). The function of subcategorization is taken care of by the notions of "completeness" and "coherence" (see Kaplan and Bresnan, 1982). These two notions basically define a biunique relation between f-structures and the predicates that they contain. Completeness requires that every grammatical function listed in the lexical form of the predicate must be contained in the f-structure and coherence requires that every grammatical function contained in the f-structure must be listed in the lexical form of the predicate. These are two of the major well-formedness conditions placed on the sentences generated by an LFG. If a string's f-structure is not complete and coherent, it is ungrammatical. Kaplan and Bresnan give the following sentences as examples of strings that are not complete and not coherent, respectively:

(8)   *The girl handed
(9)   *The girl fell the dog

One of the questions of interest to the present study is that of the generative power of theories of performance. An LFG has a (slightly modified) CF base; therefore it is reasonable to assume that the set of LFG's will generate the set of CF languages. Kaplan and Bresnan, however, offer a proof that this is not the case. First, they exclude from consideration derivations involving nonbranching dominance chains (chains such that X dominates Y which dominates Z which dominates X) or involving two or more optionality e's in a row. (An "optionality e" is introduced as an interpretation of the standard parenthesis notation; (NP) is understood as $\begin{bmatrix} NP \\ e \end{bmatrix}$ and is not to be confused with gaps introduced by rules like $NP \to e$, where e is a null element.) Derivations having either or both of these two characteristics allow infinitely many trees to correspond to one surface string. These derivations are excluded as unmotivated empirically and uninteresting intuitively. Given this, it is straightforward to show that the language generated by an LFG is recursive. Using the principle of "nearly nested correspondence" (which limits the number of times a metavariable correspondence may cross another metavariable correspondence), which is motivated syntactically, Kaplan and Bresnan claim that a nondeterministic linear bounded automaton can be constructed to accept exactly the language of any particular LFG. Thus, they claim, languages generated by LFG fall within the set of context-sensitive languages. Kaplan and Bresnan then show that the CS language $a^n b^n c^n$ can be generated by the following LFG:

$$(10) \quad S \to \quad A \qquad B \qquad C$$
$$\uparrow = \downarrow \ \ \uparrow = \downarrow \ \ \uparrow = \downarrow$$

$$A \to \left\{ \begin{array}{l} a \\ (\uparrow COUNT) = 0 \\ a \quad A \\ \qquad (\uparrow COUNT) = \downarrow \end{array} \right\}$$

$$B \to \left\{ \begin{array}{l} b \\ (\uparrow COUNT) = 0 \\ b \quad C \\ \qquad (\uparrow COUNT) = \downarrow \end{array} \right\}$$

$$C \to \left\{ \begin{array}{l} c \\ (\uparrow COUNT) = 0 \\ c \quad C \\ \qquad (\uparrow COUNT) = \downarrow \end{array} \right\}$$

The strings generated by this grammar will not have valid f-structure unless the terminal sequence is of the proper form (this is because of the S rewrite rule). That LFG should be included in the set of context-sensitive languages is a desirable result because some constructions are said to require that power (cf. the *respectively* construction in English).

A basic tenet of lexical-functional grammar is that each different syntactic frame in which a verb may appear is represented as a separate lexical entry (or lexical form) for that verb. (It is unclear whether this is necessary or if possibly these separate forms could all be listed under one entry). Relationships that could be stated in terms of transformations are instead stated as lexical redundancy rules. That is, there is in English a redundancy rule which states the passive relation that has roughly the following form:

$$(11) \quad (\uparrow \text{SUBJ}) \mapsto (\uparrow \text{BY OBJ})$$
$$(\uparrow \text{OBJ}) \ \mapsto (\uparrow \text{SUBJ})$$

This redundancy rule captures the relationship between lexical entries like the (somewhat simplified) following:

(12) *persuade*: V ($\uparrow$ PRED) = 'PERSUADE $\langle(\uparrow$ SUBJ) (OBJ)
$\qquad\qquad\qquad\qquad\qquad\qquad (\uparrow$ V COMP)$\rangle$'

(13) *persuade*: V ($\uparrow$ PRED) = 'PERSUADE $\langle(\uparrow$ BY OBJ) ($\uparrow$ SUBJ)
$\qquad\qquad\qquad\qquad\qquad\qquad (\uparrow$ V COMP)$\rangle$'

This rule is not productive in any sense. It just states a relationship that holds between many lexical entries. The various slots in which the grammatical functions are listed inform the semantic component which semantic rules to assign to which NP. That is, the first position filled by ($\uparrow$ SUBJ) in (12) and ($\uparrow$ BY OBJ) in (13) is the agent position. Therefore, in actives such as (12) the subject is interpreted as the agent of the action and in passives such as (13) the object of *by* is interpreted as the agent.

The Sausage Machine Parser*

Frazier and Fodor (1978) offer as a general motivation for a two-stage model of syntactic parsing the fact that, without some sort of chunking, the parser should be overloading at the end of a long sentence (given that the capacity of working memory is severely limited); but sentence length is not a good measure of processing complexity.

---

* Because of their relevance to parsing strategies, we repeat here some of the examples on pages 155 to 157, with some additional details.

By assuming that partial analyses are cleared out of the first-stage parser and shunted to the second-stage parser, the demand on working memory is reduced to a reasonable level. This accords with the phenomenon that the more structure the material to be stored has, the fewer demands it places on the storage space. Before proposing their own model, Frazier and Fodor discuss two two-stage parsers in the literature, that of Kimball (1973) and that of Fodor, Bever, and Garrett (1974).

In Kimball's model, the first-stage parser connects lexical items into a phrase marker for the whole sentence as they are encountered. Completed phrasal units are then taken off from the left and the second-stage parser reassembles them and relates transformationally moved constituents with their deep structure position (also possibly doing some semantic interpretation). In the Fodor, Bever, and Garrett model, the first-stage parser determines the clause boundaries in the input string and then determines the constituent structure within each clause. The second-stage parser determines the relationships among the clauses within the whole string. Frazier and Fodor claim that both these models are incorrect about the type of units the first-stage parser builds. In their *sausage machine* (SM) model, the processing unit which is the output of the first-stage parser, namely, the *preliminary phrase packager* (PPP), is determined not by syntactic status but by size. The PPP can view approximately six words (somewhat smaller than Miller's 7 plus or minus 2) which may or may not constitute a clause, a phrase, or a sentence. (Perhaps size should be defined over syllables, morphemes, or time—this is left open.) The second-stage parser, namely, the *sentence structure supervisor* (SSS) takes the packages given to it by the PPP and completes the phrase marker for the sentence. The difference between the PPP and the SSS is not in the operations they perform but rather in the units on which they operate. Frazier and Fodor call the SM model a "garden path variety" of explanation for processing difficulties. That is, the parser will choose the easiest path in parsing, then if that is correct the sentence is easy to parse; if it is not correct, the sentence must be reparsed and it is more difficult to process.

Frazier and Fodor attribute right association (Kimball, 1973) to the narrow viewing window of the PPP. They claim that the tendency to low right attachment sets in only when the element is some distance away from the other nodes to which it could be attached. In (14) there is no tendency for right association but in (15) there is.

(14) Joe bought the book for Susan
(15) Joe bought the book that I had been trying to obtain for
      Susan

This is because in (14) the PPP can still see the verb *bought* when it encounters the PP, but in (15) *bought* has already been shunted to the SSS by the time the PP is encountered. Actually, there is some confusion in this paper about right association. Frazier and Fodor modify Kimball's statement of right association and then alternate between calling it RA and "local attachment." LA accounts not only for sentences like (15), but also for (16).

(16) Though Martha claimed that she will be the first woman president, yesterday she announced that she'd rather be an astronaut.

Kimball's RA falsely predicts that *yesterday* will be associated with *claimed*, but LA insures that it is attached to *announced*. (*Will be* is not considered, probably because of tense disagreement.) LA is not stated explicitly because it is a consequence of the structure of the SM. The PPP assigns structure to adjacent words and then shunts them (up to six at a time) to the SSS which will create a PM for the whole string. In parsing (16) when the PPP reaches *yesterday, claimed* will already have been sent to the SSS and not be available to the PPP. *Will be* is not an option so *yesterday* is attached as a left sister to the new phrase. The prediction that this model makes is that longer phrases that will be attached by the SSS and not by the PPP should not show effects of LA. That is, LA is sensitive not only to the length of the intervening constituent but also to that of the constituent to be attached. Frazier and Fodor say that if the PPP is looking at six words, already lexically categorized, and has already grouped together the elements up to X, which is approximately in the middle, it will try to associate X with the elements on its left. If this is not possible, then the package being formed is closed and shunted to the SSS and X becomes the left-most element of a new package.

The human parsing mechanism not only processes what it receives, but also makes predictions about what it is about to receive (cf. Ford, Bresnan, and Kaplan, 1982, for a different model that uses this principle). That is, (17) is easily seen to be ungrammatical because the parser "expects" *meet* to have an object.

(17) John took the dean to meet

ATN models have this feature built in. This predictive ability makes it possible to locate "gaps" in sentences as well. Frazier and Fodor claim that the human parser must be purely information-driven to gain maximal reliability and efficiency, not purely top down or bottom up.

The SM model has this feature. Phrase markers can be built up in any order within the PPP or the SSS. However, all parsing obeys the general principle of *minimal attachment* (MA). This says that each lexical item (or other node) is attached into the phrase marker with the fewest possible nodes linking it to the nodes already existing in the phrase marker. This principle goes against Frazier and Fodor's claim that the SM can do parsing without recourse to "*ad hoc* parsing strategies." However, MA can be motivated in terms of the efficiency of the parser. In order to effect a simple statement of what to do if a parser does not work, we want attachment to be in some sense uniform. That is, attachment should be either maximal, so the parser will know to remove a node, or minimal, so the parser will know to insert a node. If attachments were not always the same, it would be difficult and cumbersome to state what the parser should do. Maximal attachment, although a logical possibility, is not valid because it is not a well-defined notion in a grammar with recursion (e.g., if the grammar has the rule: NP → NP S, there could always be one more NP node inserted). Therefore, minimal attachment seems not to be *ad hoc*.

Frazier and Fodor offer an explanation for the difficulty of parsing center-embedded sentences within the SM model. Center-embeddings are difficult because of the load they place on the PPP. The best option for the PPP is to package each NP or V separately, but that is inefficient and strains the SSS. If the NPs are packaged together, MA will cause them to have a conjoined structure (with the conjunction missing). This is in fact a common error. Parsing is facilitated if the first NP is long enough to be packaged alone. Then the second and third NPs and the first VP should be short so they can be packaged together. Finally, the second and third VPs should be long so they are packaged separately. Compare:

(18)   The woman the man the girl loved met died
(19)   The very beautiful young woman the man the girl loved
        met on a cruise ship in Maine died of cholera in 1962

(18) is easier to parse because it facilitates the PPP in making the right choices on the basis of length. This gives an explanation of the parsing difficulties of center-embedded sentences that does not rely on saying that two is the maximal number of S nodes that are storable in the parser or that a parsing subroutine cannot call itself more than once. The difficulty of center-embedded sentences is that of establishing the correct phrasal units.

The "Strength" of Lexical Form

In order to account for the same data that Frazier and Fodor (1978) used to motivate their "sausage machine," Ford, Bresnan, and Kaplan (1982), using lexical functional grammar, devise two principles of parsing, along with their defaults. The data to be explained involve syntactic closure biases. That is, they want to explain why it is that in (20) the preferred reading is the one in which the PP *for Susan* is attached to an NP which also dominates *the book* whereas in (21) the preferred reading is with the PP hanging directly off of the VP node.

(20)   Fred included the book for Susan
(21)   Fred carried the book for Susan

The difference between (20) and (21), Ford *et al.* claim, is counter-evidence to the predictions of the SM, as is (22).

(22)   Bill said that John died yesterday

Recall that SM is set up such that its preliminary phrase packager (PPP) assigns lexical items and phrasal nodes into groups. These packages, as they are formed, are sent off to the sentence structures supervisor (SSS) which arranges the packages into the phrase structure tree for the sentence. The limited viewing window of the PPP is to account for sentences like (20) which exhibit *late closure* (alternately called *right association*; see Kimball, 1973). In sentence (21), however, late closure does not obtain and (22) is so short that the PPP should be able to view the whole string, so there is no reason why there is a preferred reading. Both readings should be equally possible.

Ford *et al.* take the different biasing effects in (20) and (21) to be related to the different verbs used. Indeed, because this is apparently the only difference between the two sentences, it is hard to conceive of anything else to which to attribute the difference. Relying on the fact that in LFG one lexical item may have several lexical entries (or forms), Ford *et al.* postulate an explanation for the closure biases reported in the literature. Consider two of the plausible lexical entries for the verb *carry*:

(23a)   *carry* $\langle(\uparrow \text{SUBJ}) (\uparrow \text{OBJ}) (\uparrow \text{FOROBJ})\rangle$
(23b)   *carry* $\langle(\uparrow \text{SUBJ}) (\uparrow \text{OBJ})\rangle$

(23a) says that *carry* takes a subject, an object, and a *for* prepositional phrase (our extreme simplification will not effect the discussion; see Bresnan (1978) for more discussion). Suppose now that lexical form (23a) is "stronger" than (23b). (Ford *et al.* suggest that the *strength* of a lexical entry might be determined by frequency of usage, but they leave the problem to future research.) That is, if possible, the parser will assign a string containing *carry* a structure that is compatible with (23a) rather than (23b). We may then say that the parser is biased towards the strongest lexical form whenever that is possible.

There are two types of options that can arise during a parsing in the system proposed by Ford *et al.*: different constituents may be hypothesized as following and completed constituents may be attached in several ways. They postulate two principles and their defaults which govern the choice of options during the process of parsing. Both principles rely on the notion of comparative strength of lexical forms. The first principle is the principle of *lexical preference* (LP). Basically, LP states that if you have a choice between several constituents that might be hypothesized as occurring next (this is related to when there is an option in the PS rules), give priority to the options that are *coherent* with the strongest lexical entry of the predicate (that is, use [23a] rather than [23b] if possible). Coherent, as used in LFG, means that all of the grammatical functions in the f-structure must be listed in the lexical form of the local predicate (see Kaplan and Bresnan, 1982, for a precise formulation). If LP does not choose between the alternatives, that is, if none of them are coherent with the predicate, then the default comes into action. It is called *syntactic preference* (SP) and uses the strength of the alternative categories. The order of preference (strength) that is desired is lexical categories before phrasal categories and NPs before Ss. Ford *et al.* suggest two possible ways to motivate this. One is basically frequency of occurrence. Because phrasal nodes always dominate at least one lexical node (and usually more), lexical nodes are stronger. In the same way, fewer NPs dominate Ss than Ss dominate NPs; therefore, NPs are stronger. The other way to define strength is in terms of the X-bar notation (Jackendoff, 1977). The fewer the bars, the stronger the priority given to that option.

The other principle governing closure is *final arguments* (FA). FA gives low priority to attaching the last argument of the strongest lexical form of a phrase to that phrase. Assuming that *include* takes only two arguments (a subject and an object) in its strongest lexical form, then FA is part of the explanation for the preferred reading of (20) (repeated here).

(20)   Fred included the book for Susan

If *the book* were attached to the VP after it was parsed, it would be the
final argument, so it is not attached to the VP but made the head of a
complex NP. *For Susan* is then parsed and can only be attached to the
complex NP as the VP is then inaccessible. No other options are then
viable and the complex NP is attached to the VP node. If FA does not
apply, then its default, *invoked attachment* (IA) does apply. IA indicates
that if there are options for attaching a phrase, give priority to the
option of attaching it to the node that caused the phrase to be
hypothesized. FA and LA explain the basic difference between (20) and
(21). In (21) (repeated here) when the NP has been parsed it must be
attached to something.

(21)   Fred carried the book for Susan

If it is put on the VP, it will not invoke FA as it will not be the final
argument of the strongest lexical form. But it could also be attached
into a complex NP, as in (20). However, IA will cause it to be attached
to the VP since the VP is what caused it to be hypothesized. FA and its
default IA account for the tendency for right association (Kimball,
1973), or late closure, but state that only the last argument or any
subsequent to it will close late.

Taken as a whole, this system does seem to work for most of the
data, but then that is not very surprising as it is basically just a restatement
of the problem. What it all boils down to is this: Give the parser access
to which is the preferred reading ("strength" of lexical forms is
determined by which reading is preferred) and the parser will produce
a structure for it. It would seem that in order to impart to this theory
a measure of empirical content, some way of defining "strength" that
does not rely on which reading is preferred must be found. The
complexity of the notion of the strength of lexical form is further
confounded if we consider the following pair of ambiguous sentences:

(24a)   John was *shot* during a family argument at 10:45 last night
(24b)   John was *sad* during a family argument at 10:45 last night

For many speakers, *10:45* in (24a) is taken as the time of shooting,
whereas in (24b) it is taken as the time of family argument. What might
be at play here is perhaps the pragmatic strength of the impact of the
event. Ford *et al.* acknowledge that they must find a way to determine

strength, but they postpone that for future research. However, the parsing principles and notion of strength of lexical form can be put to uses that have no analogue in the SM. In general, late closure seems to be prevalent in complex sentences which are ambiguous. Ford *et al.* claim that this is additional support for their system. Consider (25):

(25) Joe carried the package that I included for Susan

The preferred analysis for this sentence attaches the PP to the lowest VP even though the strongest lexical form of *include* does not take a (FOR OBJ) and the strongest form of *carry* does. Both strong lexical forms have been rejected in favor of weaker ones. FA and SP interact to yield precisely this effect. But, this does not happen across the board, as predicted by Kimball's (1973) principle of right association, as is shown by (26):

(26) The woman positioned the dress that I wanted on that rack

In (26), the PP is associated with *position* on the preferred reading, even though it could be associated with *want*. Ford *et al.* postulate that a stronger lexical form may be rejected in favor of a weaker one just in case the meaning of the weaker form entails the meaning of the stronger form. Therefore, because *I included it for Susan* entails that *I included it*, the weaker form of *include* may supplant the stronger form. And because *I wanted it on that rack* does not entail that *I wanted it* the stronger form of *want* may not be replaced by the weaker form.

FA also explains the initial strangeness of sentences like (27). This sentence is perfectly grammatical and unambiguous,

(27) Bourbon made the candidate that we left drunk

although it sounds at first as if something were missing. This is because FA "forces material (*drunk*) into the lower VP" which would be ungrammatical, so the sentence is reanalyzed with *drunk* associated with the higher VP. These principles also account for another range of data. As Fodor (1978) points out, verbs which may occur either transitively or intransitively are not equally easy to parse on both readings. Ford *et al.* observe that (28) is easier to analyze than (29).

(28) Which student did the teacher send—to the cafeteria?
(29) Which student did the teacher send to the cafeteria for—?

This processing difference finds a simple resolution in terms of comparative strength of lexical forms. All that need be said is that *send* ⟨(SUBJ)(OBJ)(TOOBJ)⟩ is stronger than *send* ⟨(SUBJ)(TOOBJ)⟩. Also, in sentences where a gap may be in more than one place, the preferred interpretation can be predicted by this system. Consider (30):

(30)   Those are the boys that the police warned about fighting

There could be a gap following *warn* or following *fight* but because the strongest lexical form of *warn* is *warn* ⟨(SUBJ)(OBJ)(ABOUTOBJ)⟩, the preferred reading will place the gap after *warn*.

Thus, this approach accounts for not only the syntactic closure biases, but also some garden pathing effects, right association tendencies, and lexical-expectation effects in gap finding. However, this system is also not without its problems. For example, it is not always the case that a weaker form may replace a stronger form only if the weaker form entails the stronger form. Compare (31) with (26):

(26)   The woman positioned the dress that I wanted on that rack
(31)   Joe kept the gift that I wanted for Susan

Ford *et al.* explain that the stronger form of *want* in (26) may not be rejected in favor of the weaker form, but (31) shows that clearly sometimes it can. In all crucial respects (31) appears to be identical to (26), yet the opposite closure biases obtain. This will be a problem by anyone's account. There is also a problem in the opposite direction. On at least one reading *liking the things in my apartment* entails *liking the things* yet in (32), the PP *in my apartment* attaches to the higher VP and not the lower one.

(32)   I kept the things that I liked in my apartment

However, because a weaker form of the lower verb could be accepted over the strongest form, it should and the PP should be associated with the lower verb. Perhaps the wrong analysis could be precluded on semantic grounds (Ford *et al.* do this in several instances) which would force the correct analysis, but this detracts from the elegance of the system. Sentences (31) and (32) show that the weak form entailing the strong form is neither a necessary nor sufficient condition for the rejecting of a strong form in favor of the weak form.

This LFG approach will also have trouble with sentences such as (33) and (34):

> (33)   Fred reported Bill's talking to Mary
> (34)   Who(m) did Fred report Bill's talking to?

In (33) the preferred reading is with *to Mary* attached to *talking* but in (34) this is impossible so it is attached to the higher VP. The reading in (34) which is not allowed could be ruled out syntactically because gerundive NPs seem to be *wh*-islands. However, it is not clear that this restriction has a natural statement within the LFG framework (as presented in Kaplan and Bresnan, 1982, for example). Kaplan and Bresnan propose that there is a universal theory of bounding nodes (which is left to future research), which defines islands, and that specifies in which syntactic configurations bounding nodes may appear. This is opposed to Chomsky's (1977) notion that all NPs and Ss (or $\bar{\text{S}}$s) are bounding nodes. They specify certain nodes (possibly only one per rule) on the right-hand side of some of the PS rules, as bounding nodes (these are enclosed in boxes), for example,

$$(35) \quad \begin{array}{l} \text{NP} \rightarrow \boxed{\text{NP}} \ \text{N} \\ \bar{\text{S}} \rightarrow \text{(that)} \ \boxed{\text{S}} \end{array}$$

A dependency between two nodes (e.g., a *wh*-word and a gap) may not cross a bounding node unless the controlling node is a sister of that bounding node (see Kaplan and Bresnan for a more explicit explication). The rule postulated by Ford *et al.* for genitive NPs is basically: NP → NP's VP. In order to make this an island, both elements on the right-hand side of this rule must be enclosed in boxes. It is not clear that this is allowed by what Kaplan and Bresnan have in mind, but even if it is, it is just a round about way of saying that the whole NP is a bounding node. But there is no way to state in this notation that the whole genitive NP constitutes a bounding node. Also, marking a VP as a bounding node would be quite unusual. To the best of our knowledge, VPs have never been proposed as bounding nodes; only NPs, Ss, and $\bar{\text{S}}$s have.

ATN Grammars and Parsers

Augmented transition networks (ATN) were developed for processing natural languages by computers and have gone through a variety of elaborations through the works of Thorne *et al.* (1968), Bobrow and Fraser (1969), and particularly Woods and his associates (Kaplan, 1973; Woods, 1970, 1973, 1978). Bates (1978) in a tutorial on ATNs lists five advantages for the use of ATN systems: (1) perspicuity, (2) generative power, (3) efficiency of representation, (4) the ability to capture linguistic regularities, and (5) efficiency of operation.

As Wanner and Maratsos (1978) point out, the development of ATN has evolved into a generalized set of notational conventions for grammatical representations and parsing strategies which makes it possible

> to vary the system's grammatical knowledge about syntactic patterns independently of the schedule the system uses to apply that knowledge to the analysis of input sentences. . . . For example, we can construct a grammar capable of describing all interpretations of some syntactically ambiguous sentence, but arrange a schedule that causes the system to produce the psychologically popular interpretation first. This scheduling arrangement can then be tested in terms of its independent empirical consequences (p. 120).

We have noted elsewhere in this study the different approaches to natural language processing. The so-called linguistic approach requires complete syntactic analysis whereas some of the "artificial intelligence approaches" rely very little on the depth of syntactic analysis and much more on contextual, pragmatic, role of certain keywords or concepts, and other similar analyses. It is interesting that both these camps have used a variety of ATN notational conventions to develop their systems. However, Wanner and Maratsos propose that the amount of syntactic information needed for sentence comprehension varies from sentence to sentence depending on the amount of contextual or semantic information available for each sentence. Then they claim that an ATN provides a convenient and natural framework for studying variations in the depth of syntactic analysis during comprehension.

> The system is capable, in principle, of performing a complete syntactic analysis of any sentence, providing information about the contextually appropriate syntactic categorization of each word, a proper bracketing of each phrase and clause, and specification of the grammatical function of every word, phrase, and clause. These functional specifications are rich enough to provide at least as much semantically relevant grammatical information as Chomskian deep structures (Chomsky 1965). However, unlike earlier models of comprehension, which performed complete syntactic analyses, an ATN can make intermediate results available for semantic analysis in a natural way. Therefore, it provides a relatively straightforward way of simulating the case in which comprehension can be completed on the basis of partial syntactic analysis. On the other hand, unlike comprehension models that are limited to shallow syntactic analysis, an ATN sets no

prescribed limits on the depth of processing. Thus, hypotheses can be derived from any level of analysis and tested against the facts of human performance. (pp. 120, 121)

In further comparison with transformational models, Wanner and Maratsos state,

> An ATN recovers functional information directly from surface structures. No intervening transformational rules determine the deep structure of the sentence (Chomsky 1965). Models that apply such rules have proven to be both computationally unmanageable (Woods 1970; Kelly 1970) and empirically unsatisfactory (Fodor and Garrett 1966). The heuristic strategies scheme is empirically superior to transformational models precisely because it abandons the idea that comprehension involves the internal operation of transformational rules. The ATN approach maintains this advantage. (p. 121)

Statements such as the above quotations have caused some confusion in the literature about the nature of the ATN. The confusion arises from a failure to make a distinction between a grammar and its interpreter or parser—in a more general sense, to confuse competence and performance. ATN systems developed for natural language processing incorporate a grammar and an interpreter for the grammar. It should be pointed out, however, that ATN is a notational facility for writing grammars, representing semantic or knowledge structures, and so forth. It has been used for writing diverse grammars, including transformational, tagmemic, dependency, and others, for English and other languages (Bates, 1978). Thus, to say that ATNs are superior to transformational grammars is gratuitous. It all depends on what sort of transformational or other grammars one chooses to implement on an ATN system. It has been stated that the *hold* mechanism of ATN provides a psychologically real and a significant theoretical facility which is superior to the transformational treatment of the left extraposition and other similar phenomena (Wanner and Maratsos, 1978; Woods, 1973). This may be true, but again it has to do with the question of implementation, interpretation, or comprehension, which are all aspects of performance.

At this point we can look at an example of an ATN model. The following two sentences have often been used to illustrate the working of an ATN model:

(36) The old train left the station
(37) The old train the young

An elementary and highly simplified ATN grammar network that can be used for the parsing of these sentences is (for a detailed grammar see Woods *et al.*, 1972):

(38a)

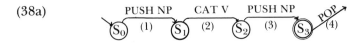

(38b)

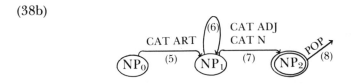

Arc	Action
1	ASSIGN SUBJECT to current phrase
2	ASSIGN ACTION to current phrase
3	ASSIGN OBJECT to current phrase
4	ASSEMBLE CLAUSE
	SEND current clause
5	ASSIGN DET to current word
6	ASSIGN MOD to current word
7	ASSIGN HEAD to current word
8	ASSEMBLE NOUN PHRASE
	SEND current phrase

It is clear that the basic device in (38a) is a finite state automaton with S_0, the initial state, and S_3, a final state. However, the PUSH arcs ([1] and [3]) transfer control or "call" a subnet (NP net in our example), and, because this call can be done recursively (for example, an NP subnet can call a relative clause subnet which can in turn call the NP subnet), the device takes the power of a pushdown automaton or context-free grammar. However, arbitrary ACTIONS can be defined in conjunction with the transition on each arc. The device then takes the power of a Turing machine.

   In our example in (38), the process starts at the initial state S_0 and the PUSH NP arc will call the NP subnet, starting at the initial state NP_0. The CAT(egory) ART(icle) on arc (5) will be satisfied by the first word in sentence (36). The action (5) would then assign the category DET (determiner) to this word and transition is made to state NP_1. By convention tests are made on arcs coming out of a node in a clockwise direction. Thus, arc (6) is tried for the CAT(egory) ADJ(ective). This is satisfied by the word *old* in (36) and the action associated with (6) assigns MOD (modifier) to *old*. A second attempt for transition on arc (6) fails because *train* is not recorded as a possible adjective in the dictionary. The transition on arc (7) is then attempted where the CAT(egory) N (noun) is expected. This is satisfied because *train* can be a noun. The action associated with (7) assigns the property HEAD (i.e., head noun) to this word. At this time transition is made to the final state of this subnet (NP_2) and the process *pops*, that is, the control is returned to the calling arc (1) with indication of success. Action (8) associated with process assembles the noun phrase and sends it to the calling routine. We are now back on arc (1) and the action associated with this arc assigns this newly found or recognized noun phrase as the SUBJECT of the sentence. Transition is made to S_1 and the CAT (egory) V (verb) is sought for the next transition. The pointer in the input sentence now points at the word *left*. This is recorded in the dictionary as a possible verb and satisfies the transition; action (2) assigns *action* or main verb to *left*. The next transition on arc (3) is another PUSH for NP and goes through the same process as in arc (1), calling the NP subnet to establish *the station* as the OBJECT noun phrase (action 3). The process at this point reaches the final state S_3 and the POP action (4) assembles the clause. The result is this analysis:

(39)

```
[CLAUSE
      SUBJECT = [NOUN PHRASE
                        DET = the
                        MOD = old
                        HEAD = train]
      ACTION = left
      OBJECT = [NOUN PHRASE
                        DET = the
                        HEAD = station]]
```

Now if we try to send the sentence in (37) through the above process, *the old train* would first be assigned as a subject noun phrase, then the word *the* is tried for a verb and fails. ATN provides for backtracking and listing other possibilities in such cases. In this case one other dictionary category entry for *train* is verb, trying this option meets with success, and for (37) we get:

$$[CLAUSE$$
$$SUBJECT = [NOUN\ PHRASE$$
$$DET = the$$
(40)
$$HEAD = old]$$
$$ACTION = train$$
$$OBJECT = [NOUN\ PHRASE$$
$$DET = the$$
$$HEAD = young]]$$

There are a number of other arcs in addition to those shown in the above example. For example, a JUMP arc will cause transition from one state to another with no action. This occurs, for example, when a sentence does not begin with a noun phrase or a noun phrase does not begin with an article. There is a WRD arc which seeks a specific word rather than a category. There are also some variations in terminology, for example, some authors use SEEK instead of PUSH.

At this point it may be useful to trace through a simplified parsing procedure for the sentences in (36) and (37) repeated below:

(36) The old train left the station
(37) The old train the young

We assume a grammar, dictionary, and syntactic nets as given below. For the simplicity of exposition, we have omitted the Action pointers on the arcs.

Grammar	Dictionary
S → NP VP	N → old, station, train, young
NP → (ART) (ADJ)* N	ART → the
VP → V NP	ADJ → old, left, young
	V → train, left

where categories in parentheses are optional and * marks possible repetitions of a category.

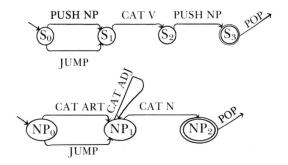

Note that the following procedures, which, incidentally, are top-down analyses, provide for alternative category selection and backtracking, for example, in Step 7 of Procedure A, when Item *3.train* is matched with category ADJ, the match fails and transition is made to the CAT N arc, where *train* is matched with category N and succeeds (Step 8) and transition is made to state NP_2. On the other hand, in Procedure B, when Item *4.the* in Step 11 fails to match with category V, the pointer is set back to Item *1.the* and transition is reverted to state NP_0 (the initial state of the NP net). In the reprocessing that ensues, the alternative categories N for *old* and V for *train* are selected. In Chapter 9, we will give more details of an algorithm for a three-stack automaton which can carry out the operations of the following procedures.

Procedure A

Step	State	Position in Sentence	Production
1	S_0	1.the	NP VP
2	S_0	1.the	NP
3	NP_0	1.the	ART ADJ* N
4	NP_0	1.the	ART (match)
5	NP_1	2.old	ADJ* N
6	NP_1	2.old	ADJ (match)
7	NP_1	3.train	ADJ (not match)
8	NP_1	3.train	N (match)
9	NP_2	4.left	pop
10	S_1	4.left	V NP
11	S_1	4.left	V (match)
12	S_2	5.the	NP
13	NP_0	5.the	ART ADJ* N
14	NP_0	5.the	ART (match)
15	NP_1	6.station	ADJ* N
16	NP_1	6.station	ADJ (not match)
17	NP_1	6.station	N (match)
18	NP_2	7..	pop
19	S_3	7..	pop

Procedure B

Step	State	Position in Sentence	Production
1	S_0	1.the	NP VP
2	S_0	1.the	NP

Procedure B (*Continued*)

Step	State	Position in Sentence	Production
3	NP_0	1.the	ART ADJ* N
4	NP_0	1.the	ART (match)
5	NP_1	2.old	ADJ* N
6	NP_1	2.old	ADJ (match)
7	NP_1	3.train	ADJ (not match)
8	NP_1	3.train	N (match)
9	NP_2	4.the	pop
10	S_1	4.the	V NP
11	S_1	4.the	V (not match)
12	NP_0	1.the	ART ADJ* N
13	NP_0	1.the	ART (match)
14	NP_1	2.old	ADJ* N
15	NP_1	2.old	N (match)
16	NP_2	3.train	pop
17	S_1	3.train	V NP
18	S_1	3.train	V (match)
19	S_2	4.the	NP
20	NP_0	4.the	ART ADJ* N
21	NP_0	4.the	ART (match)
22	NP_1	5.young	ADJ* N
23	NP_1	5.young	ADJ (match)
24	NP_1	6..	N (not match)
25	NP_1	5.young	N (match)
26	NP_2	6..	pop
27	S_3	6..	pop

A critique of ATN as a parsing or comprehension device was given by Frazier and Fodor (1978). They claimed that the sausage machine (SM), incorporating right association (RA) and minimal attachment (MA), achieved explanatory adequacy whereas ATNs did not even reach descriptive adequacy because they could not even describe RA and MA. Wanner (1980) refutes this position and claims that on closer examination the SM falls apart and ATNs remain infallible. For example, consider the following two sentences:

(41)   Tom said Bill died yesterday
(42)   Joe called everyone who smashed up

The PPP (preliminary phrase packager) of the SM could easily see the whole sentence and, according to Frazier and Fodor, can analyze it so there should be no preference. But there is a preference to low right attachment. Wanner points out that the SM has no explanation for this, but an ATN incorporating RA and MA could explain it.

Assuming an ATN like the one proposed in Wanner and Maratsos (1978), Wanner writes scheduling procedures which are equivalent to RA and MA. These are given in (43) and (44) respectively:

(43)   Order all SEND and JUMP after all other arcs
(44)   Order all CAT and WORD arcs before all SEEK arcs

(See Ford, Bresnan, and Kaplan, 1982, for a discussion of this.) Note that (43) is not ambiguous as it might appear to be at first blush, because SEND and JUMP arcs are never both options at one point. The scheduling procedures predict that there will always be low right attachment if the subnetwork is in the final state. RA does not set in, in sentences like (45), because when the NP subnetwork returns control to the VP subnetwork, it SEEKs a PP and finds it.

(45)   Joe bought the book for Susan

There is no conflict between RA and MA, because RA is not an option. Only MA can operate at that point. In this way, Wanner accounts for RA and MA, using an ATN without appeal to the geometry of the phrase marker (which Frazier and Fodor claim is important). He also accounts for their interaction without recourse to a two-stage parser with a limited viewing window.

Wanner acknowledges that there is no obvious reason why the scheduling principles are as they are and not something else. He claims that at least they give the right results and speculates that one day they may fall out of a universal theory of scheduling principles. As we will see, Fodor and Frazier (1980) criticize this view and claim that parsing schedules are a natural consequence of the SM.

Fodor and Frazier (1980) do not find Wanner's evidence that an ATN could be devised which would incorporate analogues of RA and MA without assuming a two-stage parser or a narrow viewing window, given that the ATN, in an unconstrained form, has the power equivalent to that of the Universal Turing Machine. Fodor and Frazier argue against not only the current ATNs, but ATNs in general. They claim that ATNs that do not emulate the sausage machine's two-stage parser and narrow viewing window are empirically incorrect, and that those

ATNs which do emulate the SM's structure are still less desirable than the SM because they are less constrained and offer no motivation for being as they are.

In Frazier and Fodor (1978) there was some confusion about the principles of right association (RA) and local association (LA) which Wanner (1980) seems not to have realized. Frazier and Fodor discuss Kimball's (1973) RA principle and show that it is empirically inadequate. They modify it so that it also accounts for sentences like (46) where

(46)   Though Martha claimed that she will be the first woman president yesterday she announced that she'd rather be an astronaut

RA is rejected in favor of local attachment to the left (i.e., *yesterday* associates with *announced* rather than *claimed*). Frazier and Fodor (1978) alternate between calling their new principle RA and LA, which is somewhat confusing. Also, they never state LA explicitly because it is a consequence of the structure of the SM. Fodor and Frazier (1980) point out this confusion and then begin to make the proper terminological distinction between RA and LA. They also state that although in their earlier (1978) paper they had argued against RA as a parsing strategy, Wanner's (1980) arguments that it must be included in the PPP are convincing. Also, RA must hold in the SSS, given that in sentences like (47) the PP will be attached by the SSS

(47)   John was reading a book that Mary had been reading to some very young children

(because of its length) and the preferred reading is with the PP associated with the lower verb, not the higher one. Therefore, RA must be stated as a parsing principle that both the SSS and the PPP abide by when possible.

Fodor and Frazier postulate two other parsing principles by which the SM abides. First is *revision-as-last-resort* (RALR). Basically, this states that the parser will try all other options before it modifies a structure that has already been built. This, Fodor and Frazier claim, will be needed by any parser which is to operate efficiently at decision points. The other principle is that of *minimal attachment* (MA). MA requires one to attach a lexical item (or other node) into the existing structure using as few new nodes as possible. Both the PPP and the SSS abide by these two principles. Wanner (1980) stated an analogue of MA for his ATN. He used the scheduling procedure (44) above, which gives basically the same results as MA. Fodor and Frazier point out that this *CAT-before-SEEK* approach does not choose between SEEKs. They claim that a fragment like (48) is ambiguous between a conjoined analysis and a relative analysis:

(48)   The man the girl

Both MA and CAT-before-SEEK will be garden pathed by this fragment but only MA predicts that the preferred analysis will be the conjoined one. CAT-before-SEEK makes no prediction as Wanner formulates it. However, Ford, Bresnan, and Kaplan's (1982) revision of CAT-before-SEEK (which they term *syntactic preference* or SP) will make the right prediction. Ford, Bresnan, and Kaplan motivate choosing SEEK NP before SEEK S($\overline{S}$) either in terms of number of bars in the X-bar schema (in their system NP has fewer bars than S/$\overline{S}$ and so will be chosen first, but compare Jackendoff [1977], where NP = N''' and S/$\overline{S}$ = V''') or in terms of frequency in parsing (all Ss contain NPs but not all NPs contain Ss, therefore NP is more frequent and chosen first). Fodor and Frazier offer other arguments against CAT-before-SEEK, but the SP proposal of Ford, Bresnan, and Kaplan handles all of them.

Frazier and Fodor (1978) claim that RA is a motivation for their two-stage parser. Wanner (1980) argues that it is not, and Fodor and Frazier (1980) agree, stating that what they meant in 1978 was that LA motivates postulating two stages. LA shows up in the way that the other parsing principles (RA, MA, and RALR) interact. Wanner (1980) claims that when RA and MA conflict, MA wins. Consider:

(49)   Joe bought the book for Susan

For Susan is minimally attached into the VP rather than postulating a new NP node such that *the book* and *for Susan* would be daughters of that NP. However, according to Fodor and Frazier, there is no conflict between MA and RA here. For RA to have operated, it would have violated RALR; therefore, MA had free reign. There are times when RA wins out over MA and this is orderly and predictable on the assumption of a two-stage parser. Consider sentence (50):

(50)   Joe bought the book that I had been trying to obtain for Susan

MA is inapplicable because it would take the same number of nodes to attach the PP to any one of the three verbs. Because MA does not apply, RA is free to have its way with the PP and it is associated with the lowest rightmost verb. The situation is different in (51).

(51) Joe took the book that I had wanted to include in my
       birthday gift for Susan

In this sentence, RA applies even though it violates both MA and RALR. This is because at the time the PP is to be attached, the verb (*include*) is no longer within the viewing window of the PPP and so it is not an option. Sometimes what is a last resort for the PPP will not be the last resort in terms of the whole phrase marker. That is, the PPP will give up on some analysis when it could have continued in terms of the whole sentence. When RA and MA are in conflict, RA prevails only if MA would have to use nodes that are too far removed; otherwise, MA prevails. This is the principle of local association: Associate adjacent words even if it violates other parsing principles. LA need not be stated explicitly in the parsing procedures as it is a consequence of the structure of the SM. Fodor and Frazier claim that there is no way to build this into an ATN except by making its version of RALR sensitive to the number of words being parsed by a subnetwork. Doing so would be merely a device to emulate the structure of the SM; hence, it is less desirable than the SM.

An original claim of Frazier and Fodor (1978) was that the SM could parse English without recourse to any *ad hoc* parsing strategies like ATNs. However, Fodor and Frazier (1980) have weakened this claim because they see the need for RALR, MA, and RA. They argue that these three principles can all be motivated in terms of the efficiency of the parser. But, they claim, MA, RA, and LA are only *ad hoc* scheduling procedures in an ATN. This is a definite advantage of the SM. Unless ATN theory can be constrained to make these principles equally well motivated, the SM is definitely the preferred theory.

Ford, Bresnan, and Kaplan (1982) in their discussion of the SM, and Wanner's (1980) reply to it, claim that LA does not motivate a two-stage parser as Frazier and Fodor (1978) and Fodor and Frazier (1980) claim it does. Sentences like (46) are a primary motivation for LA and hence the SM with its narrow viewing window. However, Ford, Bresnan, and Kaplan point out that the data of Frazier and Fodor are contaminated by not observing a constraint in English that S-bar complements must occur verb phrase finally. As evidence for this constraint, they give sentences like the following:

(52) I told Mary that Bill will die yesterday
(53) I told Mary yesterday that Bill will die

Sentence (52) is definitely bizarre if not ungrammatical. To associate *yesterday* with *claimed* in (46) (repeated here)

(46)   Though Martha claimed that she will be the first woman president yesterday she announced that she'd rather be an astronaut

would violate this constraint, so *yesterday* must be associated with *announced* but LA is not involved. They then claim that all other motivations of Frazier and Fodor for LA can be met with similar arguments, so they do not go through them. However, it is not clear how they will account for a sentence like:

(54)   Though John wrote the article that Mary will be reading yesterday he said that he thought that it was garbage

In this case the embedded sentence is not a complement of the verb but rather a relative clause on the NP *the article*. There does not seem to be a constraint on these comparable to the VP complements as (55) does not seem as strange as (52).

(55)   John will buy the new book that Chomsky has written tomorrow

Therefore, there is motivation for LA.

Determinism Hypothesis

Marcus (1980) has proposed a parsing strategy which is based on the hypothesis that sentences can be parsed by a deterministic procedure. Current natural language processing models in which syntactic analysis plays a major role, have nondeterministic grammars and the parsers must be provided with backtracking or some sort of parallel processing capability. The backtracking is equally applicable to top-down and bottom-up analyses. For example, in top-down hypothesis-driven parsing, we can begin by assuming that an input sentence is declarative; hence, it begins with a noun phrase. Thus, we begin by building the following initial surface structure:

(56)   Declarative: [s [NP [DET [ART

At this point we can check the input sentence for an initial article; if it is found, our hypothesis is correct so far and we can proceed with further hypotheses. If there is no article, we have to backtrack for other possible analyses. A declarative sentence may begin without an article; or we have a different type of sentence, imperative, starting with a verb:

(57) Imperative: [$_S$ [$_{VP}$ [$_V$

Similar situations arise with a left-to-right, bottom-up parsing when there are multiple productions for the rewriting of some category (S, NP, etc.).

Apart from the computational complexity (time, space) of nondeterministic parsing, there are cases, as Marcus points out, where the decision point is so far into the sentence that the parser may not be able to decide where it took the wrong path or applied the wrong production. Consider the following examples from Marcus (1980):

(58a) Is the block sitting in the box red?
(58b) Is the block sitting in the box?
(59a) Have the students who missed the exam take the makeup today
(59b) Have the students who missed the exam taken the makeup today?

Note that in both pairs of sentences in (58) and (59) the parser must scan the first seven words before it can be sure of certain structures. For example, the word *have* in (59a) is the main verb of an imperative sentence, whereas in (59b) it is an auxiliary. However, the status of *have* cannot be determined until at least past the word *exam* in both sentences. Marcus claims that these examples and many others like them notwithstanding, natural language need not be parsed by a mechanism that simulates nondeterminism, and he proposes the following hypothesis: The syntax of any natural language can be parsed by a mechanism which operates "strictly deterministically" in that it does not simulate a nondeterministic machine (p. 2).

The parser proposed by Marcus maintains two major data structures: a pushdown stack which contains the active node under consideration or construction and eventually will contain the structure of the input sentence, and a buffer which provides a look-ahead facility (cf. the "window" of the sausage machine) and also may contain completed structures, such as NPs. It is claimed that in most cases the length of the buffer need not be greater than three cells (items) and in other cases it need be extended only to five cells.

The parser has access to the bottom of the stack and the content of the buffer, and on the basis of these makes its next move. Put differently, the rules of the grammar are organized as pattern/action productions where patterns are satisfied by the "active node" in the stack and the contents of the buffer and the action or actions specify constructions, changes, and so on.

Let us at least superficially step through the parsing of the sentence in (60) to get an idea of how Marcus's proposal, implemented as PARSIFAL, works.

   (60)   John should have scheduled the meeting

(It should be pointed out that in this example we take many short cuts and represent some of the rules and procedures partially. Readers interested in the details of the process should consult Marcus [1980]).

As the process is essentially data-driven, the first word in the sentence is examined and because that marks the beginning of a noun phrase, the grammar "package" for the declarative sentence is activated. The nodes for sentence (S1) and noun phrase (NP) are inserted in the active node stack (c) and the second word (WORD2) together with its attributes from the dictionary is read into the buffer:

   (61)   The active node stack:
       c: S1 (S DECL MAJOR S)/(PARSE-AUX CPOOL)
               NP : (John)
          The buffer:
          Cell 1: WORD2 (*SHOULD VERB AUXVERB MODAL VSPL
               PAST): (should)

The two grammar packets of rules that can apply are PARSE-AUX, which contains rules for the formation of auxiliaries such as *should, will, have*, and such, and CPOOL containing rules for the formation of NPs, PPs, Ss that can be complements of verb phrases. Because the word in Cell 1 of the buffer is marked with the attribute AUXVERB, the PARSE-AUX package is activated. This will result in creating the new node AUX1 as the active node in the stack and the reading of the next word into the buffer:

   (62)   The active node stack:
          S1 (S DECL MAJOR S)/(PARSE-AUX CPOOL)
               NP: (John)

c: AUX1 (PAST VSPL AUX)/(BUILD-AUX)
The buffer
Cell 1: WORD2 (*SHOULD VERB AUXVERB MODAL
VSPL PAST): (should)
Cell 2: WORD3 (*HAVE VERB TNSLESS AUXVERB
PRES V-3S): (have)

The BUILD-AUX package becomes active. The features of number and tense are copied into the AUX1 node, and the MODAL rule will apply because WORD2 in the buffer has the MODAL attribute. As the processing of *should* is completed, it is moved into the active node and deleted from the buffer:

(63)   The active node:
S1 (S DECL MAJOR S)/(PARSE-AUX CPOOL)
NP: (John)
c: AUX1 (MODAL PAST VSPL AUX)/(BUILD-AUX)
Modal: (should)
The buffer
Cell 1: WORD3 (*HAVE VERB TNSLESS AUXVERB
PRES V-3S): (have)
Cell 2: WORD4 (*SCHEDULE COMP-OBJ VERB INF-
OBJ V-3S ED = EN EN PART PAST ED):
(scheduled)

Now the rule PERFECTIVE will apply to the attributes of WORD3; it is marked as AUX, and it is moved to the stack:

(64)   The active node:
S1 (S DECL MAJOR S)/(PARSE-AUX CPOOL)
NP: (John)
c: AUX1 (PERF MODAL PAST VSPL AUX)/(BUILD-
AUX)
MODAL: (should)
PERF: (have)
The buffer:
Cell 1: WORD4 (*SCHEDULE COMP-OBJ VERB INF-
OBJ V-3S ED = EN EN PART PAST ED):
(scheduled)

The PERFECTIVE rule is a good example of certain processes in this model. Note that after the MODAL *should* was identified and entered into the stack, the parser confronts *have* and has to decide whether it is a main verb such as in *you have the answer* or an auxiliary such as *you have given the answer*. PARSIFAL performs a simple test for this: the parser looks at the next word; if it has the attributes VERB and EN (tensed), then *have* or its other forms (*had, has*) is AUX; otherwise it is the main verb. In the present case because WORD4 in Cell 1 of the buffer has the attributes VERB and EN, *have* is marked as AUX. However, it can easily be seen that this simple approach does not work in many cases, for example,

> (65)   John had always in the past three years given a birthday present to Mary

See also the examples in (59a,b). For such sentences there are more complicated analyses including garden pathing which require going further than a three-word look ahead in the buffer.

Returning to our example in (64), note that so far AUX1 in the stack is not attached to the S (S1) node. This is done after the application of further rules, which we will skip here, and then the analysis of the auxiliary is completed (PARSE-AUX is deactivated), and the analysis of verb phrase begins (PARSE-VP is activated):

> (66)   The active node stack:
>   c: S1 (S DECL MAJOR S)/(PARSE-VP CPOOL)
>   NP: (John)
>   AUX: (should have)
>   The buffer:
>   Cell 1: WORD4 (*SCHEDULE . . .

In processing VP we again assume that the noun phrase (NP) is processed and "dropped" into the buffer as a single constituent. Thus, after applying various other rules, and a certain amount of interchanges between the stack and the buffer, the final result is something like the following:

> (67)   The active node stack:
>   S1 (S DECL MAJOR S)/(SS-FINAL CPOOL)
>   NP: (John)
>   AUX: (should have)
>   VP: ↓
>   c: VP1 (VP)/(SS-VP INF-COMP CPOOL)
>   VERB: (scheduled)
>   NP: (the meeting)
>   The buffer:
>   Cell 1: WORD7 (*.FINALPUNC PUNC): (.)

So far it has been assumed that noun phrases are apparent to the parser. They have been placed in the buffer and the stack as single constituents and they also appear in the pattern part of grammar rules. For example, the following two rules for the declarative and yes/no question sentences have NPs as part of the matching patterns.

(68) {RULE MAJOR-DECL-S IN SS-START
      [= NP] [= VERB] →
      Label c s, decl, major.
      Deactivate SS-start. Activate parse-subject}
      {RULE YES-NO-Q IN SS-START
      [= AUX VERB] [= NP] →
      Label c s, quest, ynquest, major.
      Deactivate SS-start. Activate parse-subject}

How are noun phrases recognized and assembled as constituents? Certain categories of words in the dictionary or lexicon are marked as "NP starters" (e.g., articles, certain prepositions, numerals) and there are *attention-shifting* (AS) rules that treat the buffer as a virtual device. When the beginning of an NP is sensed in the buffer, an AS rule will cause the current buffer pointer to be stored and the starting point of the noun phrase is now treated as the virtual beginning (Cell 1) of the buffer. Rule packets for constructing noun phrases are then activated, and once a noun phrase has been constructed it is stored in the buffer as a lexical unit and the buffer pointer is restored to its original position. This will virtually maintain a limit to the size of the buffer although in practice the look ahead may go far into the sentence, specially since noun phrases can have very complex structures including relative clauses and other NPs which would entail recursive call of the AS rules and shifting of the virtual start for the buffer.

The PARSIFAL model as outlined so far is theory based with a completely syntactic approach. Current linguistic and psycholinguistic theories are controversial on the level at which semantics becomes involved in syntactic analysis, but they all recognize the importance of formalized semantic procedures for any model of language understanding. Some artificial intelligence approaches, such as those specified in Schank (1975), advocate complete primacy of semantics and pragmatics over—and sometimes to the exclusion of—syntax. ATN models developed for language understanding have significant semantic components. For example, the LUNAR model of Woods *et al.* (1972) has a highly sophisticated semantic component which works in the following way: The output of syntactic analysis is sent to the syntactic component; if verified and accepted, it is directed for further analyses and final "understanding." If rejected on semantic grounds, the syntactic analysis is returned to the syntactic component for possible reanalysis. The SHRDLU model of Winograd (1972) is more interactive with semantics and pragmatics and, in that sense, it is perhaps more attuned to current linguistic thinking. It allows for noun phrases and other substructures to be tested for semantic well-formedness or acceptability before the syntactic analysis of the input sentence is completed. Marcus (1980) is well aware of the need for a significant semantic component for a natural language understanding system and devotes a chapter (Chapter 10) to discussing some interesting observations about semantics. However, as far as can be seen, the implemented PARSIFAL model as discussed in the book lacks a semantic component. Furthermore, Marcus admits that the semantic proposals contained in the chapter are somewhat speculative, and it is not clear how they can be built into an integrated system for language comprehension. It is of course well known that no one else either has a complete and well-integrated semantic component for interaction with all levels of language understanding process.

Finite-State Parsing

Chomsky (1959a,b) and Bar-Hillel *et al.* (1961) independently showed that context-free phrase-structure languages can be generated by finite-state (FS) grammars if there is no center embedding. Because it is known that for every arbitrary finite-state grammar (G) there exists a finite-state automaton (FA) as an acceptor of L(G), it follows trivially that a finite-state acceptor can be constructed for such context-free phrase-structure languages. Furthermore, Chomsky (1959a) gave an algorithm for constructing FAs for any language generated by a Chomsky normal form context-free phrase structure (cf. Chapter 1) with center embedding up to a degree n where n is fixed and finite. One problem here in a theoretical sense is that any context-free grammar which is powerful enough to generate sentences with center embedding of degree n will also generate sentences with center embedding of degree $n + 1$; hence, it will not meet the requirement for a finite-state acceptor (cf. Katz, 1966).

Recall that a parser, in addition to recognizing (or accepting) a sentence, must assign one or more structural descriptions (P-marker) to the sentence. This is, of course, essential for the comprehension of the sentence. Langendoen (1975) provides the following arguments concerning this matter. Let $L_B(G)$ be the language consisting of the set of structural descriptions that a context-free phrase-structure grammar G associates with the sentences it generates. It is easy to see that $L_B(G)$ is a context-free phrase-structure language, and a context-free phrase structure G_B that generates $L_B(G)$ can be constructed from G by replacing each production $A \rightarrow \omega$ in G by the production $A \rightarrow [A^\omega A]$. Langendoen claims that there cannot exist a finite acceptor for $L_B(G)$ unless $L_B(G)$ is finite. In other words, if the parser for G assigns an infinite number of structural descriptions to the sentences of G, then $L_B(G)$ cannot have a finite acceptor. This is stated in this theorem:

(69) If G is a context free phrase structure grammar that generates a language L(G) and associates with the sentences of L(G) a set $L_B(G)$ of P-markers, then there is a FS grammar G' that generates $L_B(G)$ if and only if $L_B(G)$ is finite. (Theorem 1 of Langendoen, p. 534).

Langendoen concludes that it is not just center embedding but the recursiveness of context-free phrase-structure grammars that causes the problem mentioned above with regard to the constraints on n degree center embedding and gives rise to the general inability of finite parsers to parse all sentences generated by context-free phrase-structure grammars. However, if we were to augment a finite-state acceptor with a counter of some sort to keep the count of the predetermined value of n, then the device would be capable of accepting sentences with center embedding up to a degree n. Chomsky (1959a) proposed such a mechanism for his algorithm. Langendoen (1975) proposes a push-down stack to keep track of each recursion for a "minimally augmented finite parser" and presents a parsing algorithm for any "normal-form" context-free phrase-structure grammar where the productions of normal-form are $A \rightarrow X$, or $A \rightarrow a$ where A is a nonterminal, X is a nonempty string of nonterminals, and a is a terminal element.

In an MIT thesis, Church (1982) provides further arguments in support of finite-state parsing and develops a model, YAP, for natural language processing. YAP is considerably based on the PARSIFAL model of Marcus, discussed in the previous section, and Church makes the following concluding remark concerning his model:

We have hypothesized that a computationally simple device is sufficient for processing natural language. By incorporating two processing constraints, FS and Marcus' Determinism, it was possible to construct a parser which approximates many competence idealizations. YAP was designed to fail precisely where the idealizations require unrealistic resources. YAP's success, as far as it goes, provides some evidence for the hypothesis. (p. 118)

The determinism actually follows as a consequence of adopting a FS device because it is well known that for any nondeterministic FSA there exists an equivalent deterministic FSA (Hopcroft and Ullman, 1979). The reference to "unrealistic resources" is also to another well-known phenomenon; that is, that the human sentence recognition device is incapable of processing sentences with greater than 3 or 4 degrees of center embedding (Langendoen, 1975). Thus, a realistic model of performance might very well be a FS device although it may be totally inadequate as a model of competence.

Exercises

1. The sentence *John said Mary arrived yesterday* is ambiguous with respect to the association of *yesterday* with *said* or *arrived*. What are some arguments about any preferred reading of this sentence?

2. The following two sentences are ambiguous with respect to the association of the time phrase *10:30 last night*:

   (a) John said Mary shot her boyfriend during a brawl at 10:30 last night
   (b) John said Mary met her boyfriend during a brawl at 10:30 last night

   For many speakers *10:30 last night* is associated with *shot* in (a) but with *brawl* in (b). How do you explain this phenomenon in terms of the "strength" of lexical form?

3. Write the following grammar in ATN and develop a procedure similar to the one given in this chapter to parse the sentences *John told Mary about the fight yesterday* for both readings of *told yesterday* and *fight yesterday*, and *The President said that he denied the rumors.*

$$\overline{S} \rightarrow S \ ADV$$
$$S \rightarrow NP \ VP$$
$$NP \rightarrow (ART) \ (ADJ)^* \ N$$
$$\overline{NP} \rightarrow NP \ C \ S$$
$$\overline{NP} \rightarrow NP \ PP$$
$$VP \rightarrow V \ (NP)$$
$$VP \rightarrow V \ PP$$
$$VP \rightarrow V \ ADV$$
$$\overline{VP} \rightarrow VP \ ADV$$
$$VP \rightarrow V \ C \ S$$
$$PP \rightarrow P \ NP$$

The above grammar is in pseudo X-bar notation. For the purposes of this exercise, you can treat the X-bar rules as indicating levels; for example, $\overline{S} \rightarrow S\ ADV$ can be treated as $\overline{S}$ being the initial symbol of the nonterminal vocabulary (the highest node) and S being an embedded sentence (descendant node). This is perhaps more clear in the rule $\overline{NP} \rightarrow NP\ C\ S$. ADV stands for adverb or adverbial, such as *yesterday*, P stands for preposition, and C stands for complementizer such as *that*. Other notations used in the grammar should be familiar.

4. Write an ATN grammar to generate the language

$$L = \{a^n b^n c^n \mid n \geq 1\}$$

5. Discuss the problems of nondeterministic parsing and contrast with deterministic parsing as proposed by Marcus.

6. Discuss the problems associated with a left-to-right parsing of the sentence *Is the man sitting in the car at the curb on Fifth Avenue blind?* (*Hint*: Consider the assignment of a category to the word *is*.)

7. A *stochastic grammar* has probabilities attached to its production rules. A stochastic grammar is called *proper* if the sum of each set of productions that expand the same constituent is 1. For example, the following is a context-free proper stochastic grammar:

> 1: $S \rightarrow NP\ VP$ .40: $DET \rightarrow ART\ ADJ$
> .75: $NP \rightarrow DET\ N$ .60: $DET \rightarrow ART$
> .25: $NP \rightarrow N$ 1: $VP \rightarrow V\ NP$

Can stochastic grammars with above characteristics be useful in designing parsers with minimum backtracking?

8. Can you think of a parsing method that can be used for the grammar in 7?

9. In case grammar (cf. Fillmore, 1968) each verb is the focal point of interpretation and its *arguments* are *deep case* relations that represent the various *cases* associated with the verb. This schema is called a *verb frame* and can be represented as in this diagram:

where *agent* indicates the doer of action, *location* indicates the location of action, *object* indicates patient or receiver of action, and so on. The cases are *semantically relevant syntactic relations* and should not be confused with the grammatical terms subject, object, and such. In the following two sentences, *John* is the agent in both but subject only in the first:

(a) John opened the door with a key
(b) The door was opened by John with a key

The *case frame* for the verb *open* is something like this:

OPEN: [object (instrument) (agent)]

indicating that object is compulsory for open, but agent and instrument are optional:

John opened the door with a key
John opened the door
The key opened the door
The door opened

Think about the utility of this approach in designing language processors. Can you write a simple case grammar and parser?

References

Bresnan (1982) is a good source for many of the topics discussed in this chapter. Halle, Bresnan, and Miller (1978) contains articles about and critical discussions of ATNs. Marcus (1980) is the best source for his approach to deterministic parsing. Claims by Gazdar (1981,1982), Pullum (1983), and Langendoen and Langsam (1984) have significant relevance to questions concerning the nature of grammar and finite-state parsing.

CHAPTER 6

Semantics

Introduction

In previous chapters we have alluded to the controversy about the primacy of syntax or semantics in language comprehension. There are also controversies about the levels of interaction between semantics and lexical accessing as well as all other aspects of language understanding. In this chapter we will give a survey of the various theories and positions on semantics for natural languages.

Putnam (1975) attempted to show why semantic theory was so much behind syntactic theory. He believed that it was because the prescientific concept that is the basis for semantics (i.e., meaning) was in much worse shape than the prescientific concept for syntax. Putnam does not say, of course, that meanings do not exist; he argues that they do not exist quite in the way we think they do.

The ordinary concept of meaning is usually taken to be ambiguous; hence we have two other terms: *extension* and *intension*. Extension is the set of truth values for a term or a proposition; it is the set of objects of which a term is true. In the extreme, an extensional viewpoint does away with propositions altogether and retains only truth values in their place. Intension, on the other hand, is the set of properties which make up the concept or meaning of a term. Intension is what it is that differs in coextensional terms that mean different things. Perhaps an example will serve to illustrate the difference between the two terms. Consider the phrase *this book*: its referent is the particular book in hand. The word *book*, however, refers to all objects for which the concept of a book is true (i.e., all books). Thus, this notion provides for the extension of the term *book*. Now, in our culture a book is "understood" to be a bound volume with separate pages, cover, and such. Consider another culture in which the concept of a book may mean a continuous long sheet of paper or other material rolled up (a scroll). The extension of the meaning of *book* into this new culture is its intension. One way of putting it is that intension is "extension in all possible worlds." (For further details and clear exposition see J. D. Fodor, 1974.) Putnam, however, disagrees with the standard definition of extension versus intension and says that *meaning* never means extension, and intension is just as vague as meaning. We will return to the views of Putnam later in this chapter.

The current literature on linguistic semantics can be roughly divided into two major theories or approaches in which research is being actively pursued. These areas can be referred to as the *sense* theory (as exemplified by Katz, 1972, 1977) and the *truth-conditional* theory (as exemplified by Davidson, 1967, 1969, and Partee, 1979). This is not to claim that other work is not being done in semantics, but rather that much of the work is done on certain specific problems without any explicit framework (see, for example, Donnellan, 1972; Kripke, 1972; and Putnam, 1975). The major diffference in these two ways of attacking the problems in semantics is that those people working within a specific theoretical framework concern themselves with the representation of semantic information and the interpretation of that representation. The work of those people outside of the two general theoretical viewpoints mentioned above tends not to be concerned with how semantic information can be encoded; rather, they discuss the proper way of approaching certain problematic areas in semantics. One purpose of this chapter is to review the two semantic frameworks that have been developed while keeping in mind that we are basically concerned with how they will fit in a model of language comprehension. We will only be concerned with the work in semantics as it relates to the claims made within the explicit frameworks.

Before proceeding to the discussion of the various semantic theories, it will be useful to discuss just what we would require of a semantic theory. As a theory, we want it to be both formal and finite. It should be finite because it is to be a description of the system internalized by speakers of natural languages. Given that the human brain is a finite system, any subsystem of it must also be finite. The requirement of formality is fairly standard for any theory. It simply means that we must be able to manipulate and define relations among the symbols used solely in terms of their form. We also require that the theory be compositional because it must in some way assign a meaning to all sentences in a natural language. Because this is an infinite set, the meanings cannot simply be listed but must rather be built up somehow. It is generally accepted that the meanings of phrase and sentences depend on the meanings of the lexical items contained in them and the syntactic configuration of the phrase or sentence. It seems that most theorists would agree that these are the minimal requirements that an adequate semantic theory must meet, whereas some (e.g., Katz, 1972; Putnam, 1975) argue that there are other requirements that we must impose.

The Truth-Conditional Theory

What we have called the truth-conditional theory is not in fact a unified theory, but actually is comprised of the work done by Montague grammarians and that done by Davidson and his followers. These differing viewpoints both draw on the work of Tarski (1956) but disagree on the way to implement Tarski's ideas. Researchers working in truth-conditional semantics would all basically agree that the goal of linguistics is a theory of truth, not a description of the competence of the native speaker. (This theory of truth would be relativized to a model, not absolute truth.) A truth definition is a formal, finite theory of how certain aspects of the "meaning" of a sentence are determined by the meanings of its parts and the syntactic operations combining them. This is done in order to coincide with Frege's principle of compositionality (or to explain how a finite mind can understand infinite number of sentences). This part of the theory draws heavily on the work of Tarski (1956) defining truth in a language. One of the general claims of this framework is that semantics is compositional but not decompositional. That is, words have meaning only in the context of the sentence and this meaning is only determinable by the way they consistently contribute to the truth conditions of the utterances in which they occur. Putnam (1975) remarks that this procedure, which is taken to be the only one possible, is the exact opposite of the procedure that has led to every success in the study of natural language.

The basic program of this approach is to translate the language being studied (the *object language*) into a *metalanguage* which is independently understood. (Usually this metalanguage ends up to be one of the formal languages of logicians.) The major difference between the approaches of Montague and Davidson is that Davidson uses an extensional metalanguage. Davidson's underlying claim is that all constructions in natural languages are extensional. Montague denies this claim. To say that a construction is extensional is to claim that the extension of the whole (a truth value in the case of sentences) can be determined by the extensions of its parts. Montague argues that many constructions in natural language (English, at least) require knowledge of the extension of the parts in all possible worlds to determine the extension of the whole.

As for Davidson's view, a semantic theory will have as consequences all sentences of the form "*s* means that *m*" where *s* is replaced by the structural description of a sentence and *m* is replaced by a singular term that refers to the meaning of that sentence. The theory must then give for every sentence *s* in the language under study a matching sentence (to replace *m*) which "gives the meaning of *s*." Davidson claims we can do this by replacing *m* with the translation of *s* in the metalanguage. However, *means that* is an intensional construction and Davidson finds this objectionable, apparently for metatheoretical reasons which are never made explicit. Therefore, he removes *means that* and defines the predicate "is T" instead. He then requires of a theory of meaning for a language L that it places, without recourse to further semantical notions like synonymy, enough restrictions on "is T" to entail all sentences that fit the schema:

$$\text{' s is T iff p '}$$

when *s* is replaced by a structural description of a sentence and *p* is replaced by that sentence. Davidson claims that this is in essence Tarski's (1956) test of the adequacy of a formal semantical definition of truth (1967, pp. 307–309).

A well-known apparent counterexample to Davidson's theory is that it could not distinguish between (1) and (2).

(1) 'Snow is white' is T iff snow is white
(2) 'Snow is white' is T iff grass is green

The argument is that if the theory entails both, then how do you tell which one gives the translation ("meaning") of "snow is white"? Davidson (1967) considers this example and concludes that it is not damaging. He says that it does not seem plausible that the optimal theory of truth-in-English would entail (2), but even assuming that it did, as long as it also entailed (1), which it must to be descriptively adequate, then (2) only provides additional information. That is, although (2) does not contain important information, it is not false, so no harm is done. Davidson (1969) changes this view somewhat and claims that we must place certain "formal and empirical constraints" on what may count as a translation of a sentence in order to avoid such possible problems. The constraint he exemplifies is called the *principle of charity*, which states: I will assume our truth-conditions to be identical until that assumption would make what you are saying bizzare. At that point, I will try to determine how our truth-theories differ and interpret your utterances accordingly. Consider Davidson's example where someone utters to you the following: "There is a hippopotamus in my refrigerator. They have strangely wrinkled skins and don't mind being touched. I squizz the juice out of two or three every morning for breakfast." At some point you must realize that the speaker is not referring to what you refer to with the term *hippopotamus*, but rather he is referring to what you would call *oranges*. Rather than thinking him crazy and uttering falsehoods, the principle of charity would require that you understand the referent of *hippo* in the speaker's terms.

Another set of apparent counterexamples to Davidsonian semantics is the intensional or *referentially oblique* constructions. A position in a sentence is referentially oblique if substitution of coextensive terms does not preserve the truth value of the sentence. Consider (3) and (4).

(3) John believes that he saw the morning star
(4) John believes that he saw the evening star

Although the *morning star* and the *evening star* refer to the same object, it is certainly possible for one of these sentences to be true while the other is false. That is to say that the verb *believe* creates a referentially oblique context. Davidson (1969) attacks this problem by offering an analysis of *oratio obliqua* (that is, indirect discourse) which he claims will tend to an analysis of psychological sentences (or *propositional attitude*, e.g., *believe*, *want*, etc.) in general. Davidson says that the problem is that we do not know the logical form of oblique contexts and that an adequate account of their logical forms should lead us to see how semantic character (truth or falsity) of sentences which contain them is determined compositionally (the truth-predicate will account for this). Indirect discourse creates somewhat of a paradox in that intuitively the embedded sentences seem to have semantically significant structure, as the theory would demand that they do; yet the failure of coextensive term substitution would invite us to treat them as semantic wholes whose truth conditions are not compositionally determined.

Davidson considers and immediately rejects the analysis of indirect discourse as containing "invisible quotation marks." The reason for this is that quotations are taken to be singular terms with no internal structure and because there are an infinite number of sentences that can follow *said that*, this claim would be equivalent to claiming that the language had an infinite number of semantically unanalyzable elements. This claim cannot be true because these semantically unanalyzable elements must somehow be represented within a finite human organism. (However, see Putnam [1975] for arguments that meanings, or truth-conditions in this case, are not a property of individuals, but rather are properties of whole speech communities. This would allow the "hidden" quotation marks analysis to go through, possibly.) The analysis that Davidson puts forth and argues for (a "paratactic" analysis) is that the *that* in this construction is actually a demonstrative pronoun referring to the utterance which follows it. That is, from a semantic point of view, (5) actually contains two sentences and is equivalent to (6).

(5)   Galileo said that the earth moves
(6)   The earth moves. Galileo said that

The idea is that the sentence following *said that* gives the content, but has no logical or semantic connection with the original attribution of saying. Davidson claims that from the semantic point of view, the embedded sentence is not contained within the whole sentence. He claims that this analysis can straightforwardly be carried over to performatives and verbs of propositional attitudes because they all serve to introduce another sentence. This analysis, according to Davidson, solves all the "standard problems" with referentially oblique contexts. Namely, the failure of extensional subsitution is explained because the referent of *that* has changed, and the problems with making valid inferences is solved because *said* is only in construction with the singular (indexical) referring term *that*. This analysis allows sentences which seem to be embedded after *said, believe*, and such to "mean" the same thing in all environments, which is a desideratum of any theory which holds to compositionality.

Foster (1976) works basically within Davidson's framework, yet he criticizes Davidson because Davidson is forced to claim that an interpreter "knows that some T-theory states that . . ." where ". . ." is filled in by that T-theory. However, *states that* is intensional (that is, not truth functional), leading Foster to claim that Davidson's theory falls apart. Davidson (1976) counters that the intensionality of *states that* is all right because he is trying to explain meaning without recourse to unexplained linguistic (semantic) concepts but can use intensional concepts like belief, intention, or states. All he must do is give an extensional semantic account for them, which he proposes. Davidson's analysis of *states that* is exactly parallel to the analysis of *oratio obliqua* already discussed. Still,

Foster claims and Davidson seems to agree that Davidson's theory is only a translational theory (that is, one must know the meaning of the metalanguage in order to get to the meaning of the object language). Davidson also indicates that

> a theory of truth, no matter how well selected, is not a theory of meaning, while the statement that a translational theory entails certain facts is not, because of irreducible indexical elements in the sentences that express it, a theory in the formal sense. (Davidson, 1976, p. 41)

Given that Davidson's theory is a translational theory of truth, he seems to be claiming that first, it is not about meaning and second, it is not even a theory. One wonders, then, if it should even be considered in discussion of theories of meaning.

A translational theory is by definition not an *interpretive* theory. An interpretive theory is, according to Foster (1976), one which if explicitly known would give mastery of a language. The reason one would study conditions on an adequate theory of meaning (like interpretiveness) is to gain philosophical insight into the nature of meaning and of language in general. We can find out what meaning is if we can determine what counts as a theory of meaning. Foster claims that to get a translational theory to be interpretive you must introduce intensionality; therefore, he must modify Davidson's theory somewhat. He requires that a theory of meaning meet at least these four conditions:

1. It must be interpretive.
2. It must have an extensional syntax.
3. Its logic must be sound.
4. Its essential vocabulary must be free from intensional concepts, especially meaning.

Foster retains, from Davidson's theory, the view that meaning depends on truth conditions and the idea of placing formal and empirical constraints on an acceptable theory (like the aforementioned and the principle of charity). However, Foster seems to be contradicting himself. Condition (1) introduces intensionality (according to Foster) because it requires interpretiveness, yet condition (2) requires an extensional syntax. Foster argues that one can have both. He devises an extensional syntax that will deal with possible worlds and involves no commitment to imaginary objects, as normal possible world semantics does.

Researchers working in the Davidsonian theory of semantics take intensionality to be undesirable and attempt to describe natural language extensionally. Montague grammarians, on the other hand, do not take an intensional analysis to be a liability but just the opposite. They hold that certain constructions in natural language are not truth-functional and therefore an intensional analysis is the only option. Although Davidson (1969, 1976) considers some intensional contexts, he by no means considers them all. Consider the following:

(7) John is looking for the morning star
(8) John is looking for the evening star

Although the parts of these sentences have the same extensions, the extension of the whole sentence (that is, its truth value) could be different. It seems as if it would not be possible to extend Davidson's paratactic analysis to this construction, or to the sentential operator *necessarily*. In order to account for these contexts and others, Montague grammarians rely on the notion of intension. The intension of a term is basically a function which maps the term onto its extension in all possible worlds. Given this notion of intension, an operator like *necessarily* can be defined as follows:

Necessarily $\emptyset$ is true in any given world iff $\emptyset$ is true in all possible worlds

Montague (1970) claimed that natural language could (and should) be studied in the same formal, rigorous way as formal languages are by logicians. Basically as a result of this, Montague grammarians define a language by means of a simultaneous recursive definition. Both Montague's syntax and semantics will be considered as they are intimately related. The categories used in this syntax are given by a categorial grammar. A categorial grammar (see Lewis, 1976) is a context-free phrase-structure grammar that has a small number of basic categories and infinitely many derived categories. This is very similar to the four possible lexical categories in the X-bar schema (Jackendoff, 1977) which allow, theoretically, an unlimited number of phrasal categories (bars). A context-free grammar is not powerful enough to describe a natural language, so the categorial base must be enriched in order to meet descriptive adequacy. We return below to exactly how Montague extended the categorial base of his grammar. Montague used two primitives, t (for truth-bearing phrases, basically sentences) and e (for expressions that denote entities) and defined an infinite set of categories in terms of these two primitives. For example, the category t (for *term*, basically an NP) is defined as t/IV (that is, what one adds to an expression of the category IV to derive an expression of the category t), IV (for intransitive verb phrases) is defined as t/e, TV (transitive verbs) are defined as IV/T, and CN (common noun) is defined as $t//e$. Note the resemblance between the definition of CN and IV. These two categories will be of the same semantic type but will differ syntactically. That is, the number of "/" only makes a difference in the syntax, not in the semantics. A Montague grammar works from the bottom up. It takes the basic expressions of the categories (given by the categorial base) and combines them into phrases and then combines the phrases until a sentence (t) is derived.

The semantic rules parallel the syntax. The syntax and semantics are generally assumed to work in tandem, but this is not necessary as long as Montague's constraints are met. The semantic rules translate words and phrases of English into an intensional logic (IL), which is intepreted by a model. As Montague formulated it, IL includes: a denumerably infinite set of types (corresponding to categories in the syntax), higher-order quantification, lambda-abstraction for all types, tenses, model operators, means for forming $\hat{}\alpha$ for any expression α ($\hat{}\alpha$ denotes the intension of α), and means for forming $v\alpha$ for any intension-denoting expression α (where $v\alpha$ denotes the extension of α). The recursive definition of types includes a clause defining for each type a a new type $\langle s,a \rangle$ for the intensions corresponding to each type a. Each type includes a denumerably infinite set of nonlogical constants and a denumerably infinite set of variables. Nonlogical constants are assigned intensions by the function f and variables are assigned extensions by the function g.

An intension is a function from possible worlds to extensions (denotations) in those possible worlds. The function takes possible worlds as arguments and gives the appropriate extensions as values. It is claimed that this intension will "do what meaning does" in that meaning determines extension. In order to teach indexicals in the same way as other expressions, we will take intension to be functions from indices to extensions, where indices are ordered pairs of a possible world and a time. The intension of a name is a function from indices to individuals, called an *individual concept*. The intension of a one-place predicate is a function from indices to sets of individuals, called a *property* and the intension of a formula is a function from indices to truth value, called a *proposition*. $[\![\alpha]\!]^{M,w,t,g}$ is the extension of α with respect to a model M,w$\in$W (the set of all possible worlds), t$\in$T (the set of times) and value assignment g. The semantic rules for IL recursively define $[\![\alpha]\!]^{M,w,t,g}$ for any α.

The relationship between the syntax and semantics of a Montague grammar is constrained by the following principles:

1. Each basic expression of English is translated into one and only one expression of IL.
2. All members of the English category x translate into IL expressions of type a.
3. For every syntactic rule of English, there is a rule of translation that specifies the translation of the output of the syntactic rule in terms of the translation of the inputs.

The rules in 3 must obey Condition 2. This gives us compositionality. The following function f maps English categories into IL types:

$$f(t) = t$$
$$f(e) = e$$
$$f(A/B) = f(A//B) = \langle\langle s, f,(B)\rangle, f(A)\rangle$$

(Note that an expression of category A/B combines with an expression of category B to give an expression of category A, whereas type $\langle a,b\rangle$ (of IL) combines with an expression of type a to give an expression of type b). The function applies to the intension of B ($\langle\langle s, f(B)\rangle\rangle$), not the extension. This is because we need intension for the referentially oblique contexts, and given the above principles we cannot sometimes have extension and sometimes intensions. Using intensions in this way makes the interpretation of transparent contexts more complicated, but possible. If extensions were used instead, oblique contexts would not be interpretable.

Syntactic rules contain three kinds of information (which is what extends Montague's syntax beyond a categorial grammar):

1. The category (or categories) of the expression(s) that serve as input
2. The category of the output (this is usually determinable from (1) given the definitions of the categories)
3. A specification of how they are combined (these are numbered independently of the number of the rule, and can be referred to in more than one rule)

The combination process may include: concatenation, with or without morphological adjustments, insertion, deletion, movement, and so forth. Consider the following rule:

S4.   If $\alpha\epsilon P_T$ (phrases of category T) and $\delta\epsilon P_{IV}$, then F_4 $(\alpha,\delta)\epsilon P_t$, where F_4 $(\alpha,\delta) = \alpha\delta'$ and δ' is the result of replacing the first verb in δ by its third-person singular present form.

Note that this rule contains all three kinds of necessary information. Each syntactic rule S_n has associated with it a translation rule T_n. For example:

T$_4$.   If $\alpha\epsilon P_T$ and $\delta\epsilon P_{IV}$, and α,δ translate into α',δ' respectively, then F_4 (α,δ) translates into $\alpha'(\hat{}\,\delta')$.

Before we can see how these rules work, we must first consider the use of *analysis trees*. Analysis trees look similar to phrase markers, but they encode the derivational history of the string. They are constructed from the bottom up and do not represent the constituent structure of the expression. All terminal nodes are members of the basic categories (B_A, for any category A). These are basically lexical items. All other nodes are strings and the numerical index of the syntactic operation by which they were derived from their daughter nodes. Consider the following analysis tree:

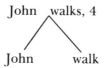

A translation of this would go as follows in the lambda conversion ($\Rightarrow$ = "translation into"):

John $\Rightarrow \lambda P[\ P\ \{j\}]$
walk $\Rightarrow$ walk$'$
John walks $\Rightarrow \lambda P[\ P\ \{j\}]\ (\hat{}\ $ walk$') \ T_4$
$\hat{}$walk$'\ \{j\}$

This says that John is in the intension of things which are walking, which is convertible into *walk$'(j)$* in the following way:

$\beta\{\alpha\} = \check{}\,\beta(\alpha)$ (called *brace convention*)
Therefore, $\hat{}$walk$'\ \{j\} = \check{}\,\hat{}$walk$'\ (j)$
$\check{}\,\hat{}\,\beta(\alpha) = \beta(\alpha)$ (called *down-up cancellation*)
Hence, $\check{}\,\hat{}$walk$'\ (j) = $ walk$'\ (j)$

Lambda conversion (referred to above) states simply that

$$\lambda x[\ldots x \ldots]\ (\alpha) = [\ldots \alpha \ldots]$$

given that x and α are of the same type.

Determining the intensions of phrases in Montague grammar is simply knowing the translation rules and having a model in which to interpret the translations. The manipulations done in order to simplify the translations are not necessary and are done only to make them more perspicuous.

It was stated that the use of intensions would complicate some translations. In order to see this, consider the following:

> (9a)   John is seeking a unicorn
> (9b)   John is kissing a unicorn

The first sentence has two readings, one in which it is true only if John is seeking some specific unicorn and the other which is true if John is unicorn-seeking, but not any one in particular. Montague analysis captures this, but also predicts that the second sentence has the same ambiguity. This is not the case. The second sentence is true only if John is kissing some particular unicorn. In order to rectify this, Montague introduced a *meaning postulate*. A meaning postualte can be understood as a constraint on possible models, for example,

$$MP1: \forall x [B (x) \rightarrow \neg M(x)]$$
$$\text{where } B = \text{is a bachelor}$$
$$\text{and } M = \text{is married}$$

MP1 tells us to consider only models where it is true and to ignore all models where it (MP1) is false. The meaning postulate that is used to deal with extensional transitive verbs says, in essence, to consider only models where the extension of both readings is the same. Katz (1972) argues that these meaning postulates are *ad hoc* and the relationship between *bachelor* and *married* is more fundamental than is captured by the meaning postulates.

Given the kinds of operations a Montague grammar can perform, one may well wonder how powerful it is in the formal mathematical sense. Cooper and Parsons (1976) give an argument that allows us to indirectly assess the power of Montague's system. Cooper and Parsons show how Montague's PTQ system, which most subsequent work has built upon, is equivalent to a transformational generative grammar.*
In particular, they use the standard theory of transformational grammar

---

* R. Montague, "The Proper Treatment of Quantification in English," in *Approaches to Natural Languages*, eds. K. J. J. Hintikka, J. M. Moravcik, and P. C. Suppes (Dordrecht: Reidel, 1973), 221–242.

with seemingly unrestricted deletion. Peters and Ritchie (1973) have claimed a formal proof that the standard theory with deletion is equivalent in power to a Turing machine and therefore not desirable as a theory of human languages. This claim created a certain amount of concern and confusion between language and grammar. Montague grammar could be subjected to the same concerns and confusions unless it is subjected to the same severe restrictions that have been proposed for TG in the extended standard theory.

The Sense Theory

The other theory of semantics that has received considerable attention is that of Katz (1972, 1977). Katz's system is interpretive and is to be hooked up with a generative grammar. The notion of *interpretive* used here is not that of Foster (1976). Katz understands interpretive to mean that it does not produce or generate as the syntax does, but rather takes as input the infinite set of phrase markers occurring at one of the well-defined levels in the syntax. In the past there has been some debate over whether semantic interpretation should come off the deep structure or the surface structure. However, given the trace theory of movement and "enriched surface-structure," there has never been shown to be any empirical difference between the two.

Katz (1972) argues that an adequate semantic theory must have at least four properties: compositionality, decompositionality, accountability for semantic properties and relations, and formality. Compositionality is basically the Fregean notion already mentioned; namely, that the meaning of a phrase depends on the meaning of the lexical items contained in that phrase and their syntactic configuration. The idea behind decompositionality is that the meanings of lexical items (words) may be complex and not just unanalyzable primitives. Thus the meaning of *bachelor* would include the meaning of *adult, unmarried, human,* and *male*. This is to account for the analyticity of sentences like (10).

(10)   Bachelors are unmarried.

Analyticity is one of the semantic properties and relations Katz claims semantic theory must explicate. Others are synonymy, antonymy, anomaly, ambiguity, meaningfulness, and so forth. The theory must account for these properties and relations formally. That is, one must be able to tell just by looking at the form of a semantic representation for a sentence whether that sentence is anomalous or meaningful. If a semantic theory meets these requirements, then we will have a description of the native speaker's semantic competence.

The semantic component of a grammar, in Katz's view, contains a finite stock of primitive symbols and a set of recursive rules for combining them. Thus there is an infinite set of primitive and derived symbols for representing the meanings in a natural language. These symbols are called *semantic markers* because they represent conceptual structure analogous to the way phrase markers represent constituent structure. Each semantic marker stands for a distinct concept. Complex semantic markers are represented as tree structures whose node labels can be either primitive or derived semantic markers. Consider the following semantic marker for the verb *chase* (taken from Katz, 1977, p. 62):

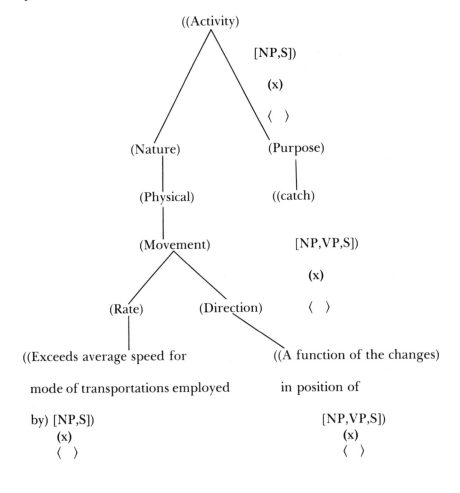

Semantic markers like the foregoing are interpreted in the following manner. First, there must be statements of correspondence which relate elementary semantic markers (those enclosed in regular parentheses) to concepts in the theory and interpretive principles which relate the formal structures built up out of the elementary symbols by the recursive rules to the internal structure of a complex concept or proposition. According to Katz, to try to formulate these principles systematically now would be premature. The dominance relation in semantic markers converts a less specified concept (the denominator) into a more specified one (the dominatee). The domination of (Purpose) by this construct:

$$((\text{Activity})[\text{NP, S}])$$
$$(x)$$
$$\langle \ \rangle$$

tells us that *chase* is a purposeful activity as opposed to a nonpurposeful activity like *wander*. Symbols like [NP, S] are called *categorized variables*.
$$(x)$$
They direct the projection rule, which combines readings to produce derived readings for the constituents of a sentence, to substitute the reading of [NP, S] into this position. Heavy parentheses around these variables, as opposed to normal parentheses, mark the position of referential terms. This is how Katz's theory would capture the specific versus the nonspecific readings of verbs like *look for*. In the specific reading the categorized variable for [NP,VP,S] would be in heavy parentheses and in the nonspecific reading it would be in normal parentheses (that is, nonreferential). Under the categorized variable is ⟨ ⟩ which would enclose selection restrictions which tell the projection rule under what conditions a reading may be substituted for the variable. The number of argument places that a predicate has is determined by the number of distinct categorized variables that are represented in that predicate's semantic marker. By inspecting the semantic marker for *chase* we can determine that it is a two-place predicate because it contains two distinct categorized variables (i.e., [NP, S] and [NP,VP,S]). Katz devised this notation to do two things.
$$(x) \qquad\qquad (x)$$
First it exhibits the "onion-like structure" of predication. That is, the domination relations provide the correct specifications on the base concept and provide a domain in which to define entailment. Second, it relates the internal predicate concepts with the proper argument places. Thus armed with the semantic markers for the lexical items in a language and a characterization of the syntactic structures of that

language, the projection rule can straightforwardly give the meaning of the sentences in that language.

This meaning so derived is what Katz terms *meaning in the null-context* (or *sentence meaning*) as opposed to *utterance meaning* in which the context plays a role. The theory of semantic competence (i.e., Katz's) gives us the sentence meaning and a theory of semantic performance (pragmatics, on Katz's view) gives us utterance meaning. Katz claims that the null-context is an idealization "in the same spirit as Chomsky's ideal speaker–hearer or the physicist's perfect vacuum or frictionless plane" (1977, p. 14). A pragmatic theory should be concerned with the mechanism used in the exploitation of context in order to produce utterances whose meaning in context differs predictably from the meaning of the sentence of which they are a token. Katz (1977, p. 16) takes a pragmatic theory to be of the following form:

> PRAG $(D(S_i), I(c(t))) = \{R_i, \dots, R_n\}$ where $(D(S_i))$ is a full grammatical description of the sentence S_i and $I(c(t))$ a specification of all information about the context c in which the token t of S_i is uttered that is relevant to t's utterance meaning, and $R_i, \dots, R_n$ is a set of readings.

The output of PRAG is a set of readings because, according to Katz, this is the simplest way of obtaining a notation for utterance meanings and allows us to capture the generalization that the utterance meaning of a token is usually the grammatical meaning of some other sentence in the language. In the null context, the output of PRAG will just replicate the semantic information in $D(S_i)$. Katz argues that grammars are theories of the structure of sentence types, which determine all the grammatical properties and relations of each sentence in a language, whereas pragmatic theories explicate the reasoning of speaker–hearers in working out the correlation in a context of a sentence token with a proposition. Following Chomsky (1965), Katz claims that a pragmatic theory should use the sound–meaning correlations given by a grammar to define new sound–meaning correlations valid only in certain contexts. That is, the theory of performance includes the theory of competence.

Speech Act

Katz's separation of semantic competence and performance (or pragmatics) brings to bear on other literature which is often brought up in discussions of semantics but which we have not mentioned so far.

This is the literature on *speech act theory* (see Austin, 1962; Grice, 1975; Searle, 1969). Speech act theory is a theory of utterances that constitute in and of themselves acts of various kinds. For example, uttering "the window is open" may, in the proper context, be an act of requesting. Given the Katzian distinction between semantics and pragmatics, it is clear that the interpretation of the above as a requestive would be dealt with by a theory of pragmatics which would construe it to have basically the sentence meaning of "I request that you close the window." Searle (1969) however, seems to disagree with this. Searle claims that the study of meaning and the study of speech acts are the same study from two different points of view. This is because every sentence in a language can, in virtue of its meaning, be used to perform a particular speech act and any speech act can in principle be given in exact formulation in a sentence in the language. Searle says (p. 19) that a complete philosophy of language has to answer both the question, How do the elements of a sentence combine to yield the meaning of the whole? and the question, What different kinds of speech acts do speakers perform in uttering sentences? He goes on to hypothesize that the speech act is the basic unit of communication. Although this hypothesis is very plausible, it is not clear that linguistics must explicate speech acts. If one takes the goal of linguistics to be a description of human communication (as speech act theorists seem to) then it must deal with speech acts. However, if the goal of linguistic theory is a description of the human language facility (as generative grammarians, Katz included, assume) then linguists need not be concerned necessarily with speech acts. However this issue is resolved within linguistics proper, it will prove useful to consider briefly the work done in speech act theory, because we are ultimately concerned with language comprehension, which is intimately tied to communication.

Searle assumes (as most researchers do) that language is a rule-governed form of behavior. To know a language is to know the rules governing it. Rules may be divided into two classes according to Searle. *Regulative* rules regulate already existing forms of behavior, and *constitutive* rules create or define new forms of behavior. The rules governing language are constitutive, not regulative. He claims that the conventions governing languages and the use of speech acts are realizations of the underlying constitutive rules. The rules Searle discusses are those governing elocutionary speech acts (those such as promising, requesting, ordering, warning, etc., that is, acts performed *in* saying something). These rules, which are stated informally, are rules which aid the flow of communication but do not contribute to grammaticality. For instance, although one of the rules governing the use of the act of promising is that the hearer would prefer that the speaker do the act promised and the speaker believes that the hearer would prefer it, the following is still a valid promise and not ungrammatical:

(11)   If you do that again, I promise I will spank you

It seems reasonable that anyone hearing this would not prefer the act to come to pass, yet it appears to be a promise none the less.

Grice (1975, pp. 41–58) has also postulated rules or principles, which help to make communication effective. He called the general principle the *cooperative principle* (CP):

> Make your conversational contribution such as is required, at the stage at which it occurs, by the accepted purpose or direction of the talk exchange in which you are engaged. (p. 45)

Grice distinguishes four categories under which the more specific principles (maxims) fall:

1. Quantity (of information to be provided)
2. Quality (truth or falsity of what is said)
3. Relation (relationship to previous context)
4. Manner (how it is said) (p. 47)

Conversational implicatures are generated by the blatant failure to fulfill a maxim. The maxim has then been "exploited." A person who has implicated q by saying that p, can be said to have conversationally implicated that q provided that:

1. He is to be presumed to be observing the maxims, or at least the CP.
2. The supposition that he is aware that q is required in order to make his saying p consistent with (1).
3. The speaker thinks that it is within the capabilities of the hearer to work out that the supposition in (2) is required.

For example, if one were to say:

(12)   Jones is meeting a woman tonight

one conversationally implies that it is something illicit. Grice says that to calculate a conversational implicature, one must already know the meaning of the utterance. This is basically what Katz (1977) says about the connection between semantic competence and performance. That is, in order to determine the pragmatic meaning of an utterance, you must first determine the meaning of the sentence of which it is a token.

It is not clear just how one could formalize Grice's maxims or Searle's rules in order to include them in a grammar. In fact, it is not even clear that one would want to. However, they do seem necessary to account for language comprehension. Thus, a model of comprehension may contain a grammar with a semantic component (whether it is of the same type, truth-conditional type, or some other type) and a theory of pragmatics which captures the insights of speech act theorists.

Bach and Harnish (1979) point out that although there has been extensive work done on grammar, meaning, and speech acts, not much has been done to try to integrate them all into an account of communication. This is the task to which they set themselves. Bach and Harnish are interested only in linguistic communication, not nonlinguistic communication as is found in Grice (1975); they take linguistic communication to depend not only on what is said (that is, structure and meaning) but also on the speaker's intention and the hearer's recognition of that intention. In their view, the speaker's communicative intention is fulfilled in the hearer's recognition of it as being intended to be recognized. This reflexive intention is one of expressing an attitude (e.g., a belief or desire). The hearer is intended to recognize the speaker's intent not only through the content and context of the utterance but also because the intent is intended to be recognized.

In the systems that have been proposed to integrate grammar and speech acts (Sadock, 1974; Searle, 1969), Bach and Harnish claim that the connection has been taken to be semantics. They reject this view and assert instead that the connection is inferential. A speaker (S) expresses an attitude by uttering a certain phonetic string and intends that the hearer (H) infer from that string and from *mutual contextual beliefs* what attitude the speaker is expressing. Mutual contextual beliefs (MCBs) are salient contextual information that S and H both have; both believe that they have them. MCBs help H infer from what S uttered what S is saying and from that to infer the content and force of S's elocutionary act. Besides the MCBs used in linguistic communication, Bach and Harnish postulate three mutual beliefs shared by all the members of a linguistic community. These also help aid the inferential acts done by the hearer. These three general mutual beliefs are called the *linguistic presumption* (LP), the *communicative presumption* (CP), and the *presumption of literalness* (PL). The LP basically states that all speakers in a linguistic community assume that they share the same language and that when one member utters something in L to another, then the hearer can identify what was said, given that he knows the meanings provided by L and the appropriate background information. The CP states that whenever there is a talk exchange between two (or more) members of the linguistic community, utterances are made with a recognizable elocutionary intent. The PL simply states that if one could be speaking literally (under the circumstances) then one is speaking literally.

Given this machinery, Bach and Harnish (pp. 76–77) then produce a *speech act schema* (SAS) which is designed to account for not only literal and nonliteral elocutionary acts but also for direct and indirect elocutionary acts. This inferential schema, if feasible, would account for a large part of what goes on in communication.

Although the three general mutual beliefs that Bach and Harnish postulate as holding within a linguistic community seem to be very reasonable and are quite specific, the same things may not be said of MCBs. It is clear what MCBs are intended to do (they fill in the gaps that the hearer might have in inferring the speaker's communicative intent) but it is not clear that they are any more than an *ad hoc* device to make the system work. They are a vital part of the SAS as is evidenced by the fact that they occur on most lines of the inference pattern. It is not clear how the speaker could determine which beliefs are mutually held or how the hearer could determine which beliefs (given that they are mutual) are relevant to the utterance currently under analysis. The question is, is it possible to determine MCBs in such a way as to make them a part of an explicit theory of language comprehension? If MCBs can be more rigorously specified, then the SAS seems to be a plausible system for how the hearer determines what elocutionary act the speaker performed.

One further aspect of the SAS is that Bach and Harnish take it to be a supplement to a linguistic parser. That is, before the inferential part of the SAS may be used, the hearer must have some way of determining the syntax and semantics of the utterance. They assume that some sort of global processing mechanism will get the hearer to a relevant point in the SAS. They have nothing to say about this processor except that they assume that it is global. Obviously, before the SAS can be tested empirically, it must incorporate a sound linguistic processor.

Sentence Verification and Propositional Models

We have frequently alluded to language comprehension in the previous sections as an aspect of semantics. This is certainly true from the viewpoint of performance and is, of course, of paramount interest for the objectives of this study. Sentence verification is an aspect of comprehension. That is, it provides a mechanism to verify the truth-value of an utterance. The work of Herbert Clark and his associates exemplifies the research in this area (see Clark, 1976; Clark, 1978; Clark and Chase, 1972; Clark and Clark, 1977). We have already discussed some of this work in Chapters 3 and 4. The following discussion is intended to make this chapter more comprehensive and to add emphasis to goals for language comprehension.

Clark (1976) states that one should not confuse the meaning of a sentence (semantics) with judgments about the sentence (true, false, nonsensical, etc.). The former deals with representation, the latter with processing. Clark believes that the sense of linguistic deep structures is what people know once they have comprehended a sentence. Consider the following sentence:

(13)   John isn't happy

It is composed of two propositions: *John is happy* and *it is false*, with the first embedded in the second:

(14)   ((John is happy) is false)

Thus, (14) is a representation of what a person knows about (13).

Clark (1976) suggests a four-stage process in sentence comprehension and verification. For example, given the sentence *A is above B* and the following picture:

A
B

We have the following procedure:

Stage 1: Represent the sentence as (A above B)
Stage 2: Represent the picture as (A above B)
Stage 3: Compare the two representations
Stage 4: Produce the response: *true*

In discussing negatives, Clark suggests that, for example, sentence (15) is represented with a positive assertion and a negative of that as in (16).

(15)   Helen isn't at home
(16)   (false (Helen at home))

Now, with a previous knowledge or presupposition that *Helen is at school*, the processing of the sentence works as follows:

Stage 1: Sentence representation (false (Helen at home))
Stage 2: Knowledge representation (Helen at school)
Stage 3: The comparison algorithm
      (a) (false (Helen at home))
      (b) ((Helen at school))
        1. Set index to T (true)
        2. Compare the inner parentheses of (a) and (b)
        3. Fails; change index to F (false)
        4. Compare the complete expressions in (a) and (b)
        5. They do not match; change index to T
Stage 4: Response generator *true*

Thus, the output of the processing for sentence (15) is *true*.

However, all terms do not have identical representations. For example, the word *absent* is considered a "full negative" compared against *present*. Hence *x is present* and *x is absent* may be represented, respectively, as

present (x)

not (present (x))

on the other hand, *absent* is treated with a positive representation and contrasted with *not present* in the pair : *Helen is not present* ; *Helen is absent*. We get the following representations:

(false (Helen present))

(Helen absent)

These representations are not arbitrary. Psychological experiments have shown different latencies for the processing of apparently synonymous concepts. For example, the processing for verifying *A above B* versus *B below A* is reported to be different, suggesting that the terms *above* and *below* have different complexity in their underlying representations.

In an analogous discussion for active versus passive sentences, Clark claims that in the following pair of sentences:

(17a)   A hit B

(17b)   B was hit by A

the active asserts what A did, whereas the passive asserts what happened to B; thus the representations would be:

(18a)   (A did (A hit B))
(18b)   ((A hit B) happened to B)

Following Clark, then, we can conclude in general that the process of comprehension goes as follows:

At Stage 1 people represent the meaning of a sentence in an abstract symbolic form.

At Stage 2 other relevant information (perception of a picture, previous knowledge, presupposition, etc.) is also represented in the same format as in Stage 1.

At Stage 3 the two representations are compared by a series of match and/or other manipulations.

At Stage 4 a response is generated or is perceived.

Stage 1 representation is in the form of some deep structural representation, and Stage 3 operations are based on the principle of congruity.

Meaning in Different Worlds

The theory of meaning, traditionally, rests on two assumptions: (a) knowing the meaning of a term is being in a certain psychological state; (b) the meaning of a term determines its extension (sameness of intension entails sameness of extension). Putnam (1975) attempts to show that these assumptions are incompatible. Assumption (a) entails the assumption of *methodological solipsism* (that is, no psychological state presupposes the existence of anyone other than the person in that state). The trouble with this position is that some psychological states, like being jealous of x, do entail that x exists. To know the meaning of a term A, it is not enough just to *grasp the intension of A*; one must also know that it is the intension of A that one has grasped. But then, if A and B are different terms, *knowing the meaning of A* and *knowing the meaning of B* must be different psychological states, so A and B cannot be synonymous (that is, there can be no synonymy). The psychological state determines intension and intension determines extension, so psychological states determine extensions. Putnam claims this follows from the two assumptions and will show later that it is false (that is, it is possible for two speakers to be in exactly the same state and yet differ in the extension assigned to the term A). He claims extension is *not* determined by psychological state.

To show that psychological state does not determine extension, Putnam gives several examples using "Twin Earth" (exactly like Earth, but the extension of certain words is different). "Water" on Twin Earth is XYZ not H_2O. Consider the period before chemists had discovered the chemical composition of water. Putnam claims it would be possible for an Earthian and a Twin Earthian to have the same beliefs about water, to be in exactly the same psychological state, and yet "water" would have different extensions. He concludes (p. 144), "Cut the pie any way you like, 'meanings' just ain't in the head!" He is claiming that meanings are not psychological states.

Putnam points out that there is a "divison of linguistic labor." We could not use the word *elm* unless someone was able to distinguish elm trees from all other trees, but not everyone to whom the distinction is important need be able to make that distinction. This division of linguistic labor presupposes division of nonlinguistic labor. It may take the collective linguistic body to fix the extension of a term.

Putnam suggest two ways for stating the meaning of a "natural-kind" term: (1) *ostensive definition* (point to an instance of it and say "this is———"); (2) give a description of the stereotype and some *markers*. He is assuming the notion *possible world* as a primitive. Putnam holds that water is H_2O in all worlds, but *water* means different things on Earth and Twin Earth. He says that natural-kind terms and the demonstrative *this* are rigid designators in Kripke's sense. A designator is rigid if it refers to the same individual in all possible worlds in which it designates. This has "startling consequences" for a theory of necessary truth. Putnam introduces *cross-world relations*. A two-place predicate is cross-world if it holds true of ordered pairs of individuals such that the first member is not in the same possible world as the second member. Because *water* is a rigid designator, it must refer to the same stuff (H_2O) in all possible worlds. Therefore, there is no possible world in which water is not H_2O, that is, *water is H_2O* is a necessary truth. Kripke splits necessary truths into two kinds: (a) epistemically necessary (these are rationally unrevisable); (b) metaphysically necessary (these are true in all possible worlds). A statement like "water is H_2O" can be metaphysically necessary and yet epistemically contingent. Putnam asserts that the extension of a term is part of that term's meaning. Therefore, a difference in extension is necessarily a difference in meaning but not a difference in the concept or psychological state associated with that term.

Finally, Putnam proposes to describe meaning by a finite sequence, or *vector*, which includes at least

> (1) the syntactic markers that apply to the word, e.g., "noun"; (2) the semantic markers that apply to the word, e.g., "animal," "period of time"; (3) a

description of the additional features of the stereotype, if any; (4) a description of the extension." (p. 190)

Exercises

1. Declarative sentences are said to be statements about events which can be true or false and, therefore, they can be accounted for by truth-conditional semantics. Philosopher J. L. Austin has noted, however, that sentences such as *I promise that I will give you the money* present certain problems in this regard. Discuss this in terms of the theories of Davidson and Montague.

2. In one TG view semantic analysis was strictly reserved for the deep structure level. It was claimed that sentence pairs such as the following:

    (a) John kissed Mary
    (b) Mary was kissed by John

had the same deep structure and, therefore, the same semantic interpretation. However, consider the following well-known pair of sentences which are clearly not equivalent:

    (c) Everyone loves somebody
    (d) Somebody is loved by everyone

Discuss some of the issues concerning surface and deep interpretations.

3. Compare and contrast *theory of truth, theory of meaning,* and *translational theory.*

4. What is *intensional logic* in the context of the Montague semantics?

5. Analyze the sentence *Mary is a spinster* in terms of Katz's semantic markers and other identifiers.

6. Frege (1892) distinguishes between reference and sense: expressions with the same sense always have the same reference, but the reverse is not true:

    (a) John's only sister
    (b) The only sister of John

(c) Morning star
(d) Evening star

(a) and (b) have the same sense and reference; (c) and (d) have the same reference but not the same sense. Discuss.

7. Carnap (1956) differs from Frege (cf. Exercise 6 above) in working with a "constructed logical language" in which he defines the two concepts of meaning as *extension* and *intension*. For Carnap two expressions have the same extension if and only if they are equivalent. On the other hand, two expressions have the same intension if and only if their equivalence can be established solely by the semantic rules of a linguistic system, irrespective of any extralinguistic facts (cf. Katz's *meaning in the null context*). Discuss and give examples.

8. Discuss the utility of speech act, if any, in language use (performance).

9. The propositional model of Clark for sentence verification provides for expressing explicit facts and additional knowledge about a sentence S in terms of a procedure. (a) Write a procedure for the propositional derivation and verification of the sentence:

The house is not near the forest

(b) Do the same as in (a) but with the prior knowledge that *the house is near a city*.

References

Wunderlich (1979) provides more details of some of the philosophical discussions in this chapter. Kempson (1977) has written a textbook for this purpose, with additional reading recommendations and examples. J. D. Fodor (1974) is perhaps the best survey of "theories of meaning in generative grammar."

#### CHAPTER 7

Natural Language Understanding
The Artificial Intelligence Approach

Introduction

Artificial intelligence (AI) covers a vast variety of subfields, each of which has become or is becoming a quasi-independent field of study and research: pattern recognition, expert systems, knowledge representation and engineering, robotics, natural language processing and understanding, automated reasoning, speech perception, problem solving and proof systems. From a more global perspective, however, AI is a discipline that identifies the common components and problems, interrelationship, and interdependence of these subfields. An analogy can be drawn with psychology and its many subdisciplines: developmental, cognitive, educational, clinical, and so forth.

To be a successful enterprise, AI must necessarily follow a two-pronged path of study and research: (a) It must study and understand the nature, sources, scope, and other aspects of human intelligence. It must also answer questions about biological endowments versus acquired knowledge (for example, as we have seen, many linguists believe that a "faculty of language," including a "universal grammar" is part of the genetic endowment of humans, whereas knowledge relevant to a particular language must be acquired), and, in general, various aspects of human behavior. (b) It must study and develop computing devices and programs which can perform tasks, similar to those done by people, which require intelligence and reasoning. A very simple and simplistic example of the latter is for the machine to solve the type of analogy problems as in (1) which are used for testing human intelligence.

(1)

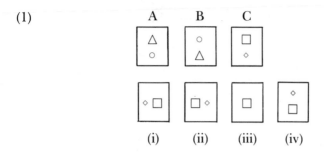

(i)   (ii)   (iii)   (iv)

The problem is this: A to B is like C to X where X is one of (i), (ii), (iii), or (iv).

Having sprung from computer science, AI research has often attempted to follow path (b) with cursory or lip service attention to path (a). This situation is rapidly changing; but even in the past, going back to the "ancient" history of computing, there have been notable and important exceptions to the trend of putting primary emphasis on path (b). It is characteristic that when Alan Turing became interested in machine intelligence, he practically gave up his research in mathematics and computer design and started studying biology (Hodges, 1983). John von Neumann, one of the architects of the original digital computers, formulated analogies between the computer and the brain and in fact wrote a book in the 1950s, *The Computer and the Brain*. Coming from the opposite direction, Pitts and McCulloch tried to show that neural nets were organized in simple *and/or* gates much in the manner of the electronic nets in computers (this has turned out not to be quite that simple). There have also been efforts in establishing interdisciplinary research projects and studies, notably between computer science and psychology at Carnegie-Mellon University where joint faculty appointments are held. Herbert Simon (1979–80) makes the following remarks concerning such cooperative efforts:

> Although their goals are quite different, the two disciplines of artificial intelligence and simulation of cognitive processes have maintained a fruitful alliance for a quarter century, and now are further associated under the label of "cognitive science." AI research is aimed at learning how to program computers to do smart things, cognitive simulation research at learning how human beings do smart things (solve problems, discover patterns, learn, reason, make decisions). But writing computer programs for performing such tasks can give us (and has given us) clues as to how humans do the same things, while discovering how people perform such tasks can provide us (and has provided us) with ideas for effective computer programs capable of doing these things. There has been a fruitful flow of ideas between these two fields in both directions during their entire histories.

There are at least two encouraging prospects currently within the AI enterprise. One is that centers, institutes, and associations are being formed in the United States and Europe for *cognitive science* with the objective of bringing together experts from various fields and of formulating joint research projects in AI in its widest sense. The people involved may include those from the fields of mathematics, computer science, linguistics, psychology, philosophy, speech and hearing sciences, neurosciences, and others. The other encouraging development is that researchers following path (b) are paying more attention to logical foundations and fundamental principles rather than engineering ingenuities. Finally, in all of these discussions, we must not lose the insight that for the foreseeable future it will be the case that computers will be better at doing some things and human beings at doing other things although all practitioners of AI may not agree with this statement.

In the remainder of this chapter, I will present additional views on the nature and practice of AI, with particular attention to natural language processing (NLP).

Views on Artificial Intelligence

An often-cited definition of AI is the one given by Minsky (1968b): *"Artificial intelligence,* the science of making machines do things that would require intelligence if done by men" (p. v). Feigenbaum and Feldman (1963) give the following example to contrast AI with research in the "simulation of cognitive processes": An AI researcher interested in programming a computer to play chess would be happy only if the program played chess well, preferably better chess than the best human player. On the other hand, the researcher interested in simulating the chess-playing behavior of a person would be unhappy if the program played chess better (or worse) than the individual, for this researcher wants the program to make exactly the same moves as the human player irrespective of the result of the game.

There never has been unanimity as to what AI is, and as recently as 1979 Martin Ringle could write: "As a newly-developed study, AI has not yet coalesced into a well-defined discipline with a universally-agreed-upon description" (Ringle, 1979b, p. 6). Ringle offers a quadripartite breakdown of AI into AI *technology,* AI *simulation,* AI *modelling,* and AI *theory.* AI technology depicts the "intelligent behavior" of a system irrespective of its data structure and tasks having any relation to human behavior. It is a purely pragmatic undertaking and must be judged on the basis of its utility. AI simulation, on the other hand, makes claims of similarity to human behavior and is often judged on the basis of the degree of similarity in the overt behavior of a computer system and a person. AI modelling

> is primarily concerned with internal components, rather than overt behavior. The finished product need resemble human behavior in none but the most abstract ways, so long as the data structures, internal states, and information processes provide a coherent and plausible model of the way people think. (p. 8)

Ringle goes further to say that AI modelling is frequently identified with cognitive psychology, and "many of the scientists who indulge in AI modelling are cognitive psychologists rather than computer scientists" (p. 9). AI theory is similar to epistemology, especially as conceived by philosophers such as Whitehead and Kant, but the fundamental principles sought by AI theorists

> are somewhat less abstract than the principles traditionally sought by philosophers. . . . AI theorists are interested in principles of knowledge and intelligence which may be used to account for concrete, physically-instantiated, time- and perspective-dependent cognition. Typically philosophers have eschewed such elements in their quest for an epistemology which was "ideal" in an almost platonic sort of way. (p. 10–11)

J. McDermott (1979) judges that "work in AI has as its goal the construction of programs which, when run on a computer, will exhibit intelligent behaviour of the same generality and power as that of human beings" (p. 110). In the same volume, K. M. Sayre (1979), a philosopher who has been close to AI for many years, states, "although artificial intelligence has its own journal, its own research institutes, and its own learned spokesmen, I do not believe it has achieved the status of a branch of knowledge" (p. 139). For Sayre, "artificial intelligence is a form of modelling" (p. 140). Contrast the following statement of Minsky's (1968b):

> While the work in artificial intelligence draws upon work in other fields, this is not a significantly interdisciplinary area; it has its own concepts, techniques, and jargon, and these are slowly growing to form an intricate, organized specialty. (p. 6)

While claiming that "whatever disclaimers their proponents make, all artificial intelligence (AI) systems in fact incorporate naive, often unexamined, psychological assumptions," Wilks (1981) affirms that "doing psychology is not the job" of AI workers (p. 337). On the other hand, in the same journal, a hundred or so pages earlier, Longuet-Higgins (1981) proposes renaming AI, and suggests "that 'theoretical psychology' is really the right heading under which to classify artificial intelligence studies of perception and cognition" (p. 200).

R. Schank, one of the central figures in AI for the last decade, begins his 1979 article as follows:

> Artificial intelligence is a field that is not exactly sure what it is about. . . . It is easy to think of problems that were considered to be part of AI ten years ago that are not now, and of parts of AI now that were part of some other field ten years ago. . . . It is even more of a problem because people in other fields cannot agree on either the relevance or the place of AI with respect to their own fields. Is AI psychology? Is it philosophy? Many computer scientists not in AI feel sure it isn't computer science. Is the work in natural language processing linguistics? Linguists feel certain that it isn't even in any way relevant. (p. 196)

Schank continues,

> What is an AI program and what is AI? . . . AI is really no more than a promise for the future. We are learning things by writing these programs. In fact, that is probably the major reason that we write them. AI represents a method by which we can learn about the nature of knowledge and the nature of man. [The programs of the Schank group] represent a sort of experimental epistemology, and viewed in that light they are significant. (p. 218)

He then supplies the definition and principles:

> *Definition*: A program is an AI program if it characterizes and uses knowledge in any way that seems to be in accord with the intuitions of people who normally do the tasks that the program is doing.

> *Principle I*: A program that solely exhibits intelligent behaviour without respect to how knowledge has been characterized and used, can be considered to be a stage 1 AI program. A Stage 1 AI program is an AI program if, and only if, that kind of intelligent behaviour has never been modelled before.

> *Principle* II: As AI program is significant only if it tells us something about the form, nature, and the use of knowledge. Such a program is a Stage 2 AI program.

> *Principle* III: The ultimate AI program that we are all aiming for is one that specifies the form in which knowledge is to be input to the program, as well as the form of the rules that use that knowledge, and produces a program that effectively models that domain. (pp. 218–220)

Schank concludes,

> The real argument to be made here is simply this: AI is the study of knowledge. It differs from epistemology in that it addresses the problem of determining the processes of applying knowledge to real-world situations as opposed to addressing more abstract questions. If AI is the study of knowledge, the only question remaining is to determine what kind of knowledge. The answer seems trivial: human knowledge of course, what

other kind is there? . . . We might at this point consider just what it is that AI has to offer the disciplines of psychology, linguistics, and philosophy. The answer is the possibility of thinking about and testing integrated processes. The above disciplines have in common a reliance on bits and pieces of unrelated evidence about the disciplines they seek to explain. An AI person looking at these same phenomena, attempts to formulate processes that mimic the behaviour of the system he seeks to find out about. The use of knowledge is a fundamental part of the process of understanding. Viewed as a method used in the process of understanding, the problems of knowledge acquisition and knowledge application are more easily attacked. (p. 221)

If the many definitions of AI provided so far suggest a field in turmoil, one which has undergone a rapid development and is still in its earliest stages, a glance at the history of AI will confirm these suggestions. The barest outline only will be supplied, however, as numerous accounts of the development of AI are available. (See references at the end of this chapter for some sources.)

As intellectual forebears of AI, few philosophers have failed to be named by one or another observer. Dreyfus (1979) swears by Plato: "It starts with Plato's separation of the intellect or rational soul from the body with its skills, emotions, and appetites" (p. 62). Haugeland (1981b) opts for Hobbes:

"Reasoning is but reckoning," said Hobbes (1651, ch. 5), in the earliest expression of the computational view of thought. Three centuries later . . . his idea . . . has become . . . the basis of an exciting new research field, called "artificial intelligence." (p. 1)

J. A. Fodor (1968, 1978) speaks of Frege and Russell as forefathers of "cognitive science" in general and therefore of AI, and thinks of adding Aristotle to the list; the predecessors of the "procedural semanticists" within AI are identified by Fodor as Locke, Hume, and the modern Verificationists. Early discussions of machine intelligence are noted by Armer (1963), who stresses the works of Samuel Butler, whose *Erewhon* (1865) concerns, among other things, the "development of mechanical consciousness." The Countess of Lovelace (1842) argued at some length concerning AI topics, in response to Babbage's mid-nineteenth-century development of an "analytical engine," a universal digital computer; her opinions are considered in detail by Turing (1950).

The second third of the twentieth century saw the development of a number of novel and interrelated disciplines whose origins are to be found, intellectually, in the nineteenth- and early twentieth-century mathematical work of Hilbert, Frege, and Russell and Whitehead. Mathematical discoveries by Turing (1937) provided a rigorous basis for the expression of intelligence as an abstract formal entity, thereby joining the vast and potent systems of logic and mathematics at one fell swoop to the meager set of tools humans had previously possessed for use in the quest to, paraphrasing Singh (1966), render human intelligence sufficiently introspective to know itself. (For a discussion of Turing's results, see Gross, 1972; Moyne, 1974; Wall, 1972, and references therein.) The importance of Turing's ideas and their relation to AI is the theme of recent comments by Pylyshyn (1979):

> The work of Turing, in a sense, marked the beginning of the study of cognitive activity from an abstract point of view, divorced in principle from both biological and phenomenological foundations. It provides a reference point for the scientific ideal of a mechanistic process which could be understood without raising the spectre of vital forces or elusive homunculi, but which at the same time was sufficiently rich to cover every conceivable formal notion of mechanism. (That the Turing formulation does cover all such notions is, of course, not provable but it has withstood all attempts to find exceptions. The belief that it does cover all possible cases of mechanism has become known as the Church-Turing thesis.) It would be difficult to overestimate the importance of this development for psychology. It represents the emergence of a new level of analysis, which is independent of physics, yet is mechanistic in spirit. It makes possible a science of structure and function divorced from material substance, while at the same time it avoids the retreat to behaviouristic peripheralism. It speaks the language of mental structures and of internal processes, thus lending itself to answering questions traditionally posed by psychologists.

> While Turing and other mathematicians, logicians, and philosophers laid the foundations for the abstract study of cognition in the 30s and 40s it was only in the last twenty or so years that this idea began to be articulated in a much more specific and detailed form: A form which lends itself more directly to attacking certain basic questions of cognitive psychology. The newer direction has grown with the continuing development of our understanding of the nature of computational processes and of the digital computer as a general, symbol-processing system. It has led to the formation of a new intellectual discipline known as artificial intelligence, which attempts to understand the nature of intelligence by designing computational systems which exhibit it. (pp. 24–25)

Pylyshyn implies that the period roughly from the end of World War II to the unfolding of the first buds of "flower power" was relatively infertile from the viewpoint of cognitive psychology and AI. This is true to the extent that developments in fields in or close to what has come to be called cognitive science (a term that is defined later) failed, with some exceptions, to give rise to traditions or subdisciplines within cognitive studies that flourish today.

Several of these fields, grouped during the period in question under the rubric *cybernetics*, which proved in subsequent years to be of minor interest and relevance to practitioners of AI, are reviewed in a readable and eloquent volume by Singh (1966). These fields are information theory, neural network theory, the theory of self-organizing systems, and cybernetic theory. A short, but more up-to-date review of some of these topics is to be found in Hunt (1975), Chapter 1; also see his references. In addition, see Minsky (1968b) and references therein; Arbib (1964, 1972); and Minsky's (1963b) extremely comprehensive bibliography. The work of Arbib (1964, 1972; Arbib and Caplan, 1979, and references therein) in cybernetics and in what he has called *brain theory* (Arbib, 1972) constitutes a major exception to the desuetude of the above-cited fields. Nonetheless, for better or for worse, the mainstream of AI has perched itself at a farther point of remove from neurophysiology than have the cyberneticians, and it is to this mainstream that attention will now be given.

References have been provided concerning AI history. No attempt will be made to rehash this or to introduce the discipline of AI more generally; this is a book-length proposition which has been taken up by such authors as Hunt (1975), Bundy (1978), and Winston (1977, 1984). Boden (1977) gives a readable popular presentation of AI. A few brief remarks might be helpful: Ringle's classification proceeds in rough chronological order historically. Although it is clear that Schank's Stage 1 AI programs must have preceded his Stage 2 programs chronologically, it is of some note that most AI researchers are no longer interested in Stage 1, but many exceptions to this claim exist. Finally, disregarding the "prehistorical" background, a rough-and-ready scheme for dividing up the evolution of AI to date might be this: The earliest period of AI research can be considered in connection with the development of the general problem solver system (GPS) started in 1957 by Newell, Shaw, and Simon (1960). This was "an attempt to synthesize in a single core of a set of concepts, methods, and strategies assumed to underlie human problem solving generally, quite apart from these features that characterize activity in any particular subject area" (Reitman, 1965, p. 4). Some of the ideas underlying GPS are discussed in Chapter 3 on psycholinguistics. A second period can be identified with Minsky (1968a). The work of this era attempted to deepen program performance by having the system digest information before attempting to use it to solve problems:

> The system does not attack problems by applying logical deduction procedures directly to a stored library of statements it has received. . . . Instead, it works by *'understanding' the statements when they are made*, consolidating this understanding by adding to or modifying the network. (Minsky, 1968b, p. 4)

In 1972, with the publication of Winograd (1972), a third period can be thought of, during which attempts to deal with inferences and deductions in natural language processing in some form or other came to the fore. Finally, the present era might be associated with Schank and Abelson (1977). Here the emphasis is on providing the deepest and broadest possible knowledge base to programs designed for full-bodied language understanding; an attempt is made to endow AI systems with representations of quasi-human goals and background knowledge. AI has found a place for itself, in this period, within an emerging discipline called *cognitive science*:

> Cognitive science includes elements of psychology, computer science, linguistics, philosophy, and education, but it is more than the intersection of these disciplines. Their integration has produced a new set of tools for dealing with a broad range of questions. In recent years, the interactions among the workers in these fields has led to exciting new developments in our understanding of intelligent systems and the development of a science of cognition. The group of workers has pursued problems that did not appear to be solvable within any single discipline. It is too early to predict the future course of this new interaction, but the work to date has been stimulating and inspiring. (Bobrow and Collins, 1975, p. x)*

In this period the programming activity of AI has come to be deemphasized by some who have preferred to characterize the field as "theoretical psychology" and "applied epistemology."

Common Sense and Learning

Perhaps the most challenging and most critical tasks for AI research at present are the problems of coping with *common sense* and *machine learning*. To illustrate these, two specific activities will be discussed. Although it may not be apparent at the outset, both examples have relevance for language understanding by machines.

The first example is the game of chess. From the very beginning, programming a computer to play chess has been a challenge. Turing, as well as the founder of information theory, Claud Shannon, and founders of AI, Allen Newell, Herbert Simon, and John Shaw, have all taken up the challenge of computer chess. Indeed, computer chess has come a long way in that there are chess programs today that can beat master players. Nevertheless, these programs have failed to meet the wishes or expectations of their AI backers. What is more important in a technical sense is that successful chess programs do not formulate their strategies in the same way that human players do.

* For a more comprehensive view of cognitive science, see Pylyshyn (1984).

At each stage in the game of chess a player can, at least theoretically, select a move from among a finite number of possible moves; then he or she can determine the best countermove that the opponent can make, and then his or her own next move and the opponent's next move, and so on to the end, at which point a move is decided upon. This strategy can be represented as a tree structure, whose nodes at each alternative level represent the possible moves of one or the other of the two players. Because the choice of moves at each position is finite (35 on average) and the game ends in a win or draw (halts), the game tree is a finite structure, and, theoretically, one can construct such a tree for the game of chess and program a computer to do an exhaustive search of the whole tree for each move. Such a program will beat any human world champion or grand master player. The problem is that as one moves from the starting point (the root) up the tree (or rather down from the root because game trees are represented upside down like P-markers of syntactic structures), the growth rate of the branches leading to new nodes is exponential. The number of branches for a chess game tree is on the order of 10^{120} (Shannon, 1950), and Waltz (1982) points out that searching such a tree only three moves down for each player in mid-game would require the examination of more than 1.8 billion moves. The time required for the exhaustive search of a chess tree is something in the order 10^{90} years if done by electronic computers which presumably have much faster speed than human brain functions.

Techniques have been developed, such as minimaxing, alpha-beta pruning, heuristic search, and others, that "prune" the game tree and cut down the number of searches based on various ways of abandoning search in branches and subtrees which are not promising for results, or on cutting the depth of the search in a tree. However, the basic strategy of such approaches is still an exhaustive search, albeit of smaller trees, and this is still unsatisfactory for large game trees, such as chess.

All this does not mean that very impressive game programs have not been written which play at the expert level. On the contrary, Hans Berliner at Carnegie-Mellon University has written a backgammon program which defeated the world champion in 1979. Two researchers at Bell Laboratories, Ken Thompson and Joe Condon, have written a chess program called *Belle* which has a rating of 2,160 where a rating of 2000 to 2199 qualifies a human player as an expert for tournaments (Waltz, 1982). But even this program examines an average of 160,000 positions per second and has to run on a specially designed computer for playing chess. One strategy in a heuristic search is to assign weights or values to pieces and also assign positive and negative values for moves that threaten or capture a piece. Then, at each level, scores are computed for each possible move and countermove, and decision is made for selecting a move without going deep into the tree.

However, human players do not seem to use any of these strategies. Human experts "know" certain opening moves, end moves, and the like, and their consequences as patterns, and make decisions about moves in ways which are not known. They use common sense to change strategies in mid-game in ways which are not fully understood. Indeed, if we knew these gestalt pattern matching and commonsensical approaches, we could not only write chess programs that worked as people do, but we could also solve the problems of language processing, translation, and so forth. Thus, we have chess programs that are at the expert level and play better than many human players, yet they do not emulate the behavior of even a novice human chess player. A crude analogy can be drawn with cars: they can go from one place to another much faster than a human can, but they cannot emulate a human's walk!

For our second example, we note that many useful and practical AI programs have been written; Patrick Winston and Karen Prendergast have even written a book about the commercial uses of AI (Winston and Prendergast, 1984). But to what extent do these programs learn and think in the way that people do? The question may be moot (and has been treated as such by AI researchers) if we are not pushing the claim of psychological reality for computer programs. We can surely program a computer to become "sad" or "happy," if sadness and happiness are two formal machine states, and if certain events, configurations, or delta-functions of a Turing machine will cause the machine to enter one of those states and emit some expected reaction by, for example, printing a message. Is this simulation or emulation of human behavior? The problem is that, for humans, happiness and sadness are not isolated states of mind in a so-called microworld; they are intimately connected with the whole process of human life and must be understood in that global context. Language learning is of particular interest to us, but this is the area where the least progress has been made in AI. This is not surprising because linguists, psychologists, and philosophers have investigated language learning, particularly child language learning for decades and centuries, and yet there does not exist a coherent and noncontroversial theory at present.

However, there are, and have been, many impressive learning programs. Arthur Samuel wrote a checkers program in the 1950s which was able to learn from its mistakes and upgrade its performance. This program as well as a number of chess programs have been able to beat their authors. Patrick Winston, nearly 20 years ago, wrote a program that could learn concepts of objects such an arch or a toy house by examining a series of examples and "near misses." One of the well-known and useful *expert systems* is the MYCIN program developed by Edward Shortliffe (1976). This is a medical diagnosis program covering cases of blood infection. It is rule-based with situation- or condition-action (C→A) rules of the Newell production type. Here is an example of a rule (Winston, 1984, p. 199):

> Rule 88:   IF the infection type is primary bacteremia
> the suspected entry point is the gastrointestinal tract
> the site of the culture is one of the sterile sites
> THEN there is evidence that the organism is bacteroides

MYCIN asks a series of pertinent questions and carries a dialogue with the doctor and arrives at a diagnosis and recommendation for treatment. It can also explain how it arrived at its conclusions. The utility of this system, which has evidently been certified by panels of qualified judges, is as follows. Often patients with blood infections arriving at an emergency room in a hospital need immediate treatment which cannot wait for the prolonged laboratory analysis of the blood sample. An experienced physician, expert in blood diseases, can guess the type of infection and prescribe a specific treatment which is always more effective than some general treatment such as antibiotics. An inexperienced doctor, such as an intern or resident, may have to depend on general treatment. MYCIN is supposed to help the latter in arriving at expert decisions. A lighthearted hypothetical conversation between a MYCIN computer (C) and a physician (P) is presented to make the point.

> C: It has been shown that while I lag behind an experienced specialist, I am better and faster in arriving at diagnosis and treatment recommendation than your young unexperienced doctors.

> P: The selection of blood infections was a clever "engineering" feat. Blood infections are rather well known and can easily be described by situation-action rules, such as you have. Many other medical procedures are far more complex, contain elements of uncertainty, and require extensive additional knowledge.

C: Tell me, how long in total number of years does it take for a person to become educated and experienced as an expert physician?

P: Oh, perhaps about 40 years.

C: Ah! People expect me to be an instant expert. Give me 40 years of experience and learning, and I will be as good as any of your experts!

P: You have put your finger exactly on the problem. Your progenitors do not know how people learn in any precise way, and could not program you to learn as people do.

C: Do you know of any fundamental principles that would prevent them from learning to do this in the future?

P: I don't know. Perhaps this is a question for psychologists and philosphers; I am neither.

There is one other aspect of learning, in contrasting man with machine which we wish to reemphasize. This is the question of *innate* versus *acquired capacity* for learning. We can teach a person to play chess or music, but we cannot teach him to become a chess or music genius. Thus, the emulation of human learning process must await the discovery of how people learn, what they are born with, and what they acquire.

Tools of Natural Language Processing

The purpose of the remainder of this chapter is to examine work in natural language understanding of the last decade or so. No claim to comprehensiveness is made; some of the more important contributions of some of the major researchers are noted. Before considering results from natural language understanding (NLU) proper, however, it is necessary to make contact with two major ideas that have had repercussions throughout the AI field: the *production system* concept of Newell and Simon and the *frame concept* of Minsky.

The literature on production system (Newell, 1973) is quite large, and if discussions of the earlier systems of Newell and Simon, which are organically related to the more current versions, are included, the volume of information available is huge. The basic operations of these systems are called *productions*, or *condition–action pairs*, or *situation–action pairs.* These are miniature programs which go into operation (*fire*) when, other things being equal, the *condition* or *situation* represented in their left-hand terms are considered by the system to obtain. In other words, each production is similar to a conditional statement whose consequent is an *action.* A production system has vast numbers of these productions, normally independent of each other, in its *long-term store* or *long-term memory.* (Sometimes no name is given to the collection of productions besides *production system.*) It also has a *short-term store* (*short-term memory,* work space) in which at any given time some number, say, between five and ten, of descriptions of situations are to be found. A production fires when its condition matches one of the descriptions in short-term memory (STM), causing a change in the contents of STM (the description is either retired, or kept active but in a holding pattern, or modified in some other way) and in the contents of long-term memory (LTM) as well. That is, the action, or right-hand term, of a fired production can have the effect of adding a new production to LTM (learning) or of deleting a production there which is discovered to be unhelpful in coping with the current situation (also learning). Other outcomes of firings include the effective restructuring of the system's representation of processes or procedures. When one firing has taken place and STM has been altered in consequence, a new *cycle* of matching and firing occurs, and so on. All of LTM is theoretically available for matching at any time. It has been stated that a match leads to a firing, other things being equal. That is to say that different versions of the production system impose different constraints on firings where more than one match occurs at one time: some schemes are based on weighting, others on recency of firing, others on specificity (i.e., the production with the greatest number of subsituations in its left-hand term gets fired). The last remark might be puzzling in light of the description of productions as primitives. In fact, the comparison is not totally apt because the right or left term, or both, can be complex, featuring conjunctions of elements and/or embeddings. (In fact, productions can even turn out to be fair-sized subroutines although this is not the norm.)

The great advantage of such a system, which is considered a tool for modelling both abstract intelligence—AI at its broadest—and human cognition (as all of the Newell/Simon systems have been), lies in the diffusion of control built into it. A production system is neither hierarchical nor modular in some halfway sense, such as relatively free passing of control to and among subroutines. It is democratic, not to say anarchic. Its control system is nominally in its interpreter, which merely sees to the matching, weighting, and firing operations; actually, however, the power lies with the data themselves, as it is quite simply the few matters to which the system (the subject) is currently paying attention (the few items in STM) that determine what actions are taken by the system as a whole (see Exercise 18).

There is much more to say about production systems but insufficient space in which to say it. Excellent treatments and interpretations of these systems are to be encountered in, for example, Pylyshyn (1979), McDermott (1979), and Winograd (1975). See Winston (1977) and Bundy (1978) for textbook accounts. All of these sources cite further references. One final observation on production systems as models of human cognition is that the notion of the production system forms a sort of link between, on the one hand, the very earliest work in AI— the work on self-organizing systems and neural nets—and pre-AI psychology such as that of Hebb (1949), and, on the other hand, the most current work in language understanding. (See Pylyshyn, 1979, for references on this work; for an excellent presentation of Hebb's ideas in their relation to information-processing psychology, see Reitman, 1965.)

With regard to his frame concept, Minsky (1975) provides the following succinct presentation:

> When one encounters a new situation (or makes a substantial change in one's view of a problem), one selects from memory a structure called a *frame*. This is a remembered framework to be adapted to fit reality by changing details as necessary.

> A *frame* is a data-structure for representing a stereotyped situation like being in a certain kind of living room or going to a child's birthday party. Attached to each frame are several kinds of information. Some of this information is about how to use the frame. Some is about what one can expect to happen next. Some is about what to do if these expectations are not confirmed.

> We can think of a frame as a network of nodes and relations. The "top levels" of a frame are fixed, and represent things that are always true about the supposed situation. The lower levels have many *terminals*—"slots" that must be filled by specific instances of data. Each terminal can specify conditions its assignments must meet. (The assignments themselves are usually smaller "subframes"). Simple conditions are specified by *markers* that might require a terminal assignment to be a person, an object of sufficient value, or a pointer to a subframe of a certain type. More complex conditions can specify relations among things assigned to several terminals.

> Collections of related frames are linked together into *frame-systems*. The effects of important actions are mirrored by *transformations* between the frames of a system. These are used to make certain kinds of calculations economical, to represent changes of emphasis and attention, and to account for the effectiveness of "imagery."

For visual-scene analysis, the different frames of a system describe the scene from different viewpoints, and the transformations between one frame and another represent the effects of moving from place to place. For nonvisual kinds of frames, the differences between the frames of a system can represent actions, cause–effect relations, or changes in conceptual viewpoints. *Different frames of a system share the same terminals*: this is the critical point that makes it possible to coordinate information gathered from different viewpoints.

Much of the phenomenological power of the theory hinges on the inclusion of expectations and other presumptions. A *frame's terminals are normally already filled with "default" assignments*. Thus, a frame may contain a great many details whose supposition is not specifically warranted by the situation. These have many uses in representing general information, most likely cases, techniques for bypassing "logic," and ways to make new generalizations.

The default assignments are attached loosely to their terminals, so that they can be easily displaced by new items that better fit the current situation. They thus can serve also as "variables" or as special cases for "reasoning by example," or as "textbook cases," and often make the use of logical quantifiers unnecessary.

The frame-systems are linked, in turn, by an *information-retrieval network*. When a proposed frame cannot be made to fit reality—when we cannot find terminal assignments that suitably match its terminal marker conditions—the network provides a replacement frame. These interframe structures make possible other ways to represent knowledge about facts, analogies, and other information useful in understanding.

Once a frame is proposed to represent a situation, a *matching* process tries to assign values to each frame's terminals, consistent with the markers at each place. The matching process is partly controlled by information associated with the frame (which includes information about how to deal with surprises) and partly by knowledge about the system's current goals. (pp. 180–181)

In addition to setting out the frame theory, Minsky sounds many notes which have proved to be keynotes in the years following the publication of his articles: the rejection of "logicism"; the encouragement of research into the structure of commonsense thinking, of stories, conversation; and the call for procedural as opposed to declarative analysis of thinking.

Models of Natural Language Processing

The work of Winograd will be considered first. Winograd (1972) is mentioned above as a watershed in the development of AI. To understand why this is the case, a look at the useful and readable literature review of the work just cited is in order. The earliest AI programs to investigate natural language were what Winograd calls *special format systems*. Examples are BASEBALL (Green *et al.*, 1963), SAD SAM (Lindsay, 1963) and STUDENT (Bobrow, 1968). The heart of these programs was a table or network arrangement of data pertaining to a specialized domain, such as baseball statistics or word problems in high school algebra. The trouble, in retrospect, was that they bypassed the problem of understanding natural language input *qua* language, and substituted a special-purpose heuristic device permitting the capturing of information relevant to its domain, but not generalizable to other domains (pp. 34–35).*

The next most advanced group of programs Winograd considers are the *text-based systems*:

> Some researchers were not satisfied with the limitations inherent in the special-format approach. They wanted systems which were not limited by their construction to a particular specialized field. Instead, they used English text, with all of its generality and diversity, as a basis for storing information. In these "text based" systems, a body of text is stored directly, under some sort of indexing scheme. An English sentence put to the understander is interpreted as a request to retrieve a relevant sentence or group of sentences from the text. Various ingenious methods were used to find possibly relevant sentences and decide which were most likely to satisfy the request. (p. 35) [Two such systems were PROTOSYNTHEX I, Simmons *et al.*, (1966) and Semantic Memory, Quillian (1968).]

> Even with complex indexing schemes, the text-based approach has a basic problem. It can only spout back specific sentences that have been stored away, and can not answer any question that demands that something be deduced from more than one piece of information. In addition, the responses often depend on the exact way the text and questions are stated in English, rather than dealing with the underlying meaning. (pp. 35–36)

What Winograd calls *limited logic systems* (e.g., Raphael [1968], Simmons [1966], and Thompson [1968]) featured two improvements over the text-based systems: Rather than storing the actual English sentences in their knowledge base, these new programs stored simple assertions in some notation other than natural language. The advantage of doing this is that it "frees simple information from being tied down to a specific way of expressing it in English" (p. 36). The second improvement is that they are able to perform limited logical inference based not only on the stored assertions, but also upon other information

---

* A heuristic device is simply an aspect of a program in which some amount of repetitious work has been avoided by fashioning the program's operations, taking into consideration the exigencies of the particular task or set of tasks it is to accomplish.

which they can deal with, amounting to translations of quantified sentences and of sentences expressing relationships such as transitivity. Disadvantages are as follows:

> All of the limited logic systems are basically similar, in that complex information is not part of the data, but is built into the system programs. Those systems which could add to their original data base by accepting English sentences could accept only simple assertions as input. The question which such systems are capable of answering could not involve complex quantified relationships (e.g., "Is there a country which is smaller than every U.S. state?" p. 38).

Next most sophisticated were the *general deduction systems*:

> The problems of limited logic systems were recognized very early . . . and people looked for a more general approach to storing and using complex information. If the knowledge could be expressed in some standard mathematical notation (such as the predicate calculus), then all the work logicians have done on theorem proving could be utilized to make a theoretically efficent deductive system. By expressing a question as a theorem to be proved, the theorem prover could actually deduce the information needed to answer any questions which could be expressed in the formalism. Complex information not easily useable in the limited logic systems could be neatly expressed in the predicate calculus, and a body of work already existed on computer theorem proving. This led to the "general deductive" approach to language-understanding programs.

> Predicate calculus seemed to be a good uniform notation, but in fact it has a serious deficiency. By putting complex information into a "neutral" logical formula, these systems ignored the fact that an important part of a person's knowledge concerns how to go about figuring things out. Our heads don't contain neat sets of logical axioms from which we can deduce everything through a "proof procedure." Instead we have a large set of heuristics and procedures for solving problems at different levels of generality. In ignoring this type of knowledge, programs run into tremendous problems of efficiency. As soon as a "uniform procedure" theorem prover gets a large set of axioms (even well below the number needed for really understanding language), it becomes bogged down in searching for a proof, since there is no easy way to guide its search according to the subject matter. In addition, a proof which takes many steps (even if they are in a sequence which can be easily predicted by the nature of the theorem) may take impossibly long since it is very difficult to describe the correct proving procedure to the system. (p. 39)

Finally come *procedural deductive systems*, of which one is incorporated in Winograd's project; these systems embody "new programming techniques capable of using procedural information, but at the same time expressing this information in ways which did not depend on the peculiarities and special structure of a particular program or subject of discussion" (p. 40). Winograd used a programming language called PLANNER to accomplish these ends:

> PLANNER (Hewitt, 1969, 1971) is a goal-oriented procedural language designed to deal with these problems. It handles simple assertions efficiently, and it can include any complex information that can be expressed in the predicate calculus. More important, complex information is expressed as procedures, and these may include knowledge of how best to go about attempting a proof. The language is "goal-oriented," in that we need not be concerned about the details of interaction among procedures. For example, theorems which may at some point ask whether an object is sturdy need not specify the program that assesses sturdiness. Instead they may say something like "Try to find an assertion that X is sturdy, or prove it using anything you can." If we know of special procedures likely to give a quick answer, we can specify that these be tried first. If at some point we add a new procedure for evaluating sturdiness, we do not need to find out which theorems use it. We need only add it to the data base, and the system will automatically try it, along with any others, whenever any theorem calls for such a test. (p. 40)

In short, "PLANNER is a uniform notation for expressing procedural knowledge just as predicate calculus is a notation for a more limited range of information" (p. 41).

The principal characteristics of Winograd's system are these: Winograd uses Halliday's *systemic grammar* (e.g., Halliday, 1966, 1967, 1970) which assigns a set of features to surface structure components of a sentence, and which enables Winograd to derive a "meaning" by parsing the surface structure alone, instead of by building a deep structure.

> PROGRAMMAR is oriented toward systemic grammar, with its identification of significant features in the consistuents being parsed. It therefore emphasizes the ability to examine the features of constituents anywhere in the parsing tree, and to manipulate the feature descriptions of nodes. (p. 45)

A second feature is that Winograd's system has the capacity to interrupt parsing when conditions arise which need to be inspected for special characteristics or problems (ill-formedness, special routines for conjoined structures and idioms). Third, Winograd rejects the use of *automatic* backup:

> It is not obvious that automatic backup is desirable in handling natural language. There are advantages instead in an intelligent parser which can understand the reason for its failure at a certain point, and can guide itself accordingly instead of backing up blindly. This is important both theoretically and as a matter of practical efficiency. (p. 46)

The particular application Winograd chose for his system is the simulated manipulation of blocks set on a table by an imaginary robot arm. The user commands and questions SHRDLU as to the status of the blocks, SHRDLU's own actions, physical properties of the "block world" and its denizens, and so on. Separate programs for syntax, semantics, and context interact in processing input, with nonsyntactic information interacting freely with syntactic parsing, questioning well-formedness, meaningfulness, and context-appropriateness as analysis goes forward. Separate components communicate the user's commands and questions to SHRDLU in the form of prescriptions for action.

Criticisms of SHRDLU have focused on the extent to which Winograd's procedures are tied to the exigencies of the blocks world (generalization of the procedures has been somewhat problematic), and on the treatment of definitions and interpretations as procedures, the latter practice having both philosophical and practical drawbacks, according to the critics.

In addition to supplying his own criticisms of SHRDLU from a phenomenological standpoint, Dreyfus (1979) cites Simon, a leading figure in AI from its inception, and Winograd himself; both men noted serious shortcomings in SHRDLU and in similar projects restricted their attention to artificial "microworlds":

> The system cannot be said to understand the meaning of "own" in any but a sophistic sense. SHRDLU's test of whether something is owned is simply whether it is tagged "owned." There is no intensional test of ownership, hence SHRDLU knows what it owns, but doesn't understand what it is to own something. SHRDLU would understand what it meant to own a box if it could, say, test its ownership by recalling how it has gained possession of the box, or by checking its possession of a receipt in payment for it; could respond differently to requests to move a box it owned from requests to move one it didn't own; and, in general, could perform those tests and actions that are generally associated with the determination and exercise of ownership in our law and culture. (Simon [1976] quoted in Dreyfus, 1979, p. 13).

> The AI programs of the late sixties and early seventies are much too literal. They deal with meaning as if it were a structure to be built up of the bricks and mortar provided by the words, rather than a design to be created based on the sketches and hints actually present in the input. This gives them a 'brittle' character, able to deal well with tightly specified areas of meaning in an artificially formal conversation. They are correspondingly weak in dealing with natural utterances, full of bits and fragments, continual (un-noticed) metaphor, and reference to much less easily formalizable areas of knowledge. (Winograd [1976] cited in Dreyfus, 1979, p. 15)

Discussions of Winograd (1972) are plentiful. Interesting accounts are to be found, for instance, in Hunt (1975), Winston (1977), Bundy (1978), Fodor (1981), Dreyfus (1979), Tennant (1980), Boden (1977), and in a more specialized context in Marcus (1980) and in Hirst (1981).

Perhaps Winograd's work since 1972 can best be understood as a struggle to work out an answer to what, in the wake of Minsky's (1975) article, has been called the *frame problem*:

> The frame problem is an abstract epistemological problem that was in effect discovered by AI thought-experimentation. When a cognitive creature, an entity with many beliefs about the world, performs an act, the world changes and many of the creature's beliefs must be revised or updated. How? It cannot be that we perceive and notice all the changes. For one thing, many of the changes we knew to occur do not occur in our perceptual fields, and hence it cannot be that we rely entirely on perceptual input to revise our beliefs. So we must have internal ways of updating our beliefs that will fill in the gaps and keep our internal model, the totality of our beliefs, roughly faithful to the world. (Dennett, 1979, p. 75)

In Bobrow and Winograd (1977) an initial attempt is made to start towards the creation of a programming language, named KRL, which will be congenial to natural language understanding work embodying some version of the approach to frame systems and the frame problem taken by Winograd and his associates. The need for such a language is expressed in these terms:

> A complete understander system demands the integration of a number of complex components, each resting on those below, as illustrated in the following table:
>
> TASK DOMAINS: Travel Arrangements, Medical Diagnosis, Story Analysis, etc.
>
> LINGUISTIC DOMAINS: Syntax and Parsing Strategies, Morphological and Lexical Analysis, Discourse Structure, Semantic Structures, etc.
>
> COMMON SENSE DOMAINS: Time, Events and States; Plans and Motivations, Actions and Causes; Knowledge and Belief Structures; Hypothetical Worlds
>
> BASIC STRATEGIES: Reasoning, Knowledge Representation, Search Strategies
>
> UNDERLYING COMPUTER PROGRAMMING LANGUAGE AND ENVIRONMENT: Representation Language: Debugging Tools; Monitoring Tools. (p. 4)

Now most researchers in NLU within AI would agree that many or even all of the factors identified in the first four lists need to be accounted for somehow in NLU systems, and to the extent that it is agreed that theories in AI are to be programmed, programming languages will be needed for this purpose. What sets Winograd apart from other groups of researchers in the field is his interpretation of the frame system notion and, more broadly, his theory of knowledge.

Bobrow and Winograd set out their beliefs ("intuitions") concerning these matters, and concerning appropriate programming devices to implement their ideas, in a "set of aphorisms":

Knowledge should be organized around conceptual entities with associated descriptions and procedures. [This is more or less Minsky's frame concept.]

A description must be able to represent partial knowledge about an entity and accommodate multiple descriptors which can describe the associated entity from different viewpoints. [This reflects the spirit of Minsky's "Frames" paper: "Thinking always begins with suggestive but imperfect plans and images; these are progressively replaced by better—but usually still imperfect—ideas."]

An important method of description is comparison with a known entity, with further specification of the described instance with respect to the prototype. [Again, there is a similarity to Minsky's remarks quoted at length earlier.]

Reasoning is dominated by a process of recognition in which new objects and events are compared to stored sets of expected prototypes, and in which specialized reasoning strategies are keyed to these prototypes. [Minsky again]

Intelligent programs will require multiple active processes with explicit user-provided scheduling and resource allocation heuristics. [This is discussed later.]

Information should be clustered to reflect use in processes whose results are affected by resource limitation and differences in information accessibility. [See later.]

A knowledge representation language must provide a flexible set of underlying tools, rather than embody specific commitments about either processing strategies or the representation of specific areas of knowledge. (p. 5)

Involved in the explication of the first aphorism is the following controversial claim that Bobrow and Winograd make (p. 6): "In general we believe that the description of a complex object or event cannot be broken down into a single set of primitives, but must be expressed through multiple views." An example of what is meant by "multiple views" is:

The description of a complex event such as *kissing* involves one viewpoint from which it is a physical event, and should be described in terms of body parts, physical motion, contact, etc. The description used from this viewpoint would have much in common with those used to describe other acts such as eating and testing someone's temperature with your lips. In the same description, we want to be able to describe kissing from a second viewpoint, as a social act involving relationships between the participants with particular combinations of motivations and emotions. Viewing kissing in this way, it would be described analogously to other social acts including hugging, caressing, and appropriate verbal communications. (p. 6)

In clarification of the third and fourth aphorisms, the following citation seems useful:

> In designing KRL we have emphasized the importance of describing an entity by comparing it to another entity described in memory. The object being used as a basis for comparison (which we call the prototype) provides a *perspective* from which to view the object being described. The details of the comparison can be thought of as a *further specification* of the prototype. Viewed very abstractly, this is a commitment to a *wholistic* as opposed to *reductionistic* view of representation. It is quite possible (and we believe natural) for an object to be represented in a knowledge system only through a set of such comparisons. There would be no simple sense in which the system contained a "definition" of the object, or a complete description in terms of its structure. However, if the set of comparisons is large and varied enough, the system can have a functionally complete representation, since it could find the answer to any question that was relevant to the reasoning processes. This represents a fundamental difference in spirit between the KRL notion of representation, and standard logical representations based on formulas built out of primitive predicates. (p. 7)

The "representations based on formulas built out of primitive predictates" will be presented shortly; they figure in the work of schools of thought different from Winograd's. Several final comments concerning KRL and Winograd's work are required. He views meaning as context-determined; an utterance on the part of a speaker occasions in the listener a set of inferences whose character is a function of the utterance's actual context at that instant. There is no invariant correspondence between words and some set of semantic primitives, Bobrow and Winograd claim, and neither is there true synonymy once context is brought in. From an organizational perspective, KRL shares with production systems, in a general way, a lack of rigid control. Also, limits are placed on processing operations in KRL so that lack of success after a while leads to a halting of the operation; this is claimed to have some psychological reality. Finally, we should point out that much of Winograd's work has not been touched upon. Further references include Bobrow and Winograd (1977), and Winograd (1974, 1975, 1976, 1980).

The primitive predicates rejected by Winograd and his co-workers are embraced by Schank and his associates. Schank (1980) discusses "some of the issues and basic philosophy that have guided [Schank's] work and that of [his] students in the last ten years" (p. 244). A dominant theme in Schank's work has been an effort to develop a theory of natural language understanding based upon the proposition that "there must be available to the mind an interlingual, i.e., language-free, representation of meaning" (p. 244). The semantics-driven parsing concept Schank has developed has been based on the effort to accurately set up "expectations about what slots needed to be filled in a conceptualization of given natural language input and the embodiment of those expectations to guide both parsing and generation" (p. 243). Semantically driven parsers were evolved, then, using "meaning-driven rules which have their basis in syntactic information about the input sentence" (p. 243). "The basic idea of slot filling and top down expectations" underlay Schank's theory of *conceptual dependency* (Schank *et al.*, 1973), and "drives [Schank's] work today (Carbonell, 1979; DeJong, 1979; Gershman, 1979; Reisbeck and Schank, 1976; Wilensky, 1978)."

Conceptual dependency is presented in Schank (1972) as a theory which (1) is conceptually based; (2) has a conceptual base that consists of a formal structure; (3) can make predictions on the basis of this conceptual structure; (4) is not limited to the understanding of isolated sentences; (5) has formal rules for analyzing natural language utterances into the conceptual base.

Conceptual dependency is a theoretical parser embodying a theory of NLU. It attempts to embrace all domains impinging on comprehension of language, from syntax through semantics to pragmatics and even to all of short- and long-term memory. The syntactic parser's first task is to try to locate the "main noun," "main verb," and "object." This accomplished, control is passed to the *verb-ACT dictionary*, whose function it is to determine the semantic representations of these three elements on the *sentential semantic level*. The sense for the verb is chosen from among the possible meanings on the basis of the meaning categories of the noun(s) co-present in the sentence; nominal categories are found by table look-up as are the full definitions of the verbs. The verb-meaning tables factor in the semantic categories of the main noun and object. So, for instance, in the sentence, *The man took the book,* the syntactic parser identifies the verb, main noun and object; *man* and *book* are looked up; because the semantic category of *man* is *human* and that of *book* is *physobj*, only one of the several senses of *take* listed in its table of meanings can be selected. Given this information about the syntax and semantics of the input on the sentential level, a conceptual processor takes control; its function is to set up a representation of the input on the *conceptual level*, which *has its own syntax and semantics*. Briefly, this means that certain predictions which ensue by virtue of the presence of a given sense, or group of senses, in the input, are formally represented. "Slots" corresponding to these predictions can then form the basis for a search through the input for material serving to specify values for these "variables." For example, in the sentence, *the man took a book*, "we know that there was a time and location of this conceptualization and furthermore that the book was taken from 'someone' or

'someplace' and is, as far as we know, in the possession of the actor."
(Schank, 1972, p. 566). A *conceptualization* is therefore set up, leaving
places for a "recipient" and an "original possessor," the first one of
which is found in the sentence (*man*), and the second is not, but is
nonetheless claimed to be plainly part of the underlying conceptual
representation of the sentence. Conceptualizations are stated partly in
terms of "semantic primitives" called ACTs. Note the similarity of the
sentence, *the man took a book*, just considered, and *I gave the man a book*.

> "Give" is like "take" in that it requires a recipient and an original possessor
> of the object. But is there any actual reason that the verbs should be
> different? Are they actually different? It would appear that conceptually the
> same underlying action has occurred. What is actually different between
> these two sentences. . . is that the initiator of the action, the actor, is
> different in each sentence. The action that was performed, namely transition
> of possession of an object, is the same for both. We thus conceptually realize
> both "give" and "take" by the ACT "trans.". . . . "Give" is then defined as
> "trans." where actor and originator are identical, while "take" is "trans."
> where actor and recipient are identical. Important here is that a great many
> other verbs besides "give" and "take" are realized as "trans." plus other
> requirements. For example, "steal," "sell," "own," "bring," "catch," and
> "want" all have senses whose complex realizates include as their ACT, the
> ACT "trans." It is this conceptual rewriting of sentences into conceptuali-
> zations with common elements that allows for recognition of similarity or
> paraphrase between utterances. The recognition of paraphrase is central to
> the problem of conceptual representation. If two sentences are agreed to
> have the same meaning, one conceptual dependency diagram must suffice
> to represent them. (pp. 567–568)

A given ACT can require one or more *conceptual cases* as dependents.
Examples of these are OBJECTIVE, RECIPIENT, DIRECTIVE, and
INSTRUMENTAL. "We use conceptual case as the basic predictive
mechanism available to the conceptual processor" (p. 568). Conceptual
relations are dependencies between conceptualizations mediated by
such relations as *causation* and *time*. Returning to ACTs, note that these
are not syntactic but semantic categories, so that a noun such as *love*,
which presupposes a lover and a love object, can be represented in
terms of ACTs. The number and names of the ACTs have changed
throughout the life of the theory: it is usually considered to be fifteen
plus a similar number of scalar-valued STATES. Some feeling for the
concepts just noted can be obtained by going through a few definitions
of ACTs from the 1972 version of the theory, the version being
presented:

PACT — Physical ACT—PACTs require an objective case and an instrumental case. They are representative of the traditional actor-action-object type constructions. Example: hit, eat, touch.

EACT — Emotional ACT—EACTs require an objective case only and are abstract in nature. Example: love.

TACT — Transfer ACT—TACTs require objective, recipient, and instrumental cases and express alienable possession. Example: trans.

DACT — Direction ACT—DACTs take derivative case, objective case, and instrumental case, and express motion of objects that are inanimate. Example: move. (p. 577, 579)

Returning to the parsing theory, it has top-down aspects in that, on the basis of an initial probe, it sets up expectations, in the form of slots to be filled, and seeks to fill them. But it has bottom-up features as well, in that when part of an input utterance cannot be fit into a slot, theoretically possible and pragmatically likely parsing are tried seriatim, until, hopefully, one is successfully integrated into the existing structure. Once the conceptual backbone of the parsing is set up, pragmatic knowledge and short- and long-term memory are searched via separate processors as processing of the input proceeds. Schank insists "that it is not possible to separate language from the rest of the intelligence mechanism of the human mind," that language "simply does not work in isolation" (p. 626). He indicates how memory and belief systems interact, in his opinion, with the linguistic mechanism.

From the beginning of Schank's theorizing, various computer implementations of his ideas have been developed. See Schank (1972) for references to the earlier ones. The best-known such implementation is MARGIE, detailed in Schank (1975a), a language understanding system including an analyzer developed by Reisbeck (1975), an inference system conceived by Reiger (1975) and a sentence generator for output from the system, the work of Goldman (1975). Unfortunately, space limitations prevent treatment of these systems here. Schank has emphasized many times, however, that his ideas are more important than his programs.

Schank's subsequent work can be seen as a working-out of the higher levels of processing alluded to above, which were present but not fully developed, in the early models of the theory: levels involving pragmatic, belief-system, and memory-organizational phenomena.

The interest of Schank and his group in inference led them to investigate text processing as a domain requiring frequent and complex inferences. A system called Script Analyzer Mechanism (SAM) was built (Cullingford, 1978) as one result of this research. This work, in turn, prompted thinking about wider domains, and the notions of *scripts, goals, plans,* and *themes* were elaborated (see Meehan, 1976; Schank and Abelson, 1975, and especially Schank and Abelson, 1977). Wilensky (1978), Cullingford (1978), Carbonell (1979), and de Jong (1979) report on working systems that embody these ideas.

Schank (1980) summarizes his current ideas as follows:

> There are a great many possible levels of description. Each of these levels is characterized by its own system of primitives and conceptual relationships. (For example, we have recently introduced a set of "basic social arts" [Schank and Carbonell, 1979] to account for actions that have social consequences). Inferences occur at each of these levels. Thus for every set of primitives, there exists a set of inferences that applies to it. (p. 254)

Scripts, goals, plans, and themes, referred to above, are data structures which can be regarded as types of frames (see the Minsky citation above for definition of *frame*). Schank has defined a *script* as

> a predetermined causal chain of conceptualizations that describe the normal sequence of things in a familiar situation. Thus there is a restaurant script, a birthday-party script, a classroom script, and so on. Each script has in it a minimum number of players and objects that assume roles within the script ... each primitive action given stands for the most important element in a standard set of actions. (Schank (1975b) quoted in Dreyfus, 1979, p. 4).

If scripts are intended to aid in the codification and prediction of normal behavior, *plans* do so for intentional behavior and *goals* and *themes* for future behavior.

A brief example of the integrated (but highly informal) use of some of these devices in text understanding is the following (from Schank, 1978): Given the simple story *John was in his history class. He knocked over the garbage can,* a Schankian interpretation in terms of the above concepts might be to

1. predict, from the odd event in the classroom script, a reaction by the teacher;
2. infer that a noisy event is a violation of the decorum goal of a teacher
3. predict a plan that will be created by the teacher that will satisfy the decorum goal;
4. understand "staying after school" as a standard scriptal way of satisfying such a goal. (1980, p. 95)

Schank sums up:

> What we are saying then is that understanding natural language sentences requires one to understand why people do what they do and to be able to place new inputs into a world model that makes sense of people's actions.

To do this requires knowledge about what people are liable to do in general, and detailed knowledge about the standard situations that predicts what they will do in a given context. (Schank, 1979b, p. 95)

Schank's approach has, of course, not been without its critics. Besides critiques from outside the immediate discipline of AI (Dresher and Hornstein, 1977; Dreyfus, 1979) there have been complaints within the discipline, both of a broad philosophical nature (Weizenbaum, 1976) and of a more technical nature. Considering the issue of determining reference of *anaphoric expressions* in discourse, certain expressions whose reference is *prima facie* unclear, as in, *They put candies in their mouths and forgot about them,* Klappholz and Lockman (1977, quoted in Hirst, 1981) find Schank and Abelson's *scripts* to be "overweight structures that inflexibly dominate processing of text." They emphasize that

the structure through which reference is resolved must be dynamically built up as the text is processed; frames or scripts could assist in this building, but cannot, however, be reliably used for reference resolution as deviations by the text from pre-defined structure will cause errors. (Hirst, 1981, p. 80)

This critique suggests a comparison of Schankian approaches to NLU with those of Winograd discussed above. Although they share a commitment to incorporating world knowledge and belief systems in their NLU theories, Winograd and Schank are diametrically opposed on the issue of how best to implement the frame concept. A fanatical Winogradian might accuse Schank of rigidity, in the spirit of Klappholz and Lockman's aforementioned remark, whereas an all-out Schankian might claim Winograd's approach is chaotic in that it is not clear just how the inferences Schank's group thinks need to be made somehow are to come about.

The preference semantics of Wilks (1973b, 1975, 1976) can be presented, in a highly simplified first approximation, as Wilks (1976) has done:

This system constructs a semantic representation for small natural language texts: the basic representation is applied directly to the text and can then be "massaged" by various forms of inference to become as deep as is necessary for tasks intended to demonstrate understanding. It is a uniform representation, in that information that might conventionally be considered as syntactic, semantic, factual or inferential is all expressed within a single type of structure. The fundamental unit of this meaning representation is the *template*, which corresponds to an intuitive notion of a *basic message* of agent-action-object form. Templates are constructed from more basic building blocks called *formulas*, which correspond to senses of individual words. In order to construct a complete text representation (called a *semantic block*) templates are bound together by two kinds of higher level structures called *paraplates* and *inference rules*. The templates themselves are built up as the construction of the representation proceeds, but the formulas, paraplates and inference rules are all present in the system at the outset, and each of these three types of pre-stored structure is ultimately constructed from an inventory of eighty *semantic primitive elements* and from functions and predicates ranging over those elements. (p. 158)

Intuitively, the formulas map words, the templates clauses, and the paraplates complex sentential structures. Wilks's NLU understanding procedure may be thought of as using, locally, the "If at first you don't succeed" method, and, globally, the "Three strikes and you're out" approach. That is, after initial *segmentation* of input along semantic lines, a *template-matching* strategy is the basic device employed as the parsing proceeds. The front line of defense is a set of *templates* expressing typical relations among abstract categories each of which figures in the definition of a large number of actual words in the language. In the more straightforward and simple cases, a segment of input can be parsed (in the sense of being mapped onto a semantic block or meaning representation—there is no syntactic parsing at all done here) simply because the salient primitive tokens of its components, when properly ordered (i.e., agent-action-object) match those of some prefabricated template—one so common that it pays to have it around to swoop up quantities of banal input fragments (Wilks calls this process *bare matching*). The second line of defense involves what Wilks calls *preferential expansion*. Here enters the concept of *semantic density*. This means that, once the *words* corresponding to subject, verb, and object have been found and mapped onto a template (simply a skeleton that has places for one subject, one verb, and one object) *preference* considerations take over. This is something like normality. That is, those combinations of the three (or two, where no object can be located) elements which make the most sense when conjoined (are *semantically* most dense) are preferred. Some of the art of setting up a preference-semantic grammar, then, is arranging the *senses* of a given word in optimal fashion in the dictionary so that the right sense will be picked most often. There is a curious principle at work here. Roughly it comes down to this: text is best parsed by maximizing the banality quotient. That is, a particular parsing—a particular hypothesis as to which of the alternate meanings a sentence can have—is the more likely to be successful, the less outlandish, the more prosaic, the thought it reflects. *Paraplate matching,* the construction of complex structures from successfully realized templates, operates on similar principles. The third line of defense involves *extraction*, or *inference making*, and *commonsense rule matching*. This process is to extract a small number of indicated inferences (not logical inferences—predictions would probably be a better term) from parsed templates for which an embodiment as a complex unit is being sought, and then to apply preformed "commonsense templates"—typical relations between clauses—and pick as the parser the one which is densest, that is, *preferred on preestablished formal grounds*. The spirit of the entire enterprise is that NLU does not really proceed on strictly logical principles. Rather, "fuzzy" concepts such as preference guide our thinking and use of language.

Wilks's attitude toward the frame idea is an interesting one. He feels it is a useful concept, but that it should be used, not to guide parsing (recall the Klappholz and Lockman criticism of Schank's use of frames as overly constraining the parsing process) but rather as one more weapon in the armamentarium of *inferential processes* which include extraction and commonsense rule matching:

> Only with their aid . . . can we make sense of the very simplest utterances that break what I would call preference restrictions. Many such phenomena would be called syntactic by others, but no matter. On my view *John ran a mile* breaks the preference of run for no object, and is, in that sense, preference-breaking or *metaphoric*. Even such trivial cases can, I believe, be profitably subsumed under a general pattern-matching algorithm . . . that matches preference-breaking items (the above as much as "my car drinks gasoline") against frame-like structures to determine an interpretation *via* what is normally the case for the mentioned entities: gas is normally *used* by a car (so that interpretation is substituted in a text representation of the sentence), just as running normally *extends* a distance such as a mile (with corresponding effects on the representation). (Wilks, 1981, p. 330)

Limitations of space preclude mentioning other important researchers in NLU. Charniak (1972, 1975, 1976, 1977a, 1977b) did work early in his career on child story comprehension (Charniak, 1972), in which one of the principal aims was the investigation of inference in discourse processing. He developed the notion of the *demon* which, in oversimplified form, is simply a miniature program set up to lie in waiting, for some short period of time after it has been activated, and actually *run* only if some condition is met which depends on the nature of the immediately following input. Charniak used demons to model the "sending out of feelers" of propositions within a discourse, for subsequent propositions with which they can combine to form inferences (see Charniak, 1976, for details). His later work has been devoted, among other concerns, to a working out of the frame idea. His model of NLU is not greatly unlike those of Schank or Wilks, but unlike them he rejects the notion of the semantic primitive:

> I believe that the apparent complexity of what we say is not a disguise for the basic simplicity of a small number of concepts arranged in a multitude of ways, but rather reflects the real complexity of the world we live in and

> our knowledge of it. We understand by relating what we are told to what we know, and given that the latter seems to be represented by concepts at all levels of generality, the former should also. (Charniak, 1977b, p. 39)

Some of the other important contributions are cited in the Bibliography.

This chapter has attempted to highlight some of the more influential research which has been done and is continuing in the field of NLU within AI. Regrettably, many important workers have not even been mentioned, and significant work of those individuals whose investigations have been discussed has had to be passed over. Nonetheless, some of the main lines of current thought have been indicated, and the interested reader can pursue the references cited.

Exercises

1. Discuss the two parallel paths for AI research, identified at the beginning of the chapter, in light of conclusions that you can draw from the study of the whole chapter.

2. In what ways does the analogy problem in (1), in the text, require machine intelligence?

3. Name five activities that computers can do better than people and five activities that people can do better.

4. What are some differences between AI theory and epistemology?

5. What are some of the competing definitions for AI?

6. What is cognitive science?

7. Here is the classic problem of missionaries and cannibals. There are three missionaries (M1, M2, M3), three cannibals (C1, C2, C3), and a boat (B) on the left bank of a river. The boat can carry a maximum of two persons in each crossing. The problem is to convey all the missionaries and cannibals from the left bank to the right bank of the river. It is important that the cannibals should not outnumber the missionaries anywhere; otherwise the cannibals will gang up on the missionaries and eat them. To solve this problem in a simple and elegant way, you must give it a proper representation. Notice, for example, that we can represent the

problem in terms of a triplet set of states. Let us assume each triplet to be (a,b,c) where a is the number of missionaries, b is the number of cannibals, and c is 1 if the boat is on the left bank and 0 if the boat is on the right bank. Thus, the initial state is $(3,3,1)$ and the goal state is $(0,0,0)$.

(a) Write an algorithm or set of instructions for solving this problem. It would be useful if you could implement this as a computer program (preferably in LISP).

(b) This problem is frequently given as an exercise for studies in AI. Why is it relevant for AI?

8. The general problem solver (GPS) was designed with the aim of enabling machines to solve problems that require intelligence. It was also intended to be a tool for studying human problem solving capabilities. It flourished for some ten years with several revisions and extensions until it was retired by its authors in 1969. The system essentially consists of two parts: the problem solver which is task-independent and the *task environment* which contains data structures. Among the latter are objects and operators. Objects may be some states, say, *A* and *B*, and operators were rules, procedures, and the like which would transform *A* to *B* or, alternatively, would diminish the difference between *A* and *B*. Thus, *A* could be an initial state and *B* could be the goal state (see Exercise 7 above). One example from Barr and Feigenbaum (1981, p. 116) is this: given the expression

$$R \wedge (\neg P \supset Q) \qquad \text{(initial state)}$$

transform it to the expression

$$(Q \vee P) \wedge R \qquad \text{(goal state)}$$

The system draws on its *operators* which are logical rules, such as $A \vee B \rightarrow B \vee A$, $A \vee B \leftrightarrow \neg(\neg A \wedge \neg B)$, and so forth and recursively sets up a series of intermediate goals until it arrives at the final goal and verifies the transformation.

(a) Think of the ways that GPS could be used for natural language understanding, for example, if the initial object is a sentence and the goal object is its semantic interpretation.

(b) Write a GPS system for travelling from one city to another, including goals for starting, uses of various modes of transportation, destination, and so forth (consult Winston, 1984, if necessary).

9. Construct arguments for the following debate:
   (a) Study of common sense should be a major component of the current AI research.
   (b) AI research can forge ahead without any serious attention to common sense.

10. Criticize the following statement:
    There are chess programs that beat expert human players, even their authors; hence, these programs have psychological reality and act just like human players.

11. Discuss the following statement (from Waltz, 1982, p. 120):

    Chess programs often play better than the people who write them, and so it is highly misleading to assert, as many people have, that computer intelligence is limited because a computer can do only what it is programmed to do.

12. Why are game programs important for AI research?

13. Criticize this statement: A good computer chess program emulates human chess players.

14. Write a set of $C{\rightarrow}A$ rules to make a computer become angry and show its anger.

15. Criticize the following statements:
    (a) Computers can learn in the same way that people do.
    (b) Computers cannot learn anything.

16. What is an expert system?

17. What is knowledge engineering?

18. Criticize this statement: Given enough time, a computer can become a musical genius.

19. Production systems were originally introduced by Post (1943), but have undergone many changes and revisions, both theoretical and

in applications. An example of a production is Rule #88 of MYCIN given in the text. An example of a complex single-line rule is the following, adopted by Moyne (1969) in the 1960s for a natural language processor.

Rm: A.B C ↑ D.E → F # ... X ... # @ ... Y ... @*** /Rn

Where Rm and Rn are, respectively, the label of the rule and pointer to the next rule to be scanned. The rule checks to see if elements $A, B,$ and C are to the left of the pointer ↑ scanning the input sentence, and elements D and E are to the right of this pointer; then if conditions within the string #... X ...# are satisfied, the elements between the dots $(B\ C\ D)$ are rewritten as F; then the actions specified within the string @ ... Y ...@ are carried out, the sentence pointer is moved to the right according to the number of *'s, and finally control is transferred to the rule labelled Rn for a similar verification and action. All the elements in the above rule need not be present in every rule of a particular processor. Notice that the above rule is a rewrite rule, similar to the productions of the formal grammars given in Chapter 1. Another example of production in the sense of Newell is the rule for the representation of the statement: You must breathe deep with lungs clogged after anesthesia.

Production:	IF	after anesthesia
	AND	lungs clogged
	THEN	breathe deep

A production system usually has three components: (a) a *rule base* consisting of a set of productions, (b) a buffer or "short-term memory" which can be regarded as context, and (c) an interpreter which controls the system and applies the rules.

With the above descriptions as your guide and additional readings, if necessary,

(a) construct a production system to identify various food items in a grocery store (see Winston, 1984);

(b) compare and contrast productions in formal languages with the productions discussed above.

20. Frame is a form of data structure or knowledge representation, still in its formative stages. It includes both declarative and procedural information in predefined structures and internal relations.

For example, the frame for a baseball game may include *slots* or "knowledge hooks" for various players: pitcher, left fielder, right fielder, catcher, batter, etc.; shape of the field; rules of the game; and so forth. Some of the information is stored in the form of static data; other information can be induced procedurally. The slots are usually filled with default data which can be replaced in particular situations; for example, a particular game of the Mets. Frames can refer to other frames and subframes recursively.

Construct a frame for a restaurant (consult the references if necessary).

21. What are text-based systems?

22. Criticize these statements:
    (a) SHRDLU is a general-purpose system for unrestricted natural language understanding.
    (b) At present it is possible to construct an unrestricted natural language understanding system.

23. What is the *frame problem*?

24. Discuss the statement: KRL language is a frame-based programming language.

25. Schank's knowledge representation for NLU is based on the theory of frames, which he calls *scripts*, for story understanding. His approach to NLU has been called *conceptual* as contrasted with *linguistic* and *perceptual* approaches. Discuss.

26. What are the role and significance of speech act in NLU?

References

The three-volume *Handbook of Artificial Intelligence* (Barr and Feigenbaum, 1981–82; Cohen and Feigenbaum, 1982) is the best ready reference for all significant present and past research in AI. Winston's textbook (1984) on AI is highly informative with many examples, algorithms, and practical models. Nilsson (1980) is more formal and more concerned with theorem-proving concepts and algorithms. For natural language processing and understanding, Winograd (1983) is the most comprehensive and up to date. Tennant (1980) contains brief descriptions and examples of many of the extant NLP projects. Sternberg (1982) is a collection of authoritative articles on aspects of human intelligence, reasoning, and other mental processes.

## PART III

Construction of a Language Processor

*Nature and Nature's laws lay hid in night: God said "Let
Newton be!" and all was light. It did not last: the Devil howling
"Ho! Let Einstein be!" restored the status quo.*

Alexander Pope and Sir John Squire

Theoretical positions and approaches toward the development of
models for language comprehension can be divided roughly into three,
some overlaps notwithstanding: *linguistic, conceptual,* and *perceptual.*
Moyne and Kaniklidis (1981) proposed a framework for the classification
of such models, and in this study we have given surveys of the underlying
theoretical views and methodologies in the three approaches. Linguistics-
based models (sometimes referred to as "theory-based") normally include a substantial grammar, together with a parser and a semantic
component. Processing within each component is usually autonomous.
Even the more recent models, which allow interaction for disambiguation, reanalysis, and backtracking, do processing in an essentially serial
manner, in that the syntactic analysis of a sentence or a clause is
complete and is then referred to the semantic/inductive component(s)
for verification and possible reanalysis. The question of autonomy has
been of interest also in theoretical linguistics and has generated debate.
The "standard theory" in generative transformational linguistics essentially advocated autonomous processing in various components of a
grammar: syntax, semantics, and so on, and Katz has continued to
support the view of semantic autonomy (cf. Jackendoff, 1981). In what
amounts to an aside comment at the end of this part, I have taken the
position that all human mental processes, particularly language perception and use, are interactive, on-line activities. As aptly stated by Kent

(1981, p. 4), at each level in the structure of the brain there are "processing elements pursuing their own jobs in parallel, while trading information with echelons above and below, and laterally with one another" (p. 14). Furthermore, I have taken the position that various relatively well defined linguistic processors, such as syntax, semantics, word accessing, and so forth, as well as pragmatics, contextual information, belief systems, the so-called knowledge of the world, and other phenomena that constitute the living experience of a human being, become active in a parallel and interactive mode at all levels of language perception, comprehension, and use.

In this final part of our study a question is posed which could be a challenge for future research: Is a realistic model of human language comprehension attainable? Chomsky (1962a) asked a similar question, with regard to language acquisition. Both these questions remain unanswered. However, models can be (and have been) constructed for what one might think of as an approximation of one aspect of the realistic model, and these models can be (and have been) useful for testing hypotheses about language comprehension and for gaining insight from the necessarily multidisciplinary nature of these models.

Classification of the current models into the above three approaches is an indication of the present state of affairs. Each approach has been pursued with tacit if not outright condemnation or attempts at falsification of the other approaches. Such controversies are futile because it is unlikely that questions about the primacy of any one of the approaches are decidable—at least at present. I believe that human language comprehension cannot be modeled on, say, merely the linguistics-based approach with the exclusion of the other approaches. We do not need empirical psychological data from laboratories for what we experience constantly in our daily experience. A human listener may perceive a whole complex idea (that from the linguistic viewpoint must be expressed by a complex sentence) by hearing one word or just a few words. On the other extreme, the so-called garden path sentences, such as the following, require one to listen carefully to the end, analyze, and reanalyze (backtrack) before understanding.

Victoria stated Walt and Beth are planning to get married next week in the church is false

I was afraid of Ali's powerful punch, especially since it had knocked out many tougher men, until I found that there was no alcohol in it

The paper cups are made of comes from trees

It is unreasonable to assume that also in the first case the listener waits to get the whole sentence, run it through a parser, semantic analyzer, and so on and then "understand" what he already knows! A realistic model must, however, recognize a cutoff point. There are sentences which are well formed with respect to a competence model (for example, a sentence with some number of center embeddings) but they are incomprehensible by a human listener; such sentences must also be rejected by a comprehension model.

Thus, the model that I present in a schema, with speculations about its functions and operations, is necessarily more complex than the extant models. It attempts to integrate the various approaches, resulting in multiplicity and redundancy which are, however, characteristics of natural systems including brain functions. A multiplicity of processors working in parallel is envisaged, as if in competition for comprehension. As soon as one of the processors succeeds, the others are disengaged. This approach needs much further research and its implementation would be far more difficult than the present models.

These speculations about the nature of human language processing aside, the main thrust of this part of the study is to provide the student with practical details for getting involved in constructing a parser, or a series of parsers with increasing degrees of complexity, for natural languages. In the first two parts of this study general theoretical issues have been discussed; in this part we come down to the business of building something tangible.

CHAPTER 8

Lexical Processing

Introduction

Any model of natural language comprehension must have a lexicon containing the vocabulary of the language. In this chapter we do not propose to construct a lexicon, but rather to discuss some of the principles and problems, to give some examples, and to leave it to the interested reader to construct his or her own lexicon. As we will see, entries in a lexicon can be quite complex and may include rules and procedures for word formation, lexical selection, insertion, and so forth. The complexity of a model lexicon and its size will depend on the approach and on the type of project one undertakes. Computer models for natural language processing (NLP) usually have a lexical processor which is an autonomous component. That is, words of the input sentences are looked up in the dictionary, and the information associated with each word, including multiple meanings of ambiguous words, are extracted for further processing and possible disambiguation at the subsequent stages of syntactic and semantic analyses.

There are theoretical issues and controversies concerning the autonomy of lexical accessing. In this chapter we will not deal with these issues except for occasional references when they become relevant to our discussions. For details of the autonomy issues and other topics concerning human lexical accessing (how humans recognize and process words in language understanding) and their implications for computer models, see Moyne (1984) and the references therein.

No attempt has been made in this chapter to follow any one particular theory of the lexicon or of its organization. We will discuss several approaches and proposals for lexical processing and the contents of a lexicon, starting with simple ones.

What Should Be in the Lexicon?

A good standard dictionary has words together with their parts of speech, their meanings, and their uses given in sample sentences or phrases. Also included may be synonyms and antonyms of the entries. As we will see, this information is not sufficient for an automated language processor. There are other facts that the human user "knows" and brings to bear on language processing. These facts must be explicitly stated in an automated dictionary. Furthermore, the definitions (meanings) of a word are often circular in standard dictionaries. For example, in the *Webster's Seventh New Collegiate Dictionary* the following sequences can be found, where words on the left are dictionary entries and words on the right are one of the meanings ascribed to them:

$$
\begin{array}{llll}
(1) & \text{corrupt} & : & \text{depraved} \\
& \text{deprave} & : & \text{corrupt} \\
& \text{depraved} & : & \text{perverted} \\
& \text{perverted} & : & \text{corrupt}
\end{array}
$$

For a standard dictionary, this arrangement may be desirable because the definitions on the right are in each case one of several definitions and the circularity points to the synonymy of these words in some sense. The trouble is, however, that even if these words had exactly the same sense (which they do not), they would be subject to different constraints in their usage; for example we can say *John corrupted the organization* but not *John perverted the organization* in the usual sense of the word.

Let us look at some examples of the *features* that must be included in a lexicon for NLP to account for the constraints on the use of the vocabulary. We will give a more rigorous presentation of these facts later in this chapter. For now, note that sentence (2a) can be transformed to sentence (2b):

(2a)   John was killed by the man
(2b)   The man killed John

But we do not want sentence (3a) to be subject to the same transformation:

(3a)   John was killed by the seashore
(3b)   *The seashore killed John

where an asterisked string is unacceptable. Thus, we must place an agent marker for certain words in the lexicon; for example, *man* will have the feature [+AGENT] but *seashore* will not. The agent marker has a distribution which is close to the feature [+HUMAN] or, perhaps [+ANIMATE], but, on the other hand, there are words which are syntactically inanimate, as evidenced by their being pronominalized by the word *it*, yet they may be subjects of passivizable sentences. Such words are *compiler, publishing house, journal,* and so forth.

> (4a) The program was processed by the compiler
> (4b) The compiler processed the program

There may be an interesting linguistic point here, but for our present purposes we can assume that the rule for marking nouns as agentive applies to the nouns which can be the logical subject of a sentence. For another example of noun marking, note that it is well known that only count nouns are subcategorized with respect to number. Thus, some provision must be made for representing other types of nouns, for example, mass nouns and abstract nouns. Hence, we will have, for example, *zinc* marked with the feature [+MASS] and knowledge with [+ABSTRACT]. Verbs, adjectives, and other categories are also subject to similar constraints. For example, distinction must be made between action and nonaction verbs: the former can occur in imperative sentences, in certain embedded sentences, and in progressive aspect; the latter cannot. Compare the verbs *kiss* and *own* in the following examples:

> (5) Kiss your mother!   *own the book!
> I told you to kiss your mother   *I told you to own the book
> The girl is kissing her mother   *The girl is owning the book

Thus, the verbs *kiss* and *own* must be subcategorized with [+ACTION] and [−ACTION], respectively. We will return to this discussion in the section, "The Lexicon in Generative Grammar."

Word Slots and Pattern Recognition

American linguist Charles Fries (1952) developed an elaborate system of frames or templates representing surface structures of English sentences. These templates had labelled slots into which words belonging to specific classes could be inserted. He divided the words of the English language into several classes which were labelled by numbers and letters.

For example, 1, 2, and 3 were roughly the labels for noun, verb, and adjective, respectively; D and f were labels for determiner and function word. Fries referred to a template as a *formula*. For example, the following is his Formula 8:

(6)
$$
\begin{array}{cccccc}
\text{D} \; 1^a & \text{D} \; 1^a & 2-\text{d} \; \text{D} & 1^b & \text{f} \; \text{D} & 1^c \\
- & - & \pm & + \; \text{E} & & + \\
\text{he} & \text{he} & & \text{it} & & \text{it}
\end{array}
$$

where the superscripts a, b, c on numbers are referential indices, that is, 1^a and 1^a are coreferential, but not 1^a, 1^b or 1^c. The markings $-$, $+$, and $\pm$ indicate number: singular, plural, and both singular and plural, respectively. That the verb is in the past or is a past participle is indicated by $2-d$; the E under the function label f stands for Group E words which are: *and, or, not, nor, but, rather than*. Finally, the pronouns *it* and *he* under the labels indicate that the labelled items can be substituted by the pronouns. An example of a sentence that can be matched by the formula in (6) is given by Fries as:

(7)   This *student* my *assistant* brought the papers and the grades

It is clear that the formulas are rather *ad hoc* and lack explanatory power in the sense of Chomsky. Furthermore, the formula in (6) does not block matching of sentences such as the following:

(8)   This student my assistant ate the house and the trees

A counter argument against this objection is to say that the formulas are a set of templates for syntactic well-formedness, and that they are not concerned with meaning, anomaly, and so forth. Taken in this sense, it is instructive to note that a relatively small number of such formulas (Fries defined ten) can account for a very large number of syntactic surface structures.

It is interesting that in later years, researchers in AI independently developed similar pattern matching procedures which were even less general than Fries's but enabled them to produce some interesting NLP systems. Raphael (1968) constructed a program called SIR which could "understand" certain set relations, part–whole relations, ownership, and spatial relations and could carry on limited conversations in English about these matters with a person. SIR contained a total of 24 patterns, such as the following:

(9)　　　———— is ————
　　　How many ———— does ———— have?
　　　———— is ———— part of ————
　　　Is ———— to the left of ————?
　　　There is one ———— on ————

where the dashes represent slots which can be filled in with words or phrases. ELIZA, a well-known early AI program (Weizenbaum, 1966), which acted as a pseudopsychiatrist and carried on a dialogue with a "patient," had pattern frames for recognition as well as generation of sentences; for example:

(10)　"Patient"　　　　　　　　　"Doctor"
　　　My $\frac{1}{\quad}$ is $\frac{2}{\quad}$　　　How long has your $\frac{1}{\quad}$ been $\frac{2}{\quad}$?
　　　$\frac{1}{\quad}$ thinks I am $\frac{2}{\quad}$　Do you think you are $\frac{2}{\quad}$?

where numbers in the slots (dashes) have been added to show the correspondence between the input (patient) and the output (doctor). On the input side, strings are matched with the fixed patterns; when a match occurs, a corresponding response is generated. If there is no match, ELIZA would respond with some innocuous remark such as "Can you be more specific?" "Can you think of another example?" Examples:

　　　Human:　My mother is hateful
　　　ELIZA:　How long has your mother been hateful?
　　　Human:　My father thinks I am crazy
　　　ELIZA:　Do you think you are crazy?

There is, of course, much more to ELIZA than these simple examples show, but the basic strategy is that of fixed patterns and pattern matching.

　　　Transition networks are essentially pattern matching devices, and more generalizable than the previous examples. Consider the following transition network:

(11)

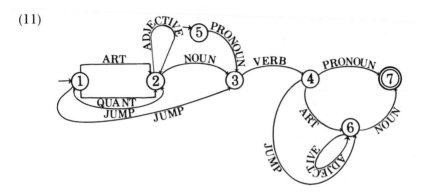

The following sentences, among many others, will match the network in (11):

> (12) The Mad Hatter touched the cake
> The Queen saw him
> She ate it
> Some queens have many friends
> Take the dirty tray!

Winograd (1983) develops systematic procedures for pattern definition and matching for some of the above examples.

Although some of the discussion in this section may appear to be more concerned with the structure of sentences than with word processing, the purpose has been to show that in this approach words and phrases can be arbitrarily inserted into patterns without much concern about selectional restrictions and subcategorization procedures which we will take up in the next section. It should be pointed out, however, that although it has been possible to build some impressive and clever models and even practical systems with this approach, the domains of their operations and applications are highly restricted, and all models built to date for NLP have broken down under any attempt at generalization.

The Lexicon in Generative Grammar

In Chapter 2, we noted that according to the TG theory, the deep structure of a sentence will contain preterminal nodes which represent lexical categories and that these preterminals are marked with dummy arguments, Δ, which can be replaced with words by the *lexical insertion rule*. As a first approximation, we can assume that this rule will select from the lexicon those lexical items (words) with appropriate category markings to replace the dummy arguments. We have already seen, however, that this procedure will not work and that the lexical insertion rule must, in addition to selecting the proper categories, match the subcategorization features and other provisions. Consider the following simple P-marker:

(13)

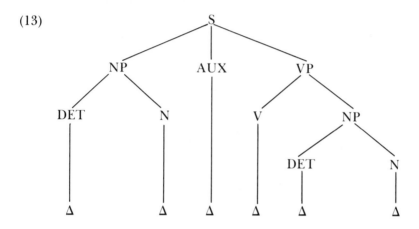

If we were to randomly select lexical items marked as determiner, noun, auxiliary, and verb and insert them in (13) above, we would get many well-formed sentences, but we would also get many bizarre strings. Thus, there must be a body of *selectional restrictions*, among other types of constraints, that block the bizarre sentences, and, as stated before, this is a part of a person's knowledge of the language. Any account intended to explain this knowledge must be able to explain why the sentence:

(14)   The plumber will fix the faucet

is a normal sentence in English, but this string is not:

(15) *The faucet has left the town

An entry in a lexicon must then consist of two parts: (a) The representation of the lexical item (word), which in TG theory is assumed to be a sequence of bundles of phonological features, but for the purposes of this study and for building computer models we can represent it as a word in its normal spelling; and (b) the set of various features which represent the meaning as well as categorial, subcategorial and contextual constraints, and other syntactic and semantic information which can be categorized as the human knowledge of language. In what follows we will give further examples to illustrate the features and rules which are mainly concerned with the structure of well-formed sentences. We will not deal here with specific semantic representations and meaning postulates; these have been dealt with in previous chapters, particularly Chapters 2 and 6. The reader should observe, however, that as the theory of lexical features develops, the demarcation between syntactic and semantic features becomes less perceptible.

We have already seen some examples of feature specifications for nouns. To give some more detail, here are two examples:

$$(16) \quad \text{man:} \begin{bmatrix} \text{MAN} \\ +\text{N} \\ [+\text{COMMON}] \\ [+\text{COUNT}] \\ \vdots \\ \vdots \\ [+\text{AGENT}] \\ [+\text{HUMAN}] \\ \vdots \end{bmatrix} \quad \text{seashore:} \begin{bmatrix} \text{SEASHORE} \\ +\text{N} \\ [+\text{COMMON}] \\ [-\text{COUNT}] \\ \vdots \\ \vdots \\ [-\text{AGENT}] \\ [-\text{ANIMATE}] \\ \vdots \end{bmatrix}$$

Notice that [+ANIMATE] is a feature that permits the sentence in (17a) but not in (17b):

(17a)   The dog chased the cat
(17b)   *The rock chased the cat

Thus, *dog* will be marked [+ANIMATE] and *rock* [−ANIMATE]. Note, however, that we have not marked *man* in (16) with [+ANIMATE]. Instead of marking all human words in the lexicon with this feature, we can have a rule for generating it:

(18)   [+HUMAN] → [+ANIMATE]

What about number? Do we have to mark *man* with [+SINGULAR]? Again, instead of entering every count noun in the lexicon for both singular and plural forms and marking them with + or − SINGULAR, we can enter only one form and have the number generated as a feature in the deep structure. Thus, the dummy place holders mentioned above or the preterminal categories can have features attached to them. Example:

(19)

$$
\begin{array}{c}
\text{S} \\
\diagup \mid \diagdown \\
\text{NP} \quad \text{AUX} \quad \text{VP} \\
\diagup \diagdown \\
\text{DET} \qquad \text{N} \\
\mid \\
\begin{bmatrix} +\text{N} \\ [+\text{SINGULAR}] \end{bmatrix}
\end{array}
$$

And we will have another rule:

$$
(20) \quad \text{N} \rightarrow \begin{bmatrix} +\text{N} \\ [\pm\text{SINGULAR}] \end{bmatrix}
$$

Now the noun phrase *the book* can be inserted in (19) but not *the books*. Note, however, that for recognition purposes in a computer model, this process must be somewhat reversed, that is, if *book* is an entry in the lexicon and *books* is an item in the input stream, then the latter must be recognized through morphological analysis, suffix matching, and so on as plural, and the feature [−SINGULAR] must be added to the developing structure.

The story is not complete as yet. Verbs must be subcategorized to provide for the selection of appropriate noun phrases to be associated with them (subject, object, etc.). Consider the verbs *eat, fall, give,* and *persuade* in the following examples:

(21a)   The cat will eat the cheese
(21b)   *The cat will fall the cheese
(21c)   *The cat will fall the balcony
(21d)   The cat will fall
(21e)   John gave the book to Mary
(21f)   John gave Mary the book
(21g)   *John ate the cheese to Mary
(21h)   John ate the cheese with Mary
(21i)   *John ate Mary the cheese
(21j)   John persuaded Mary to go home
(21k)   *John persuaded
(21l)   ?John persuaded Mary
(21m)   John ate

Traditionally, in English, verbs are classified as transitive and intransitive. Generally, transitive verbs can take object NPs and intransitive verbs cannot. In the above examples *eat* is transitive and can take a direct object (*the cheese*), but *fall* is intransitive and cannot take a direct object. However, *eat* can also be intransitive, as in (21m). Furthermore, as Radford (1981, p. 119) points out, the verbs *eat* and *devour* seem to have roughly the same meaning and have the same transitive property, but *devour* unlike *eat* cannot also be intransitive:

(22a)   I haven't eaten yet
(22b)   *I haven't devoured yet

The verb *give* takes both direct and indirect objects, whereas *eat* cannot take an indirect object. The verb *persuade* must have a direct object and also a complement. In conversation or answering a question (21l) is possible when the complement is "understood" (but if the reader does not agree, try the verb *put*). We can account for all of the above examples by the following subcategorization of the verbs:

$$(23) \quad eat : \begin{bmatrix} \text{EAT} \\ +V \\ \pm[\text{——}NP] \end{bmatrix} \qquad fall : \begin{bmatrix} \text{FALL} \\ +V \\ -[\text{——}NP] \end{bmatrix}$$

$$give : \begin{bmatrix} \text{GIVE} \\ +V \\ +[\text{——}NP \text{ to——}NP] \end{bmatrix} \qquad put : \begin{bmatrix} \text{PUT} \\ +V \\ +[\text{——}NP \text{ PP}] \end{bmatrix}$$

(see Exercise 8)   +LOC

$$persuade : \begin{bmatrix} \text{PERSUADE} \\ +V \\ +[\text{——}NP \text{ PP}] \end{bmatrix}$$

In (23) *eat* is entered as a category verb with subcategorization ±[——NP] indicating that it may be inserted into a verb phrase directly before an NP, that is, it may or may not take a direct object, whereas *fall* cannot take a direct object, and *give, put,* and *persuade* must have direct objects as well as complement phrases (PP = prepositional phrase). Notice that in the case of *give* the preposition is specified; in the case of *put* the preposition is restricted to + LOC (locative) because one can say *put the book on/upon/below the table* but not **put the book for/ from/to the table.*

Similar features and subcategorizations must be specified for adjectives, adverbs, prepositions, and other categories. For example, adjectives, like verbs, may have [+ ACTION] feature as demonstrated by the following examples:

(24a)   Be honest
(24b)   *Be short

Phrasal Dictionary

Idioms are often treated as single words, and it has been suggested that they should be entered in the lexicon as such. For example, idioms, such as *kick the bucket, shoot the breeze,* and *mark my word,* are said to be "frozen" in the sense that, among other things, they cannot become discontinuous by embedding other elements in them:

(25a)   John has kicked the bucket
(25b)   John kicked the tall bucket
(25c)   John kicked, as far as we know, the bucket

(25a) means John has died, but (25b,c) do not mean that he died. Apart from the so-called frozen idioms, there are many thousands of clichés, special phrases, and expressions that have meanings which differ from their literal meanings:

(26)   How do you do
You're welcome
Don't let me down
Palm off
Pass the buck
The buck stops here
I've washed my hands of . . .

Could these also be entered in the lexicon as a unit of some kind? Note that some of these have many variations:

(27)   I'll not let you or anybody else down
         Don't palm all your troubles off on me
         The buck does not stop here

A problem of a different sort is the following: most current parsers identify individual words, get their definitions from the dictionary, and continue with further analysis usually by a left-to-right linear scanning of the input string. The words *bucket, the,* and *kick* would be in the lexicon and would be used in other sentences, such as *John kicked the man in the stomach* or even *John kicked the bucket and made a dent in it.* When the lexical processor encounters the word *kick,* should it search for the idiom, look up *kick* alone, look ahead for more context, or what?

Moyne (1959) in an undergraduate thesis developed the following procedure and programmed it for a Russian–English machine translation project at Georgetown University (the idea was originally proposed by Michael Zarechnak). In the Russian dictionary words were marked, in addition to other features, as possible candidates for being the first, the medial, and/or the last words of Russian idioms. In "dictionary look-up" (lexical processing), then, when a word was encountered in the input string which was a possible first word of an idiom or special phrase, search would be made to see if the following words were medial and/or final words of such collocations. When an idiom or phrase was identified in this way and other contextual analyses confirmed it, the phrase would be looked up in a special idiom dictionary for its English equivalent.

Ezra Black is doing research for a "phrasal dictionary" which would contain idioms and a very large number of other phrases. Input strings can be matched against phrases by pattern matching techniques. This has an interesting theoretical implication for which we need to provide some background discussion.

There are interesting questions about the interaction of sensory perceptions with their cognitive interpretations; the former result from sensory stimuli, the latter from prestored knowledge and experience. For example, consider the well-known Müller-Lyer visual illusion: (28)

(28)

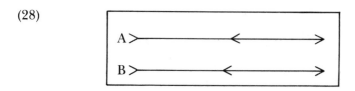

in which a person, in this case erroneously, perceives the segments in A as not equal and in B as equal. One theory about this illusion phenomenon is that the observer through previous experience (knowledge) "expects" certain shapes to be far away and others to be near.

Disregarding the role of knowledge for the present, cognitive psychologists postulate *iconic* and *echoic* stores (not to be confused with short-term memory—STM) in which chunks of visual and auditory data are respectively received and can be recalled. The amount of this data (size of the chunk), or the *perceptual span*, has been the subject of much discussion. Already in the last century, Javal (1878) had shown through experiments that reading was not done by a uniform scanning of a line in a text, but was done, rather, by jumping from one fixed point to another.

The current view is that the data collected in the iconic store is held for a very short time, much shorter than in STM. It is then passed on to the STM for cognitive processing and perception, and may subsequently be stored in the long-term memory. The iconic store then acts as a buffer to collect sufficient data to fire the cognitive processor.

On the basis of the above hypotheses, we can speculate that an entry in a phrasal dictionary is what we may call a *unit of perception* (the perceptual span, somewhat enlarged and structured in ways that remain to be explained), and that if we can develop a computer model that would do a simultaneous scanning of the entry (phrase) through parallel processing and pattern matching with the input data, then this model may perhaps represent human processing in some way. Another implication is that in the grammar, the level of syntactic analysis ends with phrasal nodes (for example, for simple sentences, S → NP VP or S → $\bar{\bar{N}}$ $\bar{\bar{V}}$), and phrasal analysis would be relegated to the lexicon. If this speculation is taken seriously, then much research remains to be done with respect, for example, to the treatment of complex NPs with sentence embedding (through on-line interaction between lexical and syntactic processors) and, more importantly, to the nature of phrasal analysis in the lexicon; for if it is the same as the current syntactic analyses, then the separation of syntactic and phrasal analyses will be reduced to an equivocation.

Exercises

1. Compare and contrast a standard dictionary, a lexicon as a component of linguistic competence (human knowledge of language), and a lexicon for a computer model of NLP.

2. Why is circularity of definitions tolerable in standard dictionaries but not desirable in a linguistic lexicon?

3. Explain the following terms; give examples:

> selectional restriction   subcategorization
> lexical insertion rule   iconic store
> perceptual span   unit of perception

4. Make up lexical entries for the following nouns and verbs:

> book   go
> sincerity   walk
> caution   introduce
> John   jump
> storm   buy
> intelligence   talk
> fish   tell

5. Can you think of cases where the gender feature (masculine/feminine) would be necessary in the subcategorization of lexical items? Give examples.

6. Why is the following sentence ungrammatical? What needs to be done to prevent the generation of the sentence:

> *I am not sure the truth

7. Rule (18) in the text, [+ HUMAN] → [+ ANIMATE], is an example of a *redundancy rule*. Write a redundancy rule that will mark *the book* in the sentence *John gave Mary the book*, or any other NP in the same environment, as OBJECT (see Exercise 8).

8. In the text the verb *give* was subcategorized as (i) + [——NP to—NP] to represent the direct and indirect objects in sentences such as *John gave the book to Mary*. But we know that this sentence can also be stated as *John gave Mary the book*, (ii) + [——NP NP]. In the standard theory it was assumed that (ii) can be derived from (i) by transformational rules. But we can assume now that (ii) can be obtained from (i) (or [i] from [ii]) by an optional redundancy rule in the lexicon:

$$+[\text{---}NP^1 \text{ to---}NP^2] \rightarrow +[\text{---}NP^2\ NP^1]$$

Subcategorize the following verbs and write redundancy rules for them: *sell, tell, present, put up* (*put up the poster* : *put the poster up*; assume *up* belongs to the category P).

9. Modify the P-marker in (13) in the text to account for the sentence *The book will be sold to Winifred.* Include necessary features in the preterminal nodes. (*Hint*: You must account for the rule VP → V PP and place *will be* under AUX.)

10. Modify the transition network in (11) in the text to process sentences with sentential complements in the VP, for example, *I saw the man climb the mountain.*

11. Winograd (1972) in SHRDLU assumes that sentences have a one-to-one correspondence with programs, that is, sentences can be translated into executable computer programs. This assumption is refuted by the speech act theory, and Winograd (1972, p. 33) himself provides the counterexample:

The city councilmen refused the women a permit because
        (a)   *they* feared violence
        (b)   *they* advocated revolution

In (a) *they* refers most likely to the city councilmen, whereas in (b) *they* refers to the women. This distinction must be judged on the basis of our knowledge of human behavior, activities of councilmen, women, and so forth. Discuss this in terms of iconic/echoic perception and cognitive processing in STM.

References

Winograd (1983) contains detailed discussions and examples for pattern matching approaches to lexical processing. Chomsky (1965) laid out the principles and outline of a linguistic lexicon. Radford (1981) has a readable and highly informative chapter on the lexicon in the extended standard theory. Aronoff (1976) provides a more advanced text on word formation. Bresnan (1982) contains advanced discussions on lexical representation, particularly on the theory of lexical-functional grammar. Peters (1983) deals with the units of language perception and acquisition along the lines discussed in the last section of this chapter.

#### CHAPTER 9

Syntactic Processing

Introduction

In this chapter we will first describe a simple parser for nondeterministic context-free phrase-structure grammars. We will then augment the parser in a couple of stages to turn it into a deterministic device with a wider scope. It is recommended that a student with no previous experience in writing parsers should start with implementing the first simple parser and do the more complex ones subsequently.

A Simple Nondeterministic Parser

The parser discussed in this section is an automaton* with a control unit (CU), three tapes, and three read, write, and erase (RWE) heads (see Chapter 1), shown in (1) below:

(1)

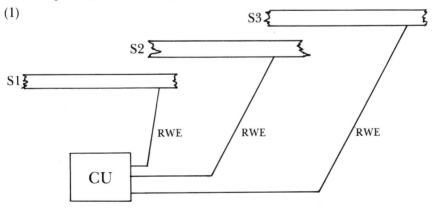

* Recall that in Chapter 1 we made a distinction between acceptor automata and parsers. Here we are presenting a somewhat unorthodox automaton which will simulate a parser. The reader can, however, think of this as a parser.

Let us call this device $\mathcal{M}_E$, with general delta-functions represented as in:

(2) $\delta(q_i, a_{mS21}, a_{nS22}) = (q_i, a_{mS3}, a_{nS21})$

We assume that the input string σ and the grammar G are both recorded on Tape S2. The device will then scan a symbol from the input and transfer control on the same tape to search for a grammar rule applicable to the symbol scanned, and continue processing in the manner shown in (5). The delta-function shown in (2) can then be interpreted as follows: If the device is in state q_i and scans the symbol a_m on Section 1 of Tape S2, which contains the input string, and if it finds a grammar rule in Section 2 of the Tape S2 which applies to the symbol scanned, the automaton will then change its state to q_j (which may be the same as q_i) and it will write the symbol a_m on Tape S3, and the symbol on the left of the arrow in the grammar rule on the top of a_m (erase a_m) on S2. If no grammar rule applies to a_m, it is written on Tape S1 (not shown in [2]). Let us take an example of the sentence in (3) and the grammar in (4) and go through the steps to see how $\mathcal{M}_E$ works.

(3)   The man left the town

(4)   S → NP VP
            NP → DET N
            NP → N
            DET → ART ADJ
            DET → ART
            VP → V NP
            ART → the; N → man, town; V → left

The procedure works as in (5). Note that in (5) we start with the input sentence in S2, and S1 and S3 are empty. In Step 2 the grammar rule ART → *the* has applied; *the* has moved to S3 and is replaced by ART in S2. In Step 3 the rule DET → ART has applied; ART has moved to S3 and is replaced by DET in S2. But no grammar rule in (4) now applies to the single symbol DET; it is, therefore, moved to S1 (Step 4). In Step 5 the sequence DET N in S1 S2 is subject to the rule NP → DET N; thus, in Step 6 N DET has moved to S3, NP is in S2, and S1 is once again empty. This process continues to the end. Notice that in the last step, the complete structural description of the sentence is in S3 recorded in a *postfix* notation.

For computer implementation, it is convenient to treat S1, S2, and S3 as pushdown stacks with pointers P1, P2, and P3 associated with

(5)

	S1	S2	S3
1.		the man left the town	
2.		ART man left the town	the
3.	DET	DET man left the town	the ART
4.	DET	man left the town	the ART
5.		N left the town	the ART man
6.		NP left the town	the ART man N DET
7.	NP	left the town	the ART man N DET
8.	NP	V the town	the ART man N DET left
9.	NP V	the town	the ART man N DET left
10.	NP V	ART town	the ART man N DET left the
11.	NP V	DET town	the ART man N DET left the ART
12.	NP V DET	town	the ART man N DET left the ART
13.	NP V DET	N	the ART man N DET left the ART town
14.	NP V	NP	the ART man N DET left the ART town N DET
15.	NP	VP	the ART man N DET left the ART town N DET NP V
16.		S	the ART man N DET left the ART town N DET NP V VP NP
17.			the ART man N DET left the ART town N DET NP V VP NP S

them, respectively. The pointers normally point to the symbol on the top of each stack except that in S3, P3 values refer to the various locations in which symbols are stored. Each stack is subdivided into four columns with the headings: *symbol, P3, Size,* and *R#*. Words from the input sentence and grammatical formatives are stored under symbol. As mentioned before, P3 refers to locations in S3. Size is indicated when more than one symbol combine to form a grammatical category, for example, the symbol NP will have Size 2 associated with it if it is made up of DET and N. On the other hand, ART will have no size indicated because it is derived from one element *the*, and so forth. The rule number which generated the particular symbol is referred to by R#. In the algorithm for this device, R# is set initially to 1 and the rules of the grammar are applied sequentially by augmenting the value of R# until one applies whose number would be in the register R#, or the search exhausts and no rule has applied. In S3, however, the column R# is used to record the alternative rules which have not been applied. For example, if the grammar contains the following two rules for DET

$$3. \ \text{DET} \rightarrow \text{ART ADJ}$$
$$4. \ \text{DET} \rightarrow \text{ART}$$

and if Rule 4 has applied in a cycle, then under R# column against the appropriate symbol in S3, Number 3 is recorded indicating that Rule 3 was an alternative which has not applied. This is useful for backtracking and for trying alternative paths when analysis has failed in one path (see below for an example and clarifications). Figure 1 shows the general schema of the three stacks.

In Figure 2, we have given an Algorithm M for the operations of M_E in the form of a flowchart for the convenience of implementation. Note that this algorithm includes provisions for backtracking when a path is blocked or analysis remains incomplete due to the selection of a wrong rule in a nondeterministic grammar.

Let us now start with a simple grammar for a fraction of the English language to see how M applies:

$$(6) \quad 1. \ \text{S} \rightarrow \text{NP VP}$$
$$2. \ \text{NP} \rightarrow \text{DET N}$$
$$3. \ \text{NP} \rightarrow \text{N}$$
$$4. \ \text{DET} \rightarrow \text{ART ADJ}$$
$$5. \ \text{DET} \rightarrow \text{ART}$$
$$6. \ \text{VP} \rightarrow \text{V NP}$$

	S1				S2				S3				
	Symbol	P3	Size	R#	Symbol	P3	Size	R#	Symbol	P3	Size	R#	
0													0
1													1
2													2
3													3
⋮													⋮
n													n

Figure 1. Structure of the three stacks of $\mathcal{M}_E$.

It is assumed that the categorial or pre-terminal symbols (N = noun, V = verb, ART = article, ADJ = adjective) are assigned to words in the dictionary. Now, given the sentence in (7), we process through our device in the manner described under (8).

(7) The old train left the station

(8) 1. Enter the sentence in S2 (see Figure 3).
  2. Access is always to the top of S1, S2, and S3.
  3. In testing for application of grammar rules in (6), first try for a match for combined elements on the top of S1 and S2. If this fails, then try for a match for the element on the top of S2.
  4. When a rule of grammar applies, the elements on the top of S2 and S1 (if applicable) are popped and are placed on the top of S3. The left hand term of the production is then pushed on S2.
  5. When no rule of grammar applies, the element on the top of S2 is popped and it is pushed on the top of S1.
  6. The above procedure for processing top elements of the stacks continues in cycles until S1 and S2 are empty and the root symbol S appears on the top of S3, in which case the sentence analysis is completed. On the other hand, if the process terminates (no rule applies) before S1 and S2 are empty (whether an S has or has not been

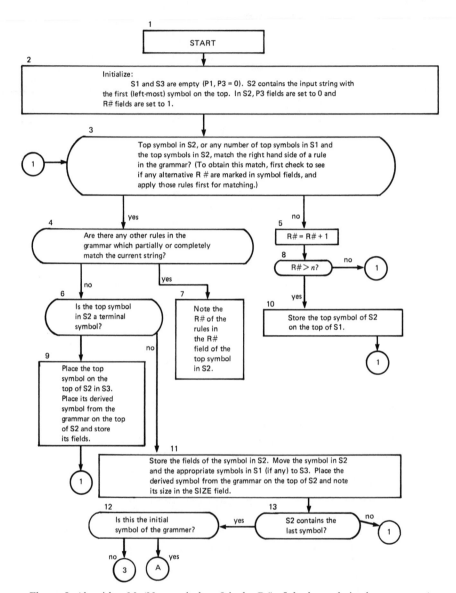

Figure 2. Algorithm M. (Note: *n* in box 8 is the R# of the last rule in the grammar.)

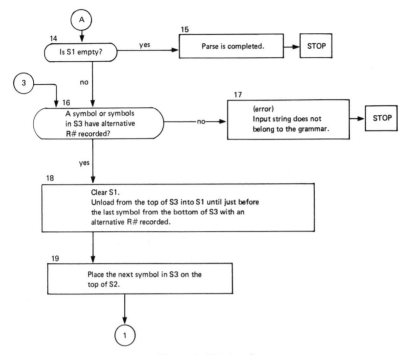

Figure 2. (*Continued*)

generated), then either the sentence is not accepted by M (the sentence is not well-formed) or the backtracking procedure must take place for an alternative analysis.

The test given in (8) is an informal description of Algorithm M. For the processing of the sentence in (7), let us first assume that the correct rules are selected and there is no backtracking. The result is shown in Figure 4.

Note that the final contents of S3 in Figure 4 represent a P-marker for the sentence. In fact the following tree structure can be constructed starting from the top of S3 where the contents of each of the registers provide for the following elements of the tree: SYMBOL = node label; SIZE = the ranking of the node (i.e., the number of branches exiting from the node —— blank (null) size has the ranking of 1 by convention); P3 = the destination of each branch (destination of the daughter nodes); and R# is ignored at this stage.

S1	S2				S3
	Symbol	P3	Size	R#	
	the	0		1	
	old .	0		1	
	train	0		1	
	left	0		1	
	the	0		1	
	station	0		1	

Figure 3. The first step in the analysis of a sentence $\mathcal{M}_E$.

	S1				S2				S3				
					Symbol	P3	Size	R#					
0									S	1	2		0
1					ART	17		5	NP	11	2	3	1
2					the				VP	3	2		2
3					ADJ	16			V	10			3
4					old				NP	5	2	3	4
5					DET	14	2	5	DET	8		4	5
6					train N	13			N	7			6
7					NP	11	2	3	station				7
8					left V	10			ART	9			8
9					DET	8		4	the				9
10					ART	9			left				10
11					the				DET	14	2	5	11
12					NP	5	2	3	N	13			12
13	DET	8		4	N	7			train				13
14	N	10			station				ART	17		5	14
15	NP	11	2	3	VP	3	2		ADJ	16			15
16	DET	14	2	5	# 8	1	2		old				16
17	ART	17		5					the				17

Figure 4. The contents of S3 showing the final analysis of the input sentence.

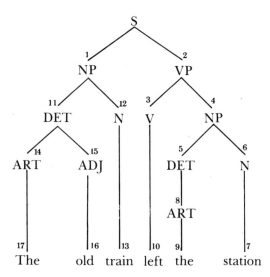

We have assumed so far that the correct rules were selected from the grammar in (6) at each cycle processing the top elements of the stacks. Suppose now that in the processing of the first determiner (DET) in sentence (7) the parser selects Rule 5 in grammar (6). Then, in S1, will be accumulated from bottom up: DET, ADJ, NP, and V. DET and ART in S3, however, would be marked with the alternative Rule 5 in their R# registers. Eventually an S will appear on the top of S3, but DET and ADJ will remain in S1 (that is, S1 will not be empty); hence, the parsing has failed. At this point, in accordance with the instructions in algorithm M, the contents of S3 are dumped to S1 up to the ART with the alternative rule marked and reprocessing is carried out, reversing the roles of S1 and S2, and this time using the alternative rule recorded in R# for the ART. This time the sentence will parse with stacks S1 and S2 empty. The reader can try this for himself by following the instructions in M (Figure 2). The same situation will apply to the following sentence:

(9)   The old train the young

If *the old train* is processed as an NP which will appear on the top of S1, then an NP for *the young* will appear on the top of S2, and no further processing would be possible. Backtracking to ART and using the alternative Rule 5 and using the alternative category N for *old* in the dictionary, we get NP for *the old*. The alternative category V for *train* is then selected from the dictionary, and the analysis will go through with success.

Recall that the processing for the ATNs was top-down, hypothesis driven. For our model, it is clear that the processing is bottom-up, data driven. This approach is more suitable and more efficient for contextual dependency of various kinds. A further observation about M to be noted is that it works most efficiently if the grammar is in Chomsky Normal Form (CNF); otherwise access must be provided to more than one element on the top of S1. Models of M have been implemented in Moyne (1977b) and in a number of student projects.

The grammar given in (6) is, of course, trivial; more comprehensive grammars can be developed for processing complex sentences. Consider the grammar G in (10).

$$
\begin{aligned}
(10) \qquad \bar{\bar{S}} &\to \text{ADV} \ \bar{S} \\
\bar{\bar{S}} &\to \text{ADV} \\
\bar{S} &\to \text{VP} \ S \\
S &\to (\bar{\text{NP}}) \ \overline{\text{VP}} \\
S &\to (\text{NP}) \ \text{VP} \\
\underline{S} &\to (\text{NP}) \ \underline{\text{VP}} \\
\overline{\text{VP}} &\to \text{VP} \ \overline{\text{ADV}} \\
\text{VP} &\to \text{V} \ \bar{\text{C}} \\
\text{VP} &\to \text{V} \ (\text{NP}) \\
\text{VP} &\to \text{V} \ \overline{\text{NP}} \\
\text{VP} &\to \text{V} \ \bar{\text{S}} \\
\text{VP} &\to \text{V} \ \text{ADV} \\
\text{VP} &\to \text{V} \ \text{PP} \\
\overline{\text{NP}} \ &\text{NP} \ \bar{\text{C}} \\
\overline{\text{NP}} &\to \text{Q} \ \text{NP} \\
\text{NP} &\to \overline{\text{DET}} \ \text{N} \\
\text{NP} &\to (\text{DET}) \ \text{N} \\
\overline{\text{DET}} &\to \left\{ \begin{matrix} \text{NP} \\ \overline{\text{NP}} \end{matrix} \right\} \text{POSS} \\
\text{DET} &\to \text{ART} \ (\text{ADJ})^* \\
\bar{\text{C}} &\to c \ \bar{\text{S}} \\
\overline{\text{C}} &\to c \ S
\end{aligned}
$$

G is a surface grammar which, in addition to the simple declarative sentences generated by the grammar in (6), can account for sentences such as:

(11a)   The President told the audience that Haig resigned
        yesterday
(11b)   The President said that he denied the rumors

(11c) The President said he denies the rumors
(11d) John heard that Bob knew that Max left yesterday
(11e) When did the President say that he thought that Haig resigned?
(11f) Why did you say that you knew that John left for Washington?
(11g) All the controllers went on strike
(11h) All the controllers' supervisors' bosses left town
(11i) Where did the President go for holiday?

We assume appropriate dictionary entries with necessary attributes (not discussed here, but obvious from the discussion of other models). The reader can run these sentences through Algorithm M for himself. For example, (11a) will produce the following P-marker.

(12)

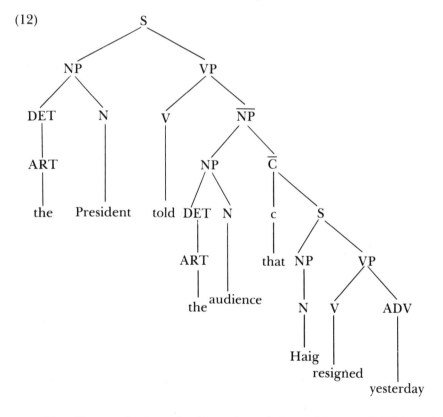

We will come back to the discussion of the ambiguity of (11a) and the possibility of the association of the time-adverbial (*yesterday*) with different constituents of the sentence. Note, however, that the surface/base nature of G allows us to represent structures such as (13) which account for the *wh*-movement transformation with the trace (e) marker

(13)

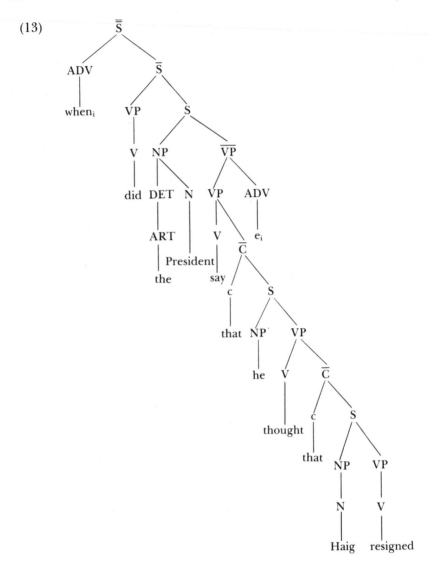

indicating the underlying attachment of the adverb. (13) represents the
sentence in (11e) with one specific interpretation of the adverb.

Some Modifications of Algorithm M and $\mathcal{M}_E$

The sample grammars for fractions of English given in (6) and
(10) are nondeterministic phrase-structure grammars and we have

provided for backtracking in Algorithm M. There are simple cases of the parser, for example, selecting between Productions 2 and 3 or between 4 and 5 in (6) where a wrong selection would require backtracking and reanalysis. There are more complex cases where the decision point is too far into the sentence and computational complexity required for keeping track of the back-up may become intolerable. For example, in the following pair of sentences:

> (14a) Is the blind man sitting in the car at the curb deaf?
> (14b) Is the blind man sitting in the car at the curb?

The word *is* must receive different categorial analysis in each sentence, but this cannot be determined until past word 11 in each case (see Chapter 5).

In addition, the parser must select from the various analyses of ambiguous sentences or produce phrase markers for all possible analyses. For example, in sentence (11d) we could have the time adverb *yesterday* associated with *John heard, Bob knew,* or *Max left.* And there are always garden path sentences such as:

> (15a) The horse raced past the barn fell
> (15b) John lifted a hundred pound bags
> (15c) I told the boy the dog bit Sue would help him

Some "engineering" and intuitive solutions can be found for some of the above problems. For example, in sentences such as (11d) there are preferred readings where listeners would tend to make one association most of the time. It may also be the case that noun phrases in English have more often than not determiners associated with them, so that in selecting between Rules 2 and 3 in grammar (6), we could have the convention that attempt must be made to apply the longest rule first. These provisions can be incorporated into our model, by writing the grammar as a *proper stochastic grammar* (cf. Gonzales and Thomason, 1978) and then having the parser apply the rules with the highest probability first. To convert the grammar, we need to write the rules with the general format of

$$(16) \qquad\qquad p : A \to \alpha$$

where p is the probability of the application of the particular rule, with the convention that the sum of the probabilities of the productions for rewriting a specific nonterminal A_i be equal to 1. Thus, if we have the set $\{A_i \rightarrow \alpha_1, A_i \rightarrow \alpha_2, \ldots, A_i \rightarrow \alpha_n\}$ or in general $A_i \rightarrow \alpha_j$ where $j = 1, 2, \ldots, n$, then

$$\sum_{j=1}^{n} p_{ij} = 1$$

When a nonterminal A has only one rewriting production, its probability is 1. Thus, the grammar in (6) can be represented as in (17) with arbitrary probabilities assigned to the rewriting rules with multiple productions:

(17)   1:   S → NP VP
                                       .75:   NP → DET N
                                       .25:   NP → N
                                       .55:   DET → ART ADJ
                                       .45:   DET → ART
                                         1:    VP → V NP

A more significant augmentation can be obtained by providing the automaton $\mathcal{M}_E$ with a look-ahead buffer. For our model, this can be done in a natural way by extending the widths of stacks S1 and S2. Suppose that we make the widths of stacks S1 and S2 into 12 registers instead of 4. This would provide for an *observation window* for the parser of seven word clusters: Three word clusters on the top of each of S1 and S2, and one on the top of S3. Each word cluster will still consist of the four fields: symbol, P3, size, and R# (Figure 5). Note, further, that lexical entries under the symbol column could be pointers to the lexicon where entries include various syntactic, semantic, and other features as well as lexical rules and procedures (see Chapter 8). Let us go through an example analogous with (5):

(18)

S1	S2	S3
1.	the old train left the station	
2.	ART A/N N/V	the old train
3. ART A/N N/V	left the station	the old train
4. ART A/N N/V	V/A ART N	the old train left the station
⋮	⋮	⋮

In (18) the input sentence *the old train left the station* is entered in S2; S1 and S3 are empty (Step 1). In Step 2, the top three accessible symbols

P3 REF.	S1												S2												S3				P3 REF.
	S11				S12				S13				S21				S22				S23				S3				
	Symbol	P3	Size	R#	Symbol	P3	Size	R#	Symbol	P3	Size	R#	Symbol	P3	Size	R#	Symbol	P3	Size	R#	Symbol	P3	Size	R#	Symbol	P3	Size	R#	
0																													0
1																													1
2																													2
3																													3
4																													4
5																													5
6																													6
7																													7
8																													8
9																													9
10																													10
11																													11
12																													12
13																													13
14																													14
15																													15
16																													16
17																													17
18																													18
19																													19
20																													20
21																													21
22																													22
23																													23
24																													24
25																													25
26																													26
27																													27
28																													28

Figure 5. The augmented version of the automaton in Figure 1.

are processed. Note that two of the three words have dual categorial markings (are ambiguous): *old* can be adjective or noun (A/N) and *train* can be noun or verb (N/V). In Step 3 the top three symbols in S2 are shifted to S1, and three more words from the input string pop to the top of S2. In Step 4 the three top words on the top of S2 are processed. The word *left* can be a verb or an adjective (V/A). At this point in Step 4 the device has access to the six symbols on the top of S1 and S2 (as well as one symbol *the* on the top of S3, but that has no bearing on the present discussion), and can disambiguate *left* as verb on the basis of the context. This in turn will allow the disambiguation of the other symbols. A complete example of running the input sentence through the revised $\mathcal{M}_E$ is in Figure 6. Sentences can, of course, be longer than six words, but often in the process of combining constituents (e.g., NP → DET N) stack positions are cleared and new words can pop on the top. Furthermore, slots can be vacated by shifting between S1 and S2. In Figure 7 an example is given for the processing of the sentence *The president told the audience that Haig resigned yesterday.* Note that in Line 8 of Figure 7, the whole string of six-word clusters is shifted, so that the leftmost NP is shifted out of the view of the observation window, but then *yesterday* pops to the surface (Line 9) and the device can process it. However, in Line 11, due to the construction of V and ADV into a VP node, an empty slot becomes available and the line is shifted back so that the leftmost NP reappears and the processing can continue to the end. In Figure 7, a convention has been adopted that suppresses the application of possible grammar rules until the whole sentence (or enough of it to account for the constituents in S2) has been observed. Recall that the rules apply to the elements in S2. In any case, the structural representation accumulated in S3 in Figure 7 is equivalent to the P-marker in (12). The procedure developed in Figure 7 will be further modified and formalized in the next section.

The Size of the Observation Window

In the previous section we suggested an extended observation window for $\mathcal{M}_E$, allowing up to seven symbol clusters to be on the top of stacks for scanning. We have not mentioned how these elements can be observed simultaneously. Before doing that, however, we will make a further change in $\mathcal{M}_E$ which will make the size of the window effectively unbounded.

Suppose that we restore the structure of S2 to its original, with one pushdown stack consisting of four subdivisions for symbol, P3, size,

P3 REF	S3 Symbol	S3 P3	S3 Size	S3 R#	S23 Symbol	S23 P3	S23 Size	S23 R#	S22 Symbol	S22 P3	S22 Size	S22 R#	S21 Symbol	S21 P3	S21 Size	S21 R#	S13 Symbol	S13 P3	S13 Size	S13 R#	S12 Symbol	S12 P3	S12 Size	S12 R#	S11 Symbol	S11 P3	S11 Size	S11 R#	P3 REF
0					train	0			old	0			the	0															0
1					N/V	23			A/N	24			ART	25															1
2					station	0			the	0			left	0			N/V	23			A/N	24			ART	25			2
3					N	20			ART	21			V/A	22			N/V	23			A/N	24			ART	25			3
4					N	20			ART	21			V	22			N	23			A	24			ART	25			4
5					N	20			DET	17			V	22			N	23			DET	18		2					5
6									NP	13		2	V	22			NP	15		2									6
7									VP	11		2	NP	15		2													7
8	S	9		2									S	9		2													8
9	NP	15		2									#																9
10	VP	11		2																									10
11	V	22																											11
12	NP	13		2																									12
13	DET	17																											13
14	N	20																											14
15	DET	18		2																									15
16	N	23																											16
17	ART	21																											17
18	ART	25																											18
19	A	24																											19
20	station																												20
21	the																												21
22	left																												22
23	train																												23
24	old																												24
25	the																												25
26																													26
27																													27
28																													28

Parse tree:

- S
  - 9 NP
    - 15 DET
      - 18 ART — 25 the
      - 19 A — 24 old
    - 16 N — 23 train
  - 10 VP
    - 11 V — 22 left
    - 12 NP
      - 13 DET — 17 ART — 21 the
      - 14 N — 20 station

Figure 6. A sentence processed by the augmented automaton.

P3 REF	S11 Symbol	S11 P3	S11 Size	S12 Symbol	S12 P3	S12 Size	S13 Symbol	S13 P3	S13 Size	S21 Symbol	S21 P3	S21 Size	S22 Symbol	S22 P3	S22 Size	S23 Symbol	S23 P3	S23 Size	S3 Symbol	S3 P3	S3 Size
0	ART	28		N	27		V	26		the	0		President	0		told	0		S	1	2
1										the	0		audience	0		that	0		NP	19	2
2										Haig	0		resigned	0		yesterday	0		VP	3	2
3	ART	28		N	27		V	26		ART	25		N	24		c	23		V	26	
4	DET	22		N	27		V	26		DET	21		N	24		c	23		N̄P	5	2
5				NP	19	2	V	26					NP	17	2	c	23		NP	17	2
6	NP	19	2	V	26		NP	17	2	c	23		Haig	0		resigned	0		C̄	7	2
7	NP	19	2	V	26		NP	17	2	c	23		N	16		V	15		c	23	
8	NP	19	2	V	26		NP	17	2	c	23		NP	14		V	15		S	9	2
9	V	26		NP	17	2	c	23		NP	14		V	15		yesterday	0		NP	14	
10	V	26		NP	17	2	c	23		NP	14		V	15		ADV	13		VP	11	
11	NP	19	2	V	26		NP	17	2	c	23		c	23		VP	11	2	V	15	
12				NP	19	2	V	26		V	17	2	NP	17	2	S	9	2	ADV	13	
13							NP	19	2	NP	19	2	V	26		C̄	7	2	yesterday	16	
14													NP	19	2	NP	5	2	N		
15																VP	3	2	resigned		
16																S	1	2	Haig		
17																			DET	21	
18																			N	24	
19																			DET	22	
20																			N	27	
21																			ART	25	
22																			ART	28	
23																			that		
24																			audience		
25																			the		
26																			told		
27																			President		
28																			the		

Figure 7. Another sentence processed by the augmented automaton.

and R#, but allow S1 to have a dynamic storage capacity which can increase or decrease according to the amount of data currently stored in it. Thus, S1 can be thought of as a tape which is infinitely extended to the left. The tape is divided into squares with each square subdivided into four parts for symbol, P3, size, and R#. The schema for this revision is outlined in the example in (20) for the input sentence in (19). The categorial markings under the words in (19) are from lexicon entries showing the ambiguity of the words; for example *old* is marked as possible adjective or noun (A/N) and *train* is marked as possible noun or verb (N/V). Initially, the input sentence is entered in S2, S3 is empty, and the size of S1 is set to null.

(19)

	the	old	train	the	young
	ART	A/N	N/V	ART	A/N

(20)

	S1				S2		S3	
01					the	01		
02				ART 20	old	02		
03			ART 20	A/N 19	train	03		
04		ART 20	A/N 19	N/V 18	the	04		
05	ART 20	A/N 19	N/V 18	ART 17	young	05	S	6 2
06	ART 20	N 19	V 18	ART 17		06	NP	14 2
07	DET 16	N 19	V 18	ART 17		07	VP	8 2
08		NP 14 2	V 18	ART 17		08	V	18
09	NP 14 2	V 18	ART 17	A/N 13		09	NP	10 2
10	NP 14 2	V 18	DET 12	N 13		10	DET	12
11		NP 14 2	V 18	NP 10 2		11	N	13
12			NP 14 2	VP 8 2		12	ART	17
13				S 6 2		13	young	
14				#		14	DET	16
15						15	N	19
16						16	ART	20
17						17	the	
18						18	train	
19						19	old	
20						20	the	

The example in (20) is organized for expositional purposes; in practice the input sentence is put in S2, as shown, and words and formatives are pushed in S3 in the manner described for the original $\mathcal{M}_E$, but in S1 there is only one line in each cycle; for example, in Cycle 1 (Line 01) there is *the* apparent in S2, and S1 and S3 are empty; in Cycle 2 (Line 02) *the* has been moved to S3, *old* is on the top of S2, and S1 contains ART 20; in Cycle 7 (Line 07) S2 is empty, ART 20 is on the top of S3, and S1 contains DET 16 N 19 V 18 ART 17. Let us call the data slots in each line *words*. Each word can then hold up to four

registers which we have labelled as ‖ Symbol │ P3 │ Size │ R# ‖

The words in Cycles 1 and 2 will then look like the following:

(21)   S1   S2   S3

| | | | | the | | | | | | | | | 01 |

| | S1 | | | | | | S2 | | | S3 | | |

| | ART | α | | | old | | | | the | | | 02 |

where α in the P3 register in S1 is the value of P3 and refers to the location of *the* in S3. Let us look at the words in Cycles 9 and 10 in (20) as they appear in the observation window of $\mathcal{M}_E$.

(22)   S1   S2   S3

| | NP 14 2 | V 18 | ART 17 | A/N 13 | | | | NP | 10 | 2 | | 09 |
| | NP 14 2 | V 18 | DET 12 | N 13 | | | | DET | 12 | | | 10 |

Comparing 09 and 10, we observe that ART has changed to DET and A/N has been disambiguated in the same cycle (or simultaneously). This is possible if we assume the following schemata for the parallel processing of all the words available in each cycle:

(23)

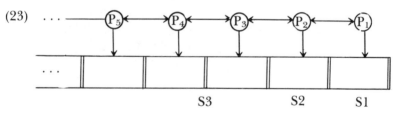

We assume that in (23) there is a processor attached to each one of the words. These processors operate in parallel with on-line communication among them. Although commercial hardware is not readily available at present for this kind of processing, such experimental designs have been studied and are being developed.

In the examples, we have seen that the dynamic storage in S1 allows us to examine any number of words necessary to disambiguate lexical items and make decisions about their categorial attributes. In English there are certain words that can indicate the boundaries of a phrase; for example, an article may mark the beginning of a noun phrase, and a noun followed by a verb may mark the end of the noun phrase. Unfortunately, all cases are not this simple, but, as stated before, our aim here is not to give details of an English grammar, but to discuss some principles and provide enough guidance for constructing a parser. It turns out, however, that our parser as discussed in the extended and revised version of $\mathcal{M}_E$ is independent of any grammar and will be as effective as is the underlying grammar. Note that in the simple model in (20) we keep shifting formatives into S1 until enough context has been accumulated to make a decision possible. For example, in Cycle 5 (line 05) the grammar "postulates" NP V NP as the structure of the input sentence. Then, in the following string:

(24)   ART 20 A/N 19 N/V 18 ART 17

the last article (ART 17) determines that N/V 18 is likely to be V 18 and A/N 19 is likely to be N 19. Such procedures can be stated in *productions*; let $j = 1, 2, 3, \ldots, n$ where n is the number of words in the input string:

$$
\begin{array}{lll}
(25a) & \text{IF} & \text{Symbol}_j = \text{ART} \\
 & & \text{Symbol}_{j+1} = \text{N/V} \\
 & \text{THEN} & \text{Symbol}_{j+1} \rightarrow \text{N} \\
(25b) & \text{IF} & \text{Symbol}_j = \text{N/V} \\
 & & \text{Symbol}_{j+1} = \text{ART} \\
 & \text{THEN} & \text{Symbol}_j \rightarrow \text{V, etc.}
\end{array}
$$

These productions are analogues of the processors indicated in (23) and can be implemented to be executed serially or in a semiparallel mode under the present state of the art.

Another characteristic to recall is that $\mathcal{M}_E$ provides surface structure analysis of the input sentence. If the semantic interpretation (not discussed here) requires a deeper analysis, then that should also be incorporated in the underlying grammar. We should also keep in mind that in the examples in this section, we have only included the categorial elements, such as N, V, NP, and so on. However, as we have seen in Chapter 8, there are many more features and subcategorial attributes which play a significant role in the syntactic and semantic analyses of the language. Again, these attributes must be included in the words entered in (20) and in the productions in (25), if we want a more realistic model of a parser.

The parser discussed in this section has certain similarities with the PARSIFAL parser discussed in Chapter 4, and with the so-called WASP parsers (Winston, 1984; see Exercise 5).

Toward a Model of Human Language Understanding

Language use and comprehension is ultimately tied with language acquisition; hence, questions concerning learnability are relevant for any study of models of human language understanding.

In the theory of formal languages, the term *grammatical inference* refers to the problem of obtaining a grammar G from a set of information consisting of sentences and nonsentences of a language L. To be more precise, suppose there is an unknown grammar G that generates the language L, but we have only two finite sets of samples R^+ and R^-. R^+ is a sample set of all and only the well-formed sentences in $L(G)$ and R^- is a sample set of all and only the strings in $\overline{L}(G)$, where $\overline{L}(G)$ is the complement set of $L(G)$. The task of grammatical inference is to obtain G from the examination of R^+ and R^-. Of course, if R^+ = $L(G)$, G can be trivially obtained by inspection, but normally $L(G)$ represents an infinite or a very large set, and R^+ and R^- are relatively small, finite sets. Furthermore, R^+ and R^- can be obtained or verified with the help of a "teacher" or an informant.

Chomsky (1962a) proposed a "language learning device" (later called *language acquisition device*—LAD) which is essentially based on the above schemata. He proposed that a child receives a set of linguistic utterances as input and produces a grammar of the language as output, using a certain innate intellectual capacity for language learning. Braine (1971) was the first psycholinguist to examine the implications of LAD and to claim that "it was intrinsically impossible as a model of language acquisition" (cf. Levelt, 1974 for further discussions).

In a well-known article in 1967 and a more technical and less fully appreciated article in 1965, E. M. Gold laid the mathematical foundations for the learnability of automata. The innovation made by Gold in the earlier paper is the extension of what may be called the finitistic decidability theory into the infinite domain. The classical approach to decidability stipulates a class of problems to be decidable if an algorithm exists which determines the question of solution for any arbitrary problem in the class in finite time. The infinitistic or *limiting decidability* extension of this is to that of classes of problems solvable by algorithms or decision procedures in infinite time; in this sense one can speak of infinite decision procedures or *limiting decision procedures*. These procedures are appropriate in situations in which (potentially) infinite information is obtainable but in which any finite subset of this information is insufficient for solving the problem. Here a limiting decision procedure examines successively larger subsets, all finite, of the total information, as the basis for solving the problem in some sense.

The parallel of this procedure for language may be as follows: The learner of a natural language, particularly a child, is presented with increasingly larger sets of primary linguistic data drawn from the (potentially) infinite set of the target language to be learned. The total information, being infinite, may determine what grammar or machine generates that total information (language), but no finite subset of it can uniquely make such a determination, because for any finite sample there are any finite number of grammars or machines capable of generating it. The learner, therefore, may be viewed as a procedure which must *infer* or induce from finite samples of an infinite language any of the grammars or machines which generate that language. This procedure itself might be viewed as discrete-deterministic in the sense that successive inferences as to the identity of the target grammars or machines are made by the procedure as it encounters successively larger samples of the language.

Now for a formal representation of the *limiting decision procedure* concept, consider a class of problems Π where Π is said to be *solvable in the limit* (limiting solvable) if (1) there exists an algorithm P such that for all $\pi_i \in \Pi$ (that is, for any problem in the class), P infers an infinite sequence of (putative) solutions, Σ, concerning the problem π_i, and (2) there exists $\sigma_\xi \in \Sigma$ (some inferred solution) such that for all $n > \xi$, σ_ξ solves the problem π_i, and (3) for all $i, j > \xi$, $\sigma_i = \sigma_j$. This essentially stipulates that some class of problems Π is limiting solvable if the following three conditions hold:

1. There exists an algorithm which, for any problem in the class, infers an infinite sequence of proposed solutions to the problem.
2. There exists some point in the sequence of inferred solutions after which the inferred solutions are correct.
3. After this point, all the inferred solutions are the same (this is in the event that there is more than one solution to the problem).

Thus, a problem is limiting solvable if there is an algorithm which infers an infinite sequence of solutions, and which, after some finite time, stabilizes on one correct solution to the problem (or to the class of problems). In a sense this algorithm solves the problem in a finite time after which it does not "change its mind."

In light of these concepts, Gold considers limiting decision procedures with respect to sets, functions, and machines; but, more importantly for our purposes, in his later paper, Gold (1967) applies these concepts to language identification *in the limit*. The most significant result of this study is the proof that only finite-state languages are learnable in the limit with positive sample sets, and that Type 0 language (equivalent to unrestricted Turing machines) are not learnable even with the help of a teacher or informant. If the claim that transformational grammars of the *Aspects* vintage precisely defined the class of Type 0 languages (Peters and Ritchie, 1973) was substantiated, then its implications should be obvious. To repeat a truism, however, children do learn languages; hence, the original question of Chomsky about the possibility of constructing a LAD remains unanswered. Studies of language learnability and acquisition have continued since Gold's fundamental papers. Levelt (1974, Vol. I) contains a survey chapter, and the most comprehensive study to date is that of Wexler and Culicover (1980). In a recent study Paul Postal and D. Terence Langendoen (Langendoen and Postal, 1984) have claimed that natural languages are not finite or infinite enumerable sets that can be specified by generative grammars. The collection of sentences in a natural language is (i) bigger than countably infinite and (ii) in fact, a "megacollection" with higher cardinality than enumerable sets. This follows from their assumption that there is no basis for imposing any size law on natural language sentences and follows from their claim that for any set of sentences in a natural language there is another sentence in that language which is the conjunction of that set. In fact, the following theorem presented by Langendoen and Postal falsifies all the extant frameworks proposed as theories of natural language, including finite-state, phrase-structure, lexical-functional, Montague, and transformational grammars with the possible exceptions of the "arc pair grammar" (Johnson and Postal, 1980; Postal, 1982).

> *Theorem*: No natural language has any constructive (i.e., proof-theoretic, generative, or Turing machine) grammar.

With regard to the question of fixed and finite sentence length, recall our comments concerning length n in Chapter 8. Katz (1966) arrived at this conclusion (quoted by Langendoen and Postal):

> Second, a syntactic theory in which some fixed N determines whether or not a string of words is well-formed would be unmotivated. It would lack

any justifiable means of choosing the N that divides the sentences from the nonsentences. Since the infinite set of strings that is considered too long is in no way structurally different from those that are granted the status of sentencehood, the length property that differentiates such strings from those that are accepted as sentences has nothing whatever to do with the structural property of syntactic well-formedness. If N is not fixed arbitrarily, the properties that fix an N are psychological properties that derive from the facts about a speaker's perceptual faculties, memory, mortality, etc. (p. 122)

What are the consequences of Langendoen and Postal's views for theories of language learning? They do not consider these as negative results, and with regard to language learning, the following is a paraphrase of what they say. Natural languages are called "natural" because human beings appear to be able to learn the one they are exposed to as children naturally—that is, spontaneously and without special instruction—unlike certain artificial languages, which children cannot learn without special instruction. The past twenty years or so has seen the development of theories of language learning which are designed to explain the specific ability of human beings to learn certain natural languages. These theories take as their starting point the observation that when children are presented with small samples of a natural language input, they develop the ability to form judgments about the grammatical status of much larger collections of sentences of that language. If limitations of time and computing space are removed, it is assumed that people would be able to judge the grammatical status of any sentence whatever of the natural languages they have learned. It is further assumed that people's ability to judge the grammatical status of the sentences of a natural language is based on their acquiring finite characterizations of that language. Given that natural languages are not finite, the same reasoning that led to the claim that theories of natural languages must be generative grammars also led to the claim that the characterization of natural language that people acquire must also be generative grammars (Chomsky, 1965; Miller and Chomsky, 1963; Wexler and Culicover 1980). But just as the first claim is false, so is the second. The finite characterization of natural languages that people acquire can just as well be nonconstructive.

Indeed, because natural languages are megacollections of sentences, they cannot be generated, and the assumption that underlies almost all serious current work in the study of human language acquisition, that to learn a natural language is to learn a grammar that generates it, is false. If human beings do learn natural languages, they must learn something other than generative grammars. One possibility is that they learn something like inductive definitions of languages (along the lines sketched in Chapter 3 of Langendoen and Postal, 1984). Thus, suppose

that all of the defining structural properties of some natural language
are exhibited by few enough relatively short sentences that children
could be exposed to them, and that children are endowed with powerful,
innately specified mechanisms to determine a collection of sentences
that is closed under a basic set of sentences drawn from among the
sentences they hear and a small set of functions. Then it seems
reasonable to suppose that they have learned that language, even
though the language comprises a megacollection of sentences.

Contrary to what is commonly supposed (cf. Wexler and Culicover,
1980), not every natural language can be learned under the ordinary
conditions of human language acquisition, for there is no guarantee
that the set of sentences that exhibit all the defining structural conditions
on sentencehood in a particular language can be presented in the
amount of time ordinarily available for language acquisition. Following
a suggestion of Wexler and Culicover (1980), we may suppose that
although natural languages are learnable in the limit (assuming an
appropriate definition of what this notion means, certainly not that of
Gold, 1967), some natural languages are not attainable. Assuming that
the attested natural languages have all been learned by those who speak
or have spoken them, all attested natural languages are attainable,
having already been attained. But the determination of the class of
attainable languages is of psycholinguistic interest only. Attainability is
not part of the definition of any natural language.

Thus the only real barrier to the learning of a natural language by
humans is the size of the set of sentences that exhibit all the defining
structural properties of that language, not the size of the language
itself. Natural languages do not have to be finite, finite-state, context-
free, context-sensitive, primitive recursive, recursive, or recursively
enumerable sets of sentences in order to be learnable. Hence the
demonstration that natural languages are megacollections does not
create any new crisis in the domain of natural language learnability.

In a footnote to the above remarks Langendoen and Postal admit
that they have not provided a full characterization of what it is to learn
a natural language nonconstructively, and they say that much more
careful work remains to be done.

We should warn the reader, however, that none of the above views
has provided a conclusive answer to the basic question of language
learning and learnability; but it will take us far afield from the scope
of this study to go into further details of these issues. The interested
reader should consult the references provided.

Another relevant concern is the question of what is the nature of
the human internal (i.e., mental) representation of information (knowl-

edge). This question has occupied a justifiably central position in virtually all attempts to construct a viable model of human information processing in general, and comprehension in particular. As such, it has always been assumed, usually tacitly, that the question is indeed an empirical one and that it is decidable at least in principle. Controversies have, however, persisted on the exact nature of the representation: procedural, propositional, semantic net, frames, and so on. These are well represented in the literature and we have discussed them in various chapters in this study. In recent studies, however, Anderson (1976, 1978) has directly challenged the validity of the decidability assumption. If Anderson is correct, then much of the research in this area will have been rendered vacuous, and, for this reason alone, it is essential to subject his arguments to extremely careful critical scrutiny (we will not do that here, however).

The earliest statement of the challenge to the empiricity of the internal representation problem occurs in Anderson (1976):

> These questions about internal representation cannot be answered. One cannot test issues just about the representation in the abstract, one must perform tests of the representation in combination with certain assumptions about the processes that use the representation. That is, one must test a representation-process pair. Given any representation-process pair it is possible to construct other pairs equivalent to it. These pairs assume different representations but make up for differences in representation by assuming compensating differences in the processes. (p. 10)

Two years later, Anderson revisits the same problem, more thoroughly armed, and since he appears to intend the latter discussion to supercede the earlier, evaluation will proceed almost entirely on the basis of the more formally rigorous work of Anderson (1978). There, Anderson clarifies to some extent why such a situation of what might be termed *representational indeterminacy* is inevitable:

> If we restrict ourselves to behavioral data, we cannot directly observe the internal processes . . . nor the internal representations. All we observe is that at various times, the stimuli . . . arrive and that sometime later response R is emitted. The question of interest is whether behavioral data . . . are adequate to constrain a theory of internal representation. Such a theory of representation will be part of a model, M, that also specifies the processes that operate on the representation . . . models with very different theories of representation can perfectly mimic the behavioral predictions of M. These alternative models will compensate for differences in the representation by different assumptions about the processes. Therefore, these models are not discriminable from M on the basis of behavioral data. And, therefore, the representation assumed by M is not discriminable from the different representations assumed by the other models. (pp. 263–64)

This passage contains Anderson's basic line of argumentation, short of formal proof, and hence some care need be exercised in attempting to understand it fully. We will reconstruct the flow of the argument as Anderson intends it and give it some added rigor in the process.

Let T be some arbitrary theory of comprehension (or more broadly still, human information processing)—at least minimally in that we require a specification of the nature of internal representation, R, and of the processes P which operate on R; thus, $T = (R, P)$. Further, let X be the set of appropriate input (stimuli, or whatever) and let Y be the set of corresponding output; for Anderson, both X and Y are constituted by (observable) behavioral data. We avoid speaking of "models" which may occasion some confusion given the word's more technical and, in this case, irrelevant meaning in metalogical discussion, and rather speak straightforwardly of theories. Then Anderson's argument is roughly:

(26a) $T = (R, P)$
(26b) $T : X \rightarrow Y$
(26c)   Both R and P are not subject to direct observation
(26d)   For any $T : X \rightarrow Y$ with $T = (R, P)$ there exists another $T' : X \rightarrow Y$ with $T' = (R', P')$
(26e)   Intermediate Conclusion: Any given Y fails to discriminate any pair of arbitrary theories T and T'
(26f)   Final Conclusion: Any given Y fails to discriminate R and R'

It should be noted that the inability to discriminate R from R'—the thesis of representational indeterminacy—stated in (26f) is deduced from the deducible inability to discriminate T from T', given some fixed Y; the latter thesis one might term that of theoretical indeterminacy. Such theoretical indeterminacy stated in (26e) is of course contingent on the provability of its formal expression found in (26d). Thus, the essential question is, What is the basis for the thesis of theoretical indeterminacy? It would appear to consist of two at least logically distinct considerations.

The first appears to be that expressed in (26c) above, namely, that neither the internal representations nor the internal processes are directly observable; in this sense, Anderson is viewing T as essentially a *black-box theory*, because its internal structure is unobservable and must be reconstructed to whatever extent possible on the basis of the (observable) input X and output Y. The second consideration which appears to motivate the claim of theoretical indeterminacy involves the putative underdetermination of theories by the nature of the output Y;

that is, there is a multiplicity of theories consonant with fixed output Y.

Both these purportive motivations for the claim of theoretical indeterminacy, and consequently for representational indeterminacy, will be explored later. At this point, it is necessary only to draw attention to the logical separability of the two theses; Thesis B: that T is essentially a black-box theory in the sense already explicated, and Thesis U: that any fixed output Y underdetermines the theories consonant with Y in the sense that no unique T is determined or fixed on the basis of Y alone (assuming that we are dealing with identical input X, as we shall continue to assume throughout).

We can appreciate this separability from the following reflections. If Thesis B were false—that is, if we could in some fashion directly examine the internal structure of the human mind as to whatever would be the correlates of R and P—then Thesis U would fail to hold in general, for if we wished to decide between two alternative theories T and T' each wholly consistent with the output Y, we could in some sense simply "look" to see what is the fact of the matter and choose the appropriate theory accordingly. Furthermore, if Thesis B were true, Thesis U is, thereby, in no way strictly entailed as there may be viable parameters of evaluation of alternative theories as to relative adequacy other than simply consistency with the output set Y.

It turns out that Anderson's "proof" of indeterminacy in this case does not hold; in fact it may prove the contrary. To show this would require development of rather abstract and formal/mathematical tools which we will not do here. The reason we do not grapple with this issue is that its outcome does not affect our view of the nature of human language comprehension whatever other implications it may have. I believe that, consistent with other aspects of comprehension, there is no unique representation. Put differently, demonstration (or adoption in model building) of any one representation does not invalidate other proposals. If this is the kind of indeterminacy that Anderson has in mind, I wholeheartedly agree with him, but his formal proofs do not show this.

We have so far discussed proposals and theoretical observations concerning the nature of natural language processors ranging from the most restricted finite-state devices to the other extreme of claiming that natural languages are transfinite collections with a cardinality higher than natural numbers and that their capacity even exceeds that of any Turing machine. Many of the arguments in support of these views are compelling, none of them are conclusive. Thus, the question we asked earlier, Is a model of human language comprehension attainable?— remains unanswered. The possible model that we outline in this section is not intended to answer this question. It is only a model in the sense that it can be described, implemented, and verified through empirical observations (all these remain to be done, of course); on the other hand, it is not a model in that it does not answer the above question in a principled way.

The conclusion we want to draw from this study is that human language comprehension cannot be modeled through any one of the approaches which have broadly been classified as linguistic, perceptual, or conceptual. Linguistic approach sometimes referred to as "theory-based" or "syntax-based," has as its primary concern syntactic analysis, design of efficient parsers, and formal semantic rules and procedures. Perceptual approach, sometimes referred to as "semantic-based," is not so much concerned with linguistic structures *per se* but with how structures are used or perceived. The approach is characterized by models of semantic memory and concepts of "teachable languages" discussed in Part II. Conceptual approach is characterized by some of the models in artificial intelligence. Its primary concern is for concepts and relationships among them, and for how linguistic objects are mapped into essentially nonlinguistic objects.

We have discussed hypotheses about and representative models of these approaches in this study. We have seen that sophisticated models in all of these approaches include important heuristic and inductive procedures and that there are overlaps between the approaches. Nevertheless, proponents of each approach have often constructed one model with specific or tacit attempts at falsification of other approaches.

Human mental processes, which include language perception, are highly complex and characteristically inefficient in the sense of theoretical computer science. Von Neumann observed early in the development of digital computers that even machines based on rigid mathematical formulations of the automata theories of the time with no redundancies and elements of chance failed to perform as expected (cf. Minsky, 1967; Moyne, 1977a).

By saying that human language comprehension cannot be modeled on any one of the extant approaches does not imply the falsification of any approach. The contention is that human language comprehension involves all of these approaches.

In Moyne (1980) a "performance" model was proposed with provisions for partitioning a sentence and simultaneous processing of various parts in an interactive design. This model is extended in Figure 8, speculating on general schemata for language comprehension.

Notice that in Figure 8 all the activity goes through the *control* device (1). This is in fact the least understood and the most difficult component to implement. All the edges connecting the components are

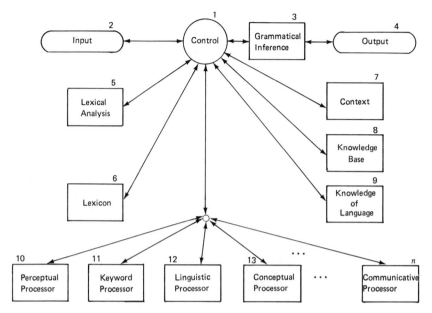

Figure 8. Schemata for language comprehension.

two-way arrows, implying highly interactive, on-line activity, including feedback from the *output* (4). Component (3), *grammatical inference*, is a learning device in the "limiting decidability" sense and it must be designed to update and augment the *knowledge of language* (9) and to ensure the extensibility of the grammar. The latter (9) contains the current knowledge of a particular language, constantly augmented as the system encounters new linguistic data (2). For the implementation of this component, one can use the grammatical notions and procedures discussed in this chapter. Components (7), (8), and (9) are data bases with dynamic storage capacity. *Context* (7) contains the current portion of the input stream already processed and is used by other components for making predictions, drawing inferences, resolving ambiguities, verifications, and so forth. *Knowledge base* (8) is the principal long-term storage data base and has similar functions as context, but in a more general and expanded sense. The nature of knowledge representation must be such as to allow for procedures and processes to take place within the data base, allowing for inductions, inferences, and so forth. For implementation, this can perhaps best be achieved by a procedural organization and production systems of the type advocated by Newell (see Chapter 7).

The *lexicon* (6) is another data base which must contain the vocabulary together with syntactic and semantic information, rules for productive processes, selectional restrictions, features of "strength" (cf. Chapter 4), and other pertinent information (see Chapters 2 and 8). The lexicon may be a collection of microglossaries with certain global rules and procedures as well as specific rules and entries for specific domains (cf. Moyne, 1977b). The *lexical analysis* (5) is the word access or recognizer.

The components (10) through (*n*) are the various processors which according to our hypotheses must be implemented to run in parallel and, in a sense, in competition with each other for the comprehension of the input data. They must contain procedures for induction, presupposition, and prediction, among others. In some cases the perception of just a few words may be enough to comprehend a sentence or whatever unit of input data is acceptable. In such cases, the *keyword processor* (11) may arrive at a relatively fast comprehension and would then signal the *control* (1) which would disengage the other processors. Models for *perceptual* (10), *conceptual* (13), and so forth have been discussed in the literature and have been surveyed in this study. Our *linguistic processor* (12) will have a central role in this model. In addition to processing sentences that must go through this processor, it must provide for any linguistic analyses required by other processors, such as recognition of phrases, clauses, and other linguistic structures. For the design of the linguistic processer (12), we propose the implementation of the augmented form of the device $\mathcal{M}_E$ discussed in this chapter.

For the design and implementation of the control (1), analogy can be drawn and insights gained from computer science in the design of operating systems with management functions for memory, processor, device, information, and resource, including interrupt, scheduling, and other well-known provisions.

Exercises

1. Using the following grammar G:

> S → (COMP) (NP) (AUX) VP
> COMP → C
> NP → (PRO) (DET) N (PP)
> VP → V (NP)* (DEG) (ADV) (PP)* (S) (ADJ)
> DET → ART (ADJ)*
> PP → P NP

the three-stack parser in Figure 1 in the text, and the Algorithm M in Figure 2, hand simulate the analysis of the following sentences as it is done in Figure 4. (In grammar G, C = complementizer, such as *that, where,* and so on; PRO = pronoun; AUX = auxiliary verbs, such as *will, should,* and so on; DEG = quantifiers, such as *very, many,* and so forth; ADV = adverb; P = preposition. Elements in parentheses are optional, and * indicates that the item can be repeated.)

   a. The fat lazy dog will chase the cat eventually
   b. He placed the very heavy lamp on the table in the study
   c. John told the man that he should go home
   d. Summer is the best time of the year

2. Given the following sentence:

   Is   the blind man sitting in the car at the curb deaf?
   AUX/V ART V/A N/V A/N/V P ART N P ART N/V A

where the markings under each word are categorial information from the lexicon and some of them indicate ambiguity, for example *sitting* could act as an adjective, a noun, or a verb (A/N/V), process through the revised and augmented $\mathcal{M}_E$ (see example [20] in the text). Your objective should be to produce the following syntactic analysis for the sentence:

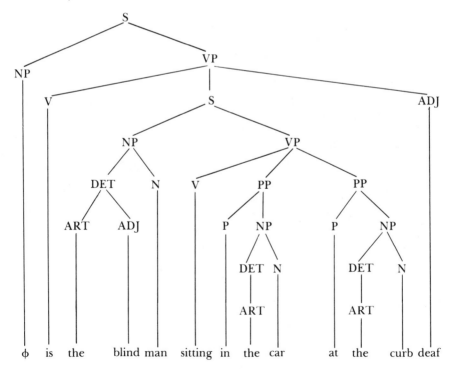

3. Write a set of productions, similar to the example (25) in the text, for the disambiguation of the ambiguous words in the sentence in Exercise 2.

4. Compare the parser depicted in Figure 2 and the revised version in example (20), and criticize the following statement: Nondeterministic parsers with backtracking are more desirable because their structures are less complex.

5. Compare and contrast the final version of $\mathcal{M}_E$ with the PARSIFAL parser discussed in Chapter 4. It would also be instructive if you were to study and compare the WASP parsers discussed in Winston (1984, pp. 309 ff.).

6. Why is a model of language learning difficult to implement?

7. Why is a model of language comprehension difficult to implement?

8. It has been claimed that transformational grammars have the power of an unrestricted Turing machine. If this were true, what sort of problems would arise for TG?

9. Integers constitute an infinite set with a cardinality (roughly, size) of $\aleph$. However, real numbers have a higher cardinality. This means that all infinite sets are not equal. Similar claims have been made about natural languages of being infinite with a cardinality of $\aleph$ or higher. We have finite brains; how is it possible to learn to use numbers or languages?

10. Several proposals have been suggested about the form of the internal (mental) representation of knowledge in humans. Anderson claims that the question of representation is undecidable. Discuss.

11. What are the three main approaches to language comprehension discussed in the literature?

12. Is parallel processing necessary for a model of language comprehension? Why?

References

Theory of parsing is discussed with many examples and exercises in Aho and Ullmann (1972, Vol. 1). Winograd (1983) contains many examples of parsers for natural language processing. Hinton and Anderson (1981) contains the so-called connectionist views on knowledge representation, parallel processing, and associative memory. Wexler and Culicover (1980) is the most comprehensive study on the theory of language acquisition.

Bibliography

Aho, A. V., and J. D. Ullman. 1972–73. *The theory of parsing, translation, and compiling.* Vols. I and II. Englewood Cliffs, N.J.: Prentice-Hall.

Akmajian, A., P. Culicover, and T. Wasow, eds. 1977. *Formal syntax.* New York: Academic Press.

Anderson, A. R., ed. 1964. *Minds and machines.* Englewood Cliffs, N.J.: Prentice-Hall.

Anderson, J. R. 1976. *Language, memory, and thought.* Hillsdale, N.J.: Erlbaum.

Anderson, J. R. 1978. Arguments concerning representations for mental imagery. *Psychological Review* 85: 249–277.

Anderson, J. R., and G. Bower. 1973. *Human associative memory.* New York: Wiley.

Anderson, S., and P. Kiparsky, eds. 1973. *A festschrift for Morris Halle.* New York: Holt, Rinehart & Winston.

Arbib, M. A. 1964. *Brains, machines and mathematics.* New York: McGraw-Hill.

Arbib, M. A. 1972. *The metaphorical brain: An introduction to cybernetics as artificial intelligence and brain theory.* New York: Wiley.

Arbib, M. A., and D. Caplan. 1979. Neurolinguistics must be computational. *Behavioral and Brain Sciences* 2: 449–483.

Armer, P. 1963. Attitudes toward intelligent machines. In *Computers and thought,* ed. E. A. Feigenbaum and J. Feldman. New York: McGraw-Hill.

Aronoff, M. 1976. *Word formation in generative grammar.* Cambridge, Mass.: MIT Press.

Atkinson, R., and R. Shiffrin. 1968. Human memory: A proposed system and its control process. In *Advances in the psychology of learning and motivation research and theory,* ed. K. Spence and J. Spence. New York: Academic Press.

Austin, J. L. 1962. *How to do things with words.* Oxford: Oxford University Press.

Bach, E. 1968. Nouns and noun phrases. In *Universals in linguistic theory,* ed. E. Bach and R. Harms. New York: Holt, Rinehart & Winston.

Bach, E. 1974. *Syntactic theory.* New York: Holt, Rinehart & Winston.

Bach, E., and R. Harms, eds. 1968. *Universals in linguistic theory.* New York: Holt, Rinehart & Winston.

Bach, E., and R. Harnish, 1979. *Linguistic communication and speech acts.* Cambridge, Mass.: MIT Press.

Barclay, J. 1973. The role of comprehension in remembering sentences. *Cognitive Psychology.* 4: 229–254.

Bar-Hillel, Y. 1964. *Language and information.* Reading, MA: Addison-Wesley.

Bar-Hillel, Y., M. Perles, and E. Shamir. 1961. On formal properties of simple phrase structure grammars. In *Zeitschrift für Phonetik, Sprachwissenschaft und Kommunikationsforschung.* Vol. 14, 143–172. Reprinted in Bar-Hillel 1964, 116–150.

Bartlett, F. 1932. *Remembering: A study in experimental and social psychology.* Cambridge: Cambridge University Press.

Barr, A., and E. A. Feigenbaum, eds. 1981. *The handbook of artificial intelligence.* Vol. 1. Los Altos, CA: Kaufmann.

Barr, A., and E. A. Feigenbaum, eds. 1982. *The handbook of artificial intelligence.* Vol. 2 Los Altos, CA: Kaufmann.

Bates, M. 1978. The theory and practice of augmented transition network grammars. In *Natural language communication with computers,* ed. L. Bolc. Berlin: Springer-Verlag.

Bever, T. 1970. The cognitive basis for linguistic structure. In *Cognition and language development,* ed. R. Hayes, 277–360. New York: Wiley.

Birkhoff, G., and T. C. Bartee. 1970. *Modern applied algebra.* New York: McGraw-Hill.

Bloomfield, L. 1933. *Language.* New York: Henry Holt.

Bobrow, D. G. 1968. Natural language input for a computer problem solving system. In *Semantic information processing,* ed. M. Minsky. Cambridge, Mass.: MIT Press.

Bobrow, D. G. 1979. KRL: Another perspective. *Cognitive Science* 3: 29–42.

Bobrow, D. G., and J. Fraser. 1969. An augmented state transition network analysis procedure. *Proc. IJCAI*: 557–567.

Bobrow, D. G., and A. Collins. 1975. *Representation and understanding: Studies in cognitive science.* New York: Academic Press.

Bobrow, D. G., and T. Winograd. 1977. An overview of KRL. *Cognitive Science* 1: 3–46.

Boden, M. 1977. *Artificial intelligence and natural man.* New York: Basic Books.

Bolc, L., ed. 1978. *Natural language communication with computers.* Berlin: Springer-Verlag.

Bower, G. 1975. Cognitive psychology. In *Handbook of learning and cognitive processes,* Vol. 1, ed. W. Estes, 25–80. Hillsdale, N.J.: Erlbaum.

Braine, M. 1971. On two types of models of the internalization of grammar. In *The ontogenesis of grammar,* ed. D. Slobin. New York: Academic Press.

Brainerd, Y. 1971. *Introduction to the mathematics of language study.* New York: Elsevier.

Bransford, J. D., and J. Franks. 1970. The abstraction of linguistic ideas. *Cognitive Psychology* 2: 331–350.

Bransford, J. D., and M. K. Johnson. 1972. Contextual prerequisites for understanding: Some investigations of comprehension and recall. *Journal of Verbal Learning and Verbal Behavior* 11: 717–726.

Bransford, J., *et al.* 1972. Sentence memory: A constructive versus interpretive approach. *Cognitive Psychology* 3: 193–209.

Bransford, J., *et al.* 1978. An analysis of memory theories from the perspective of problems of learning. In *Levels of processing and human memory,* ed. L. Cermak and F. Craik. Hillsdale, N.J.: Erlbaum.

Bresnan, J. 1978. A realistic transformational grammar. In *Linguistic theory and psychological reality,* ed. M. Halle, J. Bresnan, and G. A. Miller, 1–58. Cambridge, Mass.: MIT Press.

Bresnan, J., ed. 1982. *The mental representation of grammatical relations.* Cambridge, Mass.: MIT Press.

Brown, C., and H. Rubenstein. 1961. Test of response bias explanation of word-frequency effect. *Science* 133: 280–281.

Bundy, A., ed. 1978. *Artificial intelligence: An introductory course.* New York: North-Holland.

Butler, S. 1872. *Erewhon.* Harmondsworth, England: Penguin English Library.

Cairns, H. S. 1973. Effects of bias on processing and reprocessing of lexically ambiguous sentences. *Journal of Experimental Psychology* 97: 337–343.

Cairns, H. S. 1980. *Autonomous theories of the language processor*. Ms.

Cairns, H. S., and J. D. Kamerman. 1975. Lexical information processing during sentence comprehension. *Journal of Verbal Learning and Verbal Behavior* 14: 170–179.

Cairns, H. S., and J. R. Hsu. 1980. Effects of prior context upon lexical access during sentence comprehension: A replication and reinterpretation. *Journal of Psycholinguistic Research* 9: 1–8.

Campbell, R. N., and P. T. Smith. eds. 1978. *Recent advances in the psychology of language: Formal and experimental approaches*. New York: Plenum Press.

Carbonell, J. G. 1979. *Subjective understanding: computer models of belief systems*. Ph.D. diss., Computer Science Department, Yale University, RR–150.

Carnap, R. 1956. *Meaning and necessity*. Chicago: University of Chicago Press.

Carpenter, P., and M. Just. 1975. Sentence comprehension: A psycholinguistic processing model of verification. *Psychological Review* 83: 318–322.

Carroll, J. M., and M. K. Tanenhaus. 1975. Prologomena to a functional theory of word formation. In *Papers from the parasession functionalism*, ed. R. E. Grossman, L. J. San, and T. J. Vance. Chicago: Chicago Linguistic Society.

Carroll, J., and T. Bever. 1976. Sentence comprehension: A case study in the relation of knowledge and perception. In *Language and speech*. Vol. 7 of *Handbook of Perception*, ed. E. Carterette and M. Friedman. New York: Academic Press.

Carroll, J., M. Tanenhaus, and T. Bever. 1978. The perception of relations: The interaction of structural, functional, and contextual factors in the segmentation of sentences. In *Studies in the perception of language*, ed. W. J. M. Levelt and G. Flores d'Arcais, 187–218. New York: Wiley.

Carroll, J., T. Bever, and C. Pollack. 1981. The nonuniqueness of linguistic intuitions. *Language* 57: 368–383.

Carterette, E., and M. Friedman, eds. 1976. *Language and speech*. Vol. 7 of *Handbook of Perception*. New York: Academic Press.

Cassirer, E. 1923. *Substance and function*. New York: Dover.

Cermak, L., and F. Craik, eds. 1978. *Levels of processing and human memory*. Hillsdale, N.J.: Erlbaum.

Charniak, E. 1972. Towards a model of children's story comprehension. Memorandum. Artificial Intelligence Laboratory, MIT.

Charniak, E. 1975. A partial taxonomy of knowlege about actions. In *Advance papers of the fourth international conference on artificial intelligence*. Tbilisi. 11–98.

Charniak, E. 1976. Inference and knowledge, Part 2. In *Computational Semantics*, ed. E. Charniak and Y. Wilks. Amsterdam: North-Holland.

Charniak, E. 1977a. Ms. Malaprop: A language comprehension program. In *Advance papers of the fifth international conference on artificial intelligence*. Cambridge, Mass.: MIT Press.

Charniak, E. 1977b. A framed PAINTING: The representation of a common sense knowledge fragment. *Cognitive Science* 1: 355–394.

Charniak, E., and Y. Wilks, eds. 1976. *Computational semantics*. Amsterdam: North-Holland.

Chase, W., ed. 1973. *Visual information processing*. New York: Academic Press.

Chase, W., and H. Clark, 1972. Mental operations in the comprehension of sentences and pictures. In *Cognition in learning and memory*, ed. L. Gregg. New York: Wiley.

Cherry, E. C. 1953. On the recognition of speech with one, and with two, ears. *Journal of the Acoustic Society of America* 25.5: 975.

Chomsky, N. 1955. The logical structure of linguistic theory. M.I.T. Library. Mimeo. Also under the same title, New York: Plenum Press, 1975.

Chomsky, N. 1956. Three models for the description of language. *IRE Transactions on Information Theory*, IT-2: 113–124. Reprinted in *Readings in mathematical psychology*, Vol. 2, ed. R. D. Luce, R. Bush, and E. Galanter, 1965. New York: Wiley.

Chomsky, N. 1957. *Syntactic structures*. The Hague: Mouton.

Chomsky, N. 1959a. On certain formal properties of grammars. *Information and Control* 2: 137–167.

Chomsky, N. 1959b. A note on phrase structure grammars. *Information and Control* 2: 393–395.

Chomsky, N. 1962a. Explanatory models in linguistics. In *Logic, methodology, and philosophy of science: Proceedings of the 1960 International Congress*, ed. E. Nagel, P. Suppes, and A. Tarski. Stanford: Stanford University Press.

Chomsky, N. 1962b. Context-free grammars and pushdown storage, *QPR*, 65. Cambridge, Mass.: MIT Press.

Chomsky, N. 1963. Formal properties of grammars. In *Handbook of mathematical psychology*, Vol. 2, ed. R. D. Luce, R. Bush, and E. Galanter, 1963. New York: Wiley.

Chomsky, N. 1965. *Aspects of the theory of syntax*. Cambridge, Mass.: MIT Press.

Chomsky, N. 1970a. Remarks on nominalization. In *Readings in English transformational grammar*, ed. R. Jacobs and P. Rosenbaum. Waltham, Mass.: Ginn. Also in *Studies on semantics in generative grammar*, N. Chomsky, 1972. The Hague: Mouton.

Chomsky, N. 1970b. Deep structure, surface structure, and semantic interpretation. In *Studies in general and oriental linguistics*, ed. R. Jakobson and S. Kawamoto, 52–91, Tokyo: Tec Co.

Chomsky, N. 1972. *Studies on semantics in generative grammar*. The Hague: Mouton.

Chomsky, N. 1973. Conditions on transformations. In *A festschrift for Morris Halle*, ed. S. Anderson and P. Kiparsky. New York: Holt, Rinehart, & Winston.

Chomsky, N. 1975. *Reflections on language*. New York: Pantheon.

Chomsky, N. 1977. On *wh*-movement. In *Formal syntax*, ed. A. Akmajian, P. Culicover, and T. Wasow. New York: Academic Press.

Chomsky, N. 1981. *Lectures on government and binding*. Dordrecht: Foris.

Chomsky, N. 1982. *Some concepts and consequences of the theory of government and binding*. Cambridge, Mass.: MIT Press.

Chomsky, N., and M. Halle. 1968. *The sound pattern of English*. New York: Harper & Row.

Church, K. 1982. *On memory limitations in natural language processing*. Bloomington, Ind.: Indiana University Linguistic Club. (Reproduced from an M.I.T. thesis, 1980).

Clark, H. 1974. Semantics and comprehension. In *Linguistics and adjacent arts and sciences: Current trends in linguistics*, Vol. 12, ed. T. Sebeok. The Hague: Mouton.

Clark, H. 1976. Inferences in comprehension. In *Perception and comprehension*, ed. D. LaBerge and S. Samuels. Hillsdale, N.J.: Erlbaum.

Clark, H. 1978. Inferring what is meant. In *Studies in the perception of language*, ed. W. J. M. Levelt and G. Flores d'Arcais, 296–322. New York: Wiley.

Clark, H., and W. Chase. 1972. On the process of comparing sentences against pictures. *Cognitive Psychology* 3: 472–517.

Clark, H., and P. Lucy. 1975. Understanding what is meant from what is said: A study in conversationally conveyed requests. *Journal of Verbal Learning and Verbal Behavior* 14: 56–72.

Clark, H., and E. Clark. 1977. *Psychology and language*. New York: Harcourt Brace Jovanovich.

Clark, H., and L. Haviland. 1977. Comprehension and the given-new contract. In *Discourse, production, and comprehension,* ed. R. Freedle. Norwood, N.J.: Ablex Publishing.

Clifton, D., and P. Odom. 1966. Similarity relations among certain English sentence constructions. *Psychological Monographs* 80(5): 1–35.

Cocke, J., and J. T. Schwartz. 1970. *Programming languages and their compilers.* New York: Courant Institute, New York University.

Cohen, P., and E. Feigenbaum, eds. 1982. *The handbook of artificial intelligence.* Vol. 3. Los Altos, CA: Kaufmann.

Colby, K. M. 1975. *Artificial paranoia.* N.Y.: Pergamon Press.

Cole, R. A., ed. 1980. *Perception and production of fluent speech.* Hillsdale, N.J.: Erlbaum.

Cole, P., and J. Morgan. 1975. *Syntax and semantics.* Vol. 3 of *Speech acts.* New York: Seminar Press.

Collins, A., and A. Quillian. 1969. Retrieval time from semantic memory. *Journal of Verbal Learning and Verbal Behavior* 8: 240–247.

Collins, A., and E. Loftus. 1975. A spreading activation theory of semantic processing. *Psychological Review* 85: 407–428.

Conrad, C. 1972. Cognitive economy in semantic memory. *Journal of Experimental Psychology* 92: 149–154.

Cooper, R., and T. Parsons. 1976. Montague grammar, generative semantics, and interpretive semantics. In *Montague grammar,* ed. B. Partee, 311–362. New York: Academic Press.

Cooper, W. E., and E. C. T. Walker, eds. 1979. *Sentence Processing.* New York: Wiley.

Craik, F., and R. Lockhart. 1972. Levels of processing: A framework for memory research. *Journal of Verbal Learning and Verbal Behavior* 11: 671–684.

Craik, F., and L. Jacoby. 1975. A process view of short-term retention. In *Cognitive theory,* ed. F. Restle *et al.*, Vol. 1. Hillsdale, N.J.: Erlbaum.

Crombie, A. C. 1964. Early concepts of the sense and the mind. *Scientific American* (May 1964).

Culicover, P., J. Kimball, C. Lewis, D. Loveman, and J. Moyne, 1969. An automated recognition grammar for English. IBM Technical Report FSC 69–5007, Cambridge, Mass.

Čulik, K. 1967. On some transformations in context free grammars and languages. *Czechoslovak Mathematical Journal* 17(92), 278–310.

Cullingford, R. E. 1978. *Script application: Computer understanding of newspaper stores.* Research Report No. 116. Department of Computer Science, Yale University, New Haven.

Cutland, N. J. 1980. *Computability: An introduction to recursive function theory.* London: Cambridge University Press.

Cutler, A. 1976. Beyond parsing and lexical look-up: An enriched description of auditory sentence comprehension. In *New approaches to language mechanisms,* ed. R. Wales and E. Walker. Amsterdam: North-Holland.

Cutting, J., and B. Rosner. 1974. Categories and boundaries in speech and music. *Perception and Psychophysics* 16: 564–570.

Danks, J., and P. Sorce, 1973. Imagery and deep structure in the prompted recall of passive sentence. *Journal of Verbal Learning and Verbal Behavior* 12: 114–117.

Danks, J., and M. Schwenk. 1974. Comprehension of prenominal orders. *Memory and Cognition* 2(1A): 34–38.

Daston, P. 1957. Perception of idiosyncratically familiar words. *Perception and Motor Skills* 7: 3–6.

Davidson, D. 1967. Truth and meaning. *Synthese* 17: 304–323. Also in *Readings in the philosophy of language*, ed. J. Rosenberg and C. Travis (1971), 450–465. Englewood Cliffs, N.J.: Prentice-Hall.

Davidson, D. 1969. On saying that. In *Words and objects*, ed. D. Davidson and J. Hintikka, 158–174. New York: Humanities Press.

Davidson, D. 1976. Reply to Foster. In *Truth and meaning*, ed. G. Evans and J. McDowell, 33–41. Oxford: Clarendon Press.

Davidson, D. and J. Hintikka, eds. 1969. *Words and objects*. New York: Humanities Press.

Davidson, D., and G. Harman, eds. 1972. *Semantics of natural language*. Dordrecht: Reidel.

de Beaugrande, R., and B. N. Colby. 1979. Narrative models of action and interaction. *Cognitive Science* 3: 43–66.

de Jong, G. F. 1979. *Skimming stories in real time: An experiment in integrated understanding.* Ph.D. diss., Computer Science Department, Research Report No. 158, Yale University.

Dennett, D. C. 1979. Artificial intelligence as philosophy and psychology. In *Philosophical perspectives in artificial intelligence*, ed. M. Ringle, 57–80. Atlantic Highlands, N.J.: Humanities Press.

Dewer, K., *et al.* 1977. Recognition memory for single tones with and without context. *Journal of Experimental Psychology: Human Learning and Memory* 3: 60–67.

Dinneen, F. P., ed. 1966. *Monograph series on language and linguistics*, No. 19. Washington, D.C.: Georgetown University Press.

Donnellan, K. 1972. Proper names and identifying descriptions. In *Semantics of natural language*, ed. D. Davidson and G. Harman, 356–379. Dordrecht: Reidel.

Dresher, B. E., and N. Hornstein. 1977. On some supposed contribution of artificial intelligence to the scientific study of language. *Cognition.* 4: 4.

Dreyfus, H. 1979. *What computers can't do: The limits of artificial intelligence.* rev. ed. New York: Harper & Row.

Epstein, W. 1961. The influence of syntactic structure on learning. *American Journal of Psychology* 74: 80–85.

Erdelyi, M. 1974. A new look at the new look: Perceptual defence and vigilance. *Psychological Review* 81: 1–25.

Eriksen, C. 1963. Perception and personality. In *Concepts of personality*, ed. J. Wepman and R. Heine. Chicago: Aldine.

Ervin, S. 1961. Changes with age in the verbal determinants of word-association. *American Journal of Psychology* 74: 361–372.

Estes, W., ed. 1975. *Handbook of learning and cognitive processes.* Vol. 1. Hillsdale, N.J.: Erlbaum.

Estes, W., ed. 1978a. *Handbook of learning and cognitive processes.* Vol. 5. Hillsdale, N.J.: Erlbaum.

Estes, W., ed. 1978b. *Handbook of learning and cognitive processes.* Vol. 6. Hillsdale, N.J.: Erlbaum.

Evans, G., and J. McDowell, eds. 1976. *Truth and meaning.* Oxford: Clarendon Press.

Evey, J. 1963. *The theory and applications of pushdown store machines.* Ph.D. diss., Department of Applied Mathematics, Harvard University.

Feigenbaum, E. A., and J. Feldman. 1963. *Computers and thought.* New York: McGraw-Hill.

Fiengo, R. 1974. *Semantic conditions on surface structure.* Ph.D. diss., Department of Linguistics, Massachusetts Institute of Technology.

Fiengo, R. 1977. On trace theory. *Linguistic Inquiry* 8: 35–61.

Fiengo, R. 1980. *Surface structure.* Cambridge, Mass.: Harvard University Press.

Fillenbaum, S. 1966. Memory for gist: Some relevant variables. *Language and Speech* 9: 217–227.

Fillenbaum, S. 1968. Sentence similarity determined by semantic relations: The learning of converses. *Proceedings of the 76th Annual Convention of the American Psychological Association* 3: 9–10.

Fillenbaum, S. 1974. Pragmatic normalization: Further results for some conjunctive and disjunctive sentences. *Journal of Experimental Psychology* 102: 574–578.

Fillmore, C. 1966. A proposal concerning English prepositions. In *Monograph series on language and linguistics*, No. 19, ed. F. P. Dinneen, 19–33. Washington, D.C.: George-town University Press.

Fillmore, C. 1968. The case for case. In *Universals in linguistic theory*, ed. E. Bach and R. Harms, 1–90. New York: Holt, Rinehart & Winston.

Fillmore, C. 1969. Towards a modern theory of case. In *Modern studies in English*, ed. D. Riebel and S. Schane, 361–375. Englewood Cliffs, N.J.: Prentice-Hall.

Findler, N., ed. 1979. *Associative networks: Representation and use of knowledge by computers.* New York: Academic Press.

Fodor, J. A. 1968. *Psychological explanation.* New York: Random House.

Fodor, J. A. 1978. Tom Swift and his procedural grandmother. *Cognition* 6: 229–247. Also in *Representations: Philosophical essays on the foundations of cognitive science*, J. A. Fodor (1981). Cambridge, Mass.: MIT Press.

Fodor, J. A. 1981. *Representations: Philosophical essays on the foundations of cognitive science.* Cambridge, Mass.: MIT Press.

Fodor, J. A., and J. J. Katz, eds. 1964. *The structure of language: Readings in the philosophy of language.* Englewood Cliffs, N.J.: Prentice-Hall.

Fodor, J. A., and T. Bever. 1965. The psychological reality of linguistic segments. *Journal of Verbal Learning and Verbal Behavior* 4: 414–420.

Fodor, J. A., and M. Garrett. 1966. Some reflections on competence and performance. In *Psycholinguistic papers*, ed. J. Lyons and R. Wales, 135–154. Edinburgh: University, of Edinburgh Press.

Fodor, J. A., and M. Garrett. 1967. Some syntactic determinants of sentential complexity. *Perception and Psychophysics* 2: 289–296.

Fodor, J. A., M. Garrett, and T. Bever. 1968. Some syntactic determinants of sentential complexity II. *Perception and Psychophysics* 3: 453–461.

Fodor, J. A., T. G. Bever, and M. F. Garrett. 1974. *The psychology of language: An introduction to psycholinguistics and generative grammar.* New York: McGraw-Hill.

Fodor, J. D. 1974. *Semantics: Theories of meaning in generative grammar.* New York: Thomas Crowell. Reprinted Cambridge, Mass.: Harvard University Press.

Fodor, J. D. 1978. Parsing strategies and constraints on transformation. *Linguistic Inquiry* 9: 427–473.

Fodor, J. D., and L. Frazier. 1980. Is the human sentence parsing mechanism an ATN? *Cognition* 8: 417–459.

Ford, M., J. Bresnan, and R. Kaplan. 1982. A competence-based theory of syntactic closure. In *The mental representation of grammatical relations*, ed. J. Bresnan. Cambridge, Mass.: MIT. Press.

Forster, K. I. 1979. Levels of processing and the structure of the language processor. In *Sentence processing*, ed. W. E. Cooper and E. C. T. Walker., New York: Wiley.

Forster, K., and I. Olbrei. 1973. Semantic heuristics and syntactic analysis. *Cognition* 2: 319–347.

Forster, K., and E. Bednall. 1976. Terminating and exhaustive search in lexical access. *Memory and Cognition* 4: 53–61.

Foss, D. 1969. Decision processes during sentence comprehension: Effects of lexical item difficulty and position upon decision times. *Journal of Verbal Learning and Verbal Behavior* 8: 457–462.

Foss, D. 1970. Some effects of ambiguity upon sentence comprehension. *Journal of Verbal Learning and Verbal Behavior* 9: 457–462.

Foss, D., and R. Lynch. 1969. Decision processes during sentence comprehension: Effects of surface structure on decision times. *Perception and Psychophysics* 5: 145–148.

Foss, D., and C. Jenkins. 1973. Some effects of context on the comprehension of ambiguous sentences. *Journal of Verbal Learning and Verbal Behavior* 12: 577–589.

Foss, D., and D. Hakes. 1978. *Psycholinguistics.* Englewood Cliffs, N.J.: Prentice-Hall.

Foster, J. 1976. Meaning and truth theory. In *Truth and meaning,* ed. G. Evans and J. McDowell, 1–32. Oxford: Clarendon.

Franks, J. 1974. Toward understanding understanding. In *Cognition and symbolic processes,* ed. W. Weimer and D. Palermo. Hillsdale, N.J.: Erlbaum.

Franks, J., and J. Bransford. 1972. The acquisition of abstract ideas. *Journal of Verbal Learning and Verbal Behavior* 11: 311–315.

Frantz, G. 1974. *Generative semantics: An introduction with bibliography.* Bloomington, Ind.: Bloomington Linguitics Club.

Frazier, L., and J. D. Fodor. 1978. The sausage machine: A new two-stage parsing model. *Cognition* 6: 291–325.

Frederiksen, C. 1975. Representing logical and semantic structure of knowledge acquired from discourse. *Cognitive Psychology* 7: 371–458.

Freedle, R., ed. 1977. *Discourse, production, and comprehension.* Norwood, N.J.: Ablex Publishing.

Frege, G. 1892. Uber Sinn und Bedeutung. *Zeitschrift für Philosophie und philosophische Kritik* 100: 25–50.

French, P., T. Ueheling, and H. K. Wettstein, eds. 1979. *Contemporary perspectives in the philosophy of language.* Minneapolis: University of Minnesota Press.

Fries, C. C. 1952. *The structure of English.* New York: Harcourt, Brace.

Gazdar, G. 1981. Unbounded dependencies and coordinated structures. *Linguistic Inquiry* 12: 155–184.

Gazdar, G. 1982. Phrase structure grammar. In *The nature of syntactic representation,* ed. P. Jacobson and G. Pullum. Dordrecht: Reidel.

Gershman, A. V. 1979. *Knowledge-based parsing.* Ph.D. diss., Computer Science Department, Research Report 156, Yale University.

Ginsburg, S., and Partee, B. 1969. A mathematical model of transformational grammar. *Information and Control* 15: 297–334.

Gold, E. M. 1965. Limiting recursion. *Journal of Symbolic Logic* 30: 1.

Gold, E. M. 1967. Language identification in the limit. *Information and Control* 10: 447–474.

Goldman, N. 1975. Conceptual generation. In *Conceptual information processing,* ed. R. C. Schank. Amsterdam: North-Holland.

Gonzales, R. C., and M. G. Thomason. 1978. *Syntactic pattern recognition.* Reading, Mass.: Addison-Wesley.

Gough, P. 1965. Grammatical transformations and speed of understanding. *Journal of Verbal Learning and Verbal Behavior* 4: 107–111.

Gough, P. 1966. The verification of sentences: The effects of delay of evidence and sentence length. *Journal of Verbal Learning and Verbal Behavior* 4: 492–496.

Green, P. E., A. R. Wolf, C. Chomsky, and K. Laughery. 1963. BASEBALL: An automatic question answerer. In *Computers and thought,* ed. E. A. Feigenbaum and J. Feldman. New York: McGraw-Hill.

Gregg, L., ed. 1972. *Cognition in learning and memory.* New York: Wiley.

Grice, H. 1975. Logic and conversation. In *Syntax and semantics,* ed. P. Cole and J. Morgan, 41–58. Vol. 3 of *Speech Acts.* New York: Seminar Press.

Gross, M. 1972. *Mathematical models in linguistics.* Englewood Cliffs, N.J.: Prentice-Hall.

Gross, M., M. Halle, and M. P. Schutzenberger, eds. 1972. *Formal language analysis.* The Hague: Mouton.

Gross, M., and A. Lentin. 1967. *Notions sur les grammaires formelles.* Paris: Gauthier-Villars. English translation (1970). *Introduction to formal grammars.* Heidelberg-New York: Springer-Verlag.

Grossman, R. E., L. J. San, and T. J. Vance, eds. 1975. *Papers from the parasession functionalism.* Chicago: Chicago Linguistic Society.

Gunderson, K., ed. 1975. *Minnesota studies in the philosophy of science.* Vol. 7. Minneapolis, Minn.: University of Minnesota Press.

Hakes, D. 1971. Does verb structure affect sentence comprehension? *Perception and Psychophysics* 10: 229–232.

Hakes, D. 1972. Effects of reducing complement constructions on sentence comprehension. *Journal of Verbal Learning and Verbal Behavior* 11: 278–286.

Hakes, D., and H. Cairns. 1970. Sentence comprehension and relative pronouns. *Perception and Psychophysics* 8: 5–8.

Hakes, D., and D. Foss. 1970. Decision processes during sentence comprehension: Effects of surface structure reconsidered. *Perception and Psychophysics* 8: 413–416.

Halle, M., J. Bresnan, and G. A. Miller, ed. 1978. *Linguistic theory and psychological reality.* Cambridge, Mass.: MIT Press.

Halliday, M. A. K. 1966. *The English verbal group: A specimen of a manual of analysis.* Nuffield Programme in Linguistics and English Teaching.

Halliday, M. A. K. 1967–68. Notes on transitivity and theme in English. *English Journal of Linguistics,* 3: 37–81; 4: 179–215.

Halliday, M. A. K. 1970. Functional diversity in language as seen from the consideration of modality and mood in English. *Foundations of Language* 6: 322–361.

Harris, Z. S. 1951. *Methods in structural linguistics.* Chicago: University of Chicago Press.

Harris, Z. S. 1952. Discourse analysis. *Language* 28: 1–30.

Harris, Z. S. 1957. Co-occurrence and transformation in linguistic structure. *Language* 33: 283–340.

Harrison, M. 1978. *Introduction to formal language theory.* Reading, Mass.: Addison-Wesley.

Haugeland, J. ed. 1981a. *Mind design: Philosophy, psychology, artificial intelligence.* Cambridge, Mass.: MIT Press.

Haugeland, J. 1981b. Semantic engines: An introduction to mind design. In *Mind design,* ed. J. Haugeland. Cambridge, Mass.: MIT Press.

Haviland, S., and H. Clark. 1974. What is new? Acquiring new information as a process in comprehension. *Journal of Verbal Learning and Verbal Behavior* 13: 512–521.

Hayes, R., ed. 1970. *Cognition and language development.* New York: Wiley.

Hebb, D. O. 1949. *The organization of behavior.* New York: Wiley.

Held, R., and W. Richards. 1972. *Perception: Mechanisms and models.* San Francisco: Freeman.

Herriot, P. 1969. The comprehension of active and passive sentences as a function of pragmatic expectations. *Journal of Verbal Learning and Verbal Behavior* 8: 166–169.

Hewitt, C. 1969. PLANNER: A language for proving theorems in robots. *Proceedings of the International Joint Conference on Artificial Intelligence*, 295–301 Bedford, Mass.: Mitre Corp.

Hewitt, C. 1971. *Description and theoretical analysis (using schemes) of PLANNER: A language for proving theorems and manipulating models in robots*. Ph.D. diss., Massachusetts Institute of Technology.

Hinton, G. E., and J. A. Anderson, eds. 1981. *Parallel models of associative memory*. Hillsdale, N.J.: Erlbaum.

Hirst, G. 1981. *Anaphora in natural language understanding: A survey*. Berlin: Springer-Verlag.

Hobbes, T. [1651] 1947. *Leviathan*. Reprint edited by Michael Oakeshott. Oxford: Oxford University Press.

Hockett, C. F. 1955. A manual of phonology. *International Journal of American Linguistics* 11.

Hockett, C. F. 1970. *A Leonard Bloomfield anthology*. Bloomington: Indiana University Press.

Hodges, A. 1983. *Alan Turing: The enigma*. New York: Simon & Schuster.

Holmes, V. M., and K. Forster. 1972. Perceptual complexity and understanding sentence structure. *Journal of Verbal Learning and Verbal Behavior* 11: 148–156.

Holmes, V. M., R. Arwas, and M. F. Garrett. 1977. Prior context and the perception of lexically ambiguous sentences. *Memory and Cognition* 5: 103–110.

Hopcroft, J., and J. Ullman. 1979. *Introduction to automata theory, languages, and computations*. Reading, Mass.: Addison-Wesley.

Hornby, P. 1974. Surface structure and presupposition. *Journal of Verbal Learning and Verbal Behavior* 13: 530–538.

Howes, D. 1954. On the recognition of word frequency as a variable affecting speed of recognition. *Journal of Experimental Psychology* 48: 106–112.

Howes, D., and R. Solomon. 1951. Visual duration threshold as a function of word-probability. *Journal of Experimental Psychology* 41: 401–410.

Hunt, E. B. 1975. *Artificial intelligence*. New York: Academic Press.

Jackendoff, R. 1969. An interpretive theory of negation. *Foundations of Language* 5(2): 218–241.

Jackendoff, R. 1977. $\overline{X}$ *Syntax: A study of phrase structure*. Cambridge, Mass.: MIT Press.

Jackendoff, R. 1981. On Katz's autonomous semantics. *Language* 57:425–435.

Jacobs, R., and P. Rosenbaum, eds. 1970. *Readings in English transformational grammar*. Waltham, Mass.: Ginn.

Jacobsen, B. 1977. *Transformational generative grammar*. Amsterdam: North-Holland.

Jacobson, P., and G. Pullum, eds. 1982. *The nature of syntactic representation*. Dordrecht: Reidel.

Jacoby, L., and F. Craik. 1978. Effects of elaboration of processing at encoding and retrieval: Trace distinctiveness and recovery of initial context. In *Levels of processing and human memory*, ed. L. Cermak and F. Craik. Hillsdale, N.J.: Erlbaum.

Jakobson, R., and S. Kawamoto, eds. 1970. *Studies in general and oriental linguistics*. Tokyo: Tec Co.

Jarvella, R. J., and J. G. Collas. 1974. Memory for the intentions of sentences. *Memory and Cognition* 2: 185–188.

Javal, L. E. 1876. Essai sur la physiologie de la lecture. *Annales d'Oculistique* 82: 242–253.

Johnson, D., and P. Postal. 1980. *Arc pair grammar*. Princeton, N.J.: Princeton University Press.

Johnson, M., J. Bransford, and S. Solomon. 1973. Memory for tacit implications of sentences. *Journal of Experimental Psychology* 98: 203–205.

Johnson-Laird, P. N. 1974. Experimental psycholinguistics. *Annual Review of Psychology* 25: 135–160.

Joshi, A. K. 1972. A class of transformational grammars. In *Formal language analysis,* ed. M. Gross, M. Halle, M. P. Schutzenberger. The Hague: Mouton.

Kaplan, R. 1972. Augmented transition networks as psychological models of sentence comprehension. *Artificial Intelligence* 3: 77–100.

Kaplan, R. 1973. A general syntactic processor. In *Natural language processing,* ed. R. Rustin. Amsterdam: North-Holland.

Kaplan, R. 1975. On process models for sentence analysis. In *Explorations in cognition,* ed. D. Norman and D. Rumelhart. San Francisco: Freeman.

Kaplan, R., and J. Bresnan. 1982. Lexical functional grammar. In *The mental representation of grammatical relations,* ed. J. Bresnan. Cambridge, Mass.: MIT Press.

Katz, J. 1966. *The philosophy of language.* New York: Harper & Row.

Katz, J. 1972. *Semantic theory.* New York: Harper & Row.

Katz, J. 1977. *Propositional structure and elocutionary force.* New York: Harper & Row.

Katz, J. 1981. *Language and other abstract objects.* Totowa, N.J.: Rowman & Littlefield.

Katz, J., and J. A. Fodor. 1963. The structure of a semantic theory. *Language* 39: 170–210. (Reprinted in *The structure of language,* ed. J. A. Fodor and J. J. Katz, 479–518. Englewood Cliffs, N.J.: Prentice-Hall.

Katz, J., and P. Postal. 1964. *An integrated theory of linguistic descriptions.* Cambridge, Mass.: MIT Press.

Keenan, E., ed. 1973. *Formal semantics of natural languages.* Cambridge: Cambridge University Press.

Keenan, J. M., B. MacWhinney, and D. Mayhew. 1977. Pragmatics in memory: A study of natural conversation. *Journal of Verbal Learning and Verbal Behavior* 16: 549–560.

Kelly, E. 1970. *A dictionary based approach to lexical disambiguation.* Ph.D. diss., Harvard University.

Kempson, R. M. 1977. *Semantic theory.* Cambridge: Cambridge University Press.

Kent, E. W. 1981. *The brains of men and machines.* New York: McGraw-Hill.

Kiefer, F. 1975. Coordination within sentences and sentence combinability within "texts." In *Style and text: Studies presented to Nils Enkvist,* 349–359. Stockholm: Skriptor.

Kiefer, F. 1977. Some observations concerning the differences between sentence and text. In *Linguistic structures processing,* ed. A. Zampoli, 235–254. Amsterdam: North-Holland.

Kimball, J. 1967. Predicates definable over transformational derivations by intersection with regular languages. *Information and Control* 11: 177–195.

Kimball, J. 1973. Seven principles of surface structure parsing in natural language. *Cognition* 2: 15–47.

Kintsch, W. 1974. *The representation of meaning in memory.* Hillsdale, N.J.: Erlbaum.

Kintsch, W. 1978. Comprehension and memory of text. In *Handbook of learning and cognitive processes,* Vol. 5, ed. W. Estes. Hillsdale, N.J.: Erlbaum.

Kintsch, W., and J. Keenan. 1973. Reading rate and retention as a function of the number of propositions in the base structure of sentences. *Cognitive Psychology* 5: 257–274.

Kintsch, W., and T. A. Van Dijk. 1978. Toward a model of text comprehension and production. *Psychological Review* 85: 363–394.

Klappholz, A. D., and A. D. Lockman. 1977. *The use of dynamically extracted context for anaphoric reference resolution.* Unpublished manuscript, Department of Electrical Engineering and Computer Science, Columbia University.

Knuth, D. E. 1968. *The art of computer programming.* Vol. 1. Reading, Mass.: Addison-Wesley.

Kripke, S. 1972. Naming and necessity. In *Semantics of natural language,* ed. D. Davidson and G. Harman. Dordrecht: Reidel.

Kuroda, S-Y. 1964. Classes of languages and linear-bounded automata. *Information and Control* 7: 207–223.

LaBerge, D., and S. Samuels, eds. 1976. *Perception and Comprehension.* Hillsdale, N.J.: Erlbaum.

Lachman, R., J. L. Lachman, and E. C. Butterfield. 1979. *Cognitive psychology and information processing: An introduction.* Hillsdale, N.J.: Erlbaum.

Ladefoged, P., and D. Broadbent. 1960. Perception of sequence in auditory events. *Quarterly Journal of Experimental Psychology* 13: 162–170.

Lakoff, G. 1970a. *Irregularity in syntax.* New York: Holt, Rinehart & Winston.

Lakoff, G. 1970b. Global rules. *Language:* 46, 627–639.

Lakoff, G., and J. Ross. 1976. Is deep structure necessary? In *Syntax and semantics,* ed. J. McCawley, 159–164. New York: Academic Press.

Landweber, P. 1963. Three theorems on phrase structure grammars of type 1. *Information and Control* 6: 131–136.

Lang, E. 1973. *Studien zur semantik der koordinativen Vernupfung.* Diss. Berlin.

Langendoen, D. T. 1975. Finite-state parsing of phrase-structure languages and the status of readjustment rules in grammar. *Linguistic Inquiry* 6: 533–554.

Langendoen, D. T., and J. A. Moyne. 1981. On the form of the output of the human sentence parsing mechanism. Paper presented at the annual meeting of the Association for Computational Linguistics, New York, Dec. 1981.

Langendoen, D. T., and Y. Langsam. 1984. The representation of constituent structures for finite-state parsing. In *Proceedings of the 10th International Conference on Computational Linguistics,* 24–27. Stanford: Stanford University Press.

Langendoen, D. T., and P. M. Postal. 1984. *The vastness of natural languages.* Oxford: Blackwells.

Levelt, W. J. M. 1966. Generative grammatica en psycholinguistiek. II Psycholinguistis-chonderzoek. *Nederlands Tijdschrift voor de Psychologie* 21: 367–400.

Levelt, W. J. M. 1974. *Formal grammars in linguistics and psycholinguistics.* Vols. 1–3. The Hague: Mouton.

Levelt, W. J. M. 1978. A survey of studies in sentence perception: 1970–1976. In *Studies in the perception of language,* ed. W. J. M. Levelt and G. Flores d'Arcais. New York: Wiley.

Levelt W. J. M., and G. Flores d'Arcais, eds. 1978. *Studies in the perception of language.* New York: Wiley.

Levin, S. R. 1965. *Langue* and *parole* in American linguistics. *Foundations of Language* 1: 83–94.

Lewis, D. 1976. General semantics. In *Montague grammar,* ed. B. Partee, 1–50. New York: Academic Press.

Liberman, A. 1970. The grammars of language and speech. *Cognitive Psychology* 1: 301–323.

Lindsay, R. K. 1963. Inferential memory on the basis of machines which understand natural languages. In *Computers and thought,* ed. E. A. Feigenbaum and J. Feldman. New York: McGraw Hill.

Longuet-Higgins, H. C. 1981. Artificial intelligence—a new theoretical psychology? *Cognition* 10: 197–200.

Lovelace, M. C., Countess of. 1842. *Scientific memoirs.* Edited by R. Taylor. London: Johnson Reprint.

Luce, R. D., R. Bush, and E. Galanter, eds. 1963. *Handbook of mathematical psychology.* Vols. 1–3. New York: Wiley.

Luce, R. D., R. Bush, and E. Galanter, eds. 1965. *Readings in mathematical psychology.* Vols. 1–2. New York: Wiley.

Lyons, J. 1968. *Introduction to theoretical linguistics.* Cambridge: Cambridge University Press.

Lyons, J., and R. Wales, eds. 1966. *Psycholinguistic papers.* Edinburgh: University of Edinburgh Press.

McCawley, J. 1968a. The role of semantics in grammar. In *Universals in linguistic theory,* ed. E. Bach and R. Harms, 125–170. New York: Holt, Rinehart & Winston.

McCawley, J. 1968b. Concerning the base component of a transformational grammar. *Foundations of Language* 4: 243–269.

McCawley, J. ed. 1976. *Syntax and semantics,* 7. New York: Academic Press.

McDermott, J. 1979. Representing knowledge in intelligent systems. In *Philosophical perspectives in artifical intelligence,* ed. M. Ringle. Atlantic Highlands, N.J.: Humanities Press.

Machtey, M., and P. Young. 1978. *An introduction to the general theory of algorithms.* New York: North-Holland.

Marcus, M. P. 1980. *A theory of syntactic recognition for natural language.* Cambridge, Mass.: MIT Press.

Marks, L., and G. Miller. 1964. The role of semantic and syntactic constraints in the memorization of English sentences. *Journal of Verbal Learning and Verbal Behavior* 3: 1–5.

Marslen-Wilson, W. 1973. Linguistic structure and speech shadowing at very short latencies. *Nature* 244: 522–523.

Marslen-Wilson, W. 1975. Sentence perception as an interactive parallel process. *Science* 189: 226–228.

Marslen-Wilson, W. 1976. Linguistic descriptions and psychological assumptions in the study of sentence perception. In *New approaches to language mechanisms,* ed. R. Wales and E. Walker. Amsterdam: North-Holland.

Marslen-Wilson, W., and L. Tyler. 1975. Processing structure of sentence perception. *Nature* 257: 785–786.

Marslen-Wilson, W., and L. Tyler. 1976. Memory and levels of processing in a psycholinguistic context. *Journal of Experimental Psychology: Human Learning Memory* 2: 112–119.

Marslen-Wilson, W., and A. Welsh. 1978. Processing interactions and lexical access during word-recognition in continuous speech. *Cognitive Psychology* 10: 29–63.

Marslen-Wilson, W., L. Tyler, and M. Seidenberg. 1978. Sentence processing and the clause boundary. In *Studies in the perception of language,* ed. W. J. M. Levelt and G. Flores d'Arcais, 219–246. New York: Wiley.

Marslen-Wilson, W., and L. Tyler. 1980. The temporal structure of spoken language understanding. *Cognition* 8: 1–71.

Martin, J. 1972. Rhythmic (hierarchical) versus serial structure in speech and other behavior. *Psychological Review* 79: 487–509.

Martinet, A. 1953. Structural linguistics. In *Anthropology Today,* ed. A. L. Kroeber, 576–586. Chicago: University of Chicago Press.

Mead, M. 1952. Review of *Methods in Structural Linguistics,* by Zelig Harris. *International Journal of American Linguistics* 18: 257–260.

Meehan, J. 1976. *The metanovel: Writing stories by computer.* Ph.D. diss., Department of Computer Science, Research Report 74, Yale University.

Meyer, B. J. F. 1975. *The organization of prose and its effects on memory.* Amsterdam: North-Holland.

Miller, G. 1956. The magical number seven, plus or minus two: Some limits on our capacity for processing information. *Psychological Review* 60: 81–97.

Miller, G. 1962a. Some psychological studies of grammar. *American Psychologist* 17: 748–762.

Miller, G. 1962b. Decision units in the perception of speech. *IRE Transactions in Information Theory* IT-8: 81–83.

Miller, G., E. Galanter, and K. Pibram. 1960. *Plans and structures of behavior.* New York: Holt, Rinehart & Winston.

Miller, G., and N. Chomsky. 1963. Finitary models of language users. In *Handbook of mathematical psychology,* ed. R. D. Luce et al. Vol. 2, 419–491. New York: Wiley.

Miller, G., and S. Isard. 1963. Some perceptual consequences of linguistic rules. *Journal of Verbal Learning and Verbal Behavior* 2: 217–228.

Miller, G., and P. Johnson-Laird. 1976. *Language and perception.* Cambridge, Mass.: Harvard University Press.

Minsky, M. 1963a. Steps toward artificial intelligence. In *Computers and thought,* ed. E. A. Feigenbaum and J. Feldman. New York: McGraw-Hill.

Minsky, M. 1963b. A selected descriptor-indexed bibliography to the literature on artificial intelligence. In *Computers and thought,* E. A. Feigenbaum and J. Feldman. New York: McGraw-Hill.

Minsky, M. 1967. *Computation: Finite and infinite machines.* Englewood Cliffs, N.J.: Prentice-Hall.

Minsky, M., ed. 1968a. *Semantic information processing.* Cambridge, Mass.: MIT Press.

Minsky, M. 1968b. "Introduction." In *Semantic information processing,* ed. M. Minsky. Cambridge, Mass.: MIT Press.

Minsky, M. 1975. Framework for representing knowledge. In *The psychology of computer vision,* ed. P. H. Winston. New York: McGraw-Hill.

Mistler-Lachman, J. 1972. Levels of comprehension in processing of normal and ambiguous sentences. *Journal of Verbal Learning and Verbal Behavior* 13: 98–106.

Montague, R. 1970. English as a formal language I. *Linguaggi nella societa e nella tecnica.* Milan: Edizioni di Communita.

Moore, E., ed. 1964. *Sequential machines: Selected papers.* Reading, Mass.: Addison-Wesley.

Moore, T., ed. 1973. *Cognitive development and the acquisition of language.* New York: Academic Press.

Morton, J. 1979. Word recognition. In *Psycholinguistics 2,* ed. J. Morton and J. Marshall. Cambridge, Mass.: MIT Press.

Morton, J., and J. Marshall, eds. 1979. *Psycholinguistics 2: Structures and processes.* Cambridge, Mass.: MIT Press.

Moyne, J. A. 1959. Idiomatic structures in machine translation. Washington, D.C.: Georgetown University.

Moyne, J. A. 1962. Proto-RELADES: A restrictive natural language system. *IBM Technical Report,* BPC, 3, Cambridge, Mass.

Moyne, J. A., ed. 1969. An automated recognition grammar for English. *IBM Technical Report,* FSC, 69-5007, Cambridge, Mass.

Moyne, J. A. 1974. Some grammars and recognizers for formal and natural languages. In *Advances in information systems science,* ed. J. T. Tou, Vol. 5, 263–333. New York: Plenum Press.

Moyne, J. A. 1975. Relevance of computer science to linguistics and vice versa. *International Journal of Computer and Information Sciences* 4: 265–279.

Moyne, J. A. 1977a. The computability of competence. *Journal of Linguistics* 188: 5–9.

Moyne, J. A. 1977b. Simple-English for data base communication. *International Journal of Computer and Information Sciences* 6: 327–342.

Moyne, J. A. 1980. Language use: A performance model. *International Journal of Computer and Information Sciences* 9: 459–481.

Moyne, J. A. 1981. Controversies about autonomous lexical accessing. Paper read at IBM Systems Research Institute, New York, Dec. 11, 1981.

Moyne, J. A. 1984. Problems of lexical accessing for models of language comprehension. *Cognition and Brain Theory* 7: 181–197.

Moyne, J. A., and C. Kaniklidis. 1981. Models of language comprehension. *Cognition and Brain Theory* 4: 265–284.

Nagel, E., P. Suppes, and A. Tarski, eds. 1962. *Logic, methodology, and philosophy of science: Proceedings of the 1960 International Congress.* Stanford: Stanford University Press.

Newell, A. 1973. Production systems: Models of control systems. In *Visual information processing,* ed. W. Chase. New York: Academic Press.

Newell, A., J. C. Shaw, and H. A. Simon. 1960. Report on a general problem solving program. *Proceedings of the international conference on information processing.* Paris: UNESCO.

Newell, A., and H. Simon. 1972. *Human problem solving.* Englewood Cliffs, N.J.: Prentice-Hall.

Newmeyer, B. 1980. *Linguistic theory in America.* New York: Academic Press.

Nilsson, N. J. 1980. *Principles of artificial intelligence.* Palo Alto, CA: Tioga.

Norman, D., and D. Rumelhart, eds. 1975. *Explorations in cognition.* San Francisco: Freeman.

Olson, D., and N. Filby. 1972. On the comprehension of active and passive sentences. *Cognitive Psychology* 3: 241–263.

Paivio, A. 1969. Mental imagery in associative learning and memory. *Psychological Review* 76: 241–263.

Paivio, A. 1971. *Imagery and verbal processes.* New York: Holt, Rinehart, & Winston.

Partee, B., ed. 1976. *Montague grammar.* NY: Academic Press.

Partee, B. 1979. Montague grammar: Mental representation and reality. In *Contemporary perspectives in the philosophy of language,* ed. P. French *et al.* Minneapolis: University of Minnesota Press.

Pastore, R., *et al.* 1976. Categorical perception of both simple auditory and visual stimuli. Paper presented at the meeting of the Acoustic Society of America, Washington, D.C.

Peters, A. M. 1983. *The units of language acquisition.* Cambridge: Cambridge University Press.

Peters, S., ed. 1972. *Goals of linguistic theory.* Englewood Cliffs, N.J.: Prentice-Hall.

Peters, S., and R. W. Ritchie. 1969. A note on the universal base hypothesis. *Journal of Linguistics* 5: 150–152.

Peters, S., and R. W. Ritchie. 1971. On restricting the base component of transformational grammars. *Information and Control* 18: 483–501.

Peters, S., and R. W. Ritchie. 1973. On the generative power of transformational grammar. *Information Sciences* 6: 49–83.

Petrick, S. R. 1976. On natural language based computer systems. *IBM Journal of Research and Development* 20: 314–325.

Posner, M. 1973. Coordination of internal codes. In *Visual information processing,* ed. W. Chase. New York: Academic Press.

Posner, M., and M. Rogers. 1978. Chronometric analysis of abstraction and recognition. In *Handbook of learning and cognitive processes*, Vol. 5, ed. W. Estes. Hillsdale, N.J.: Erlbaum.

Post, E. 1943. Formal reductions of the general combinatorial problem. *American Journal of Mathematics* 65: 197–268.

Postal, P. 1964a. Limitations of phrase structure grammars. In *The structure of language: Readings in the philosophy of language*, ed. J. A. Fodor and J. J. Katz. Englewood Cliffs, N.J.: Prentice-Hall.

Postal, P. 1964b. Constituent structure: A study of contemporary models of syntactic description. *International Journal of American Linguistics* (Part 3) 30(1): 1–122.

Postal, P. 1972. The best theory. In *Goals of linguistic theory*, ed., S. Peters, 131–170. Englewood Cliffs, N.J.: Prentice-Hall.

Postal, P. 1982. Some arc pair grammar descriptions. In *The nature of syntactic representation*, ed. P. Jacobson and G. Pullum. Dordrecht: Reidel.

Premack, D. 1976. Language and intelligence in ape and man. *American Scientist* 64: 674–683.

Pullum, G. K. 1983. Context freeness and the computer processing of human languages. *Proceedings of the 21st Annual Meeting of the Association for Computational Linguistics*, Cambridge, Mass., 1–6.

Putnam, H. 1975. The meaning of meaning. In *Minnesota studies in the philosophy of science*, Vol. 7, ed. K. Gunderson, 131–193. Minneapolis: University of Minnesota Press.

Pylyshyn, Z. 1973. What the mind's eye tells the mind's brain: A critique of mental imagery. *Psychological Bulletin* 80: 1–24.

Pylyshyn, Z. 1979. Complexity and the study of artificial and human intelligence. In *Philosophical perspectives in artificial intelligence*, ed. M. Ringle. Atlantic Highlands, N.J.: Humanities Press.

Pylyshyn, Z. 1984. *Computation and cognition: Toward a foundation for cognitive science*. Cambridge, Mass.: MIT Press.

Quillian, R. 1968. Semantic memory. In *Semantic information processing*, ed. M. Minsky. Cambridge, Mass.: MIT Press.

Quillian, R. 1969. The teachable language comprehender: A simulation program and theory of language. *Communications of the ACM* 12: 459–476.

Rabin, M., and D. Scott. 1959. Finite automata and their decision problems. *IBM Journal of Research and Development* 3: 114–125. Also in *Sequential machines: Selected papers*, ed. E. Moore, 63–91. Reading, Mass.: Addison-Wesley.

Radford, A. 1981. *Transformational syntax*. Cambridge: Cambridge University Press.

Raphael, B. 1968. SIR: A computer program for semantic information retrieval. In *Semantic information processing*, ed. M. Minsky, 33–145. Cambridge, Mass.: MIT Press.

Reiger, C. J. 1975. Conceptual memory and inferences. In *Conceptual information processing*, ed. R. C. Schank. Amsterdam: North-Holland.

Reisbeck, C. K. 1975. Conceptual analysis. In *Conceptual information processing*, ed. R. C. Schank. Amsterdam: North-Holland.

Reisbeck, C. K., and R. C. Schank. 1976. Comprehension by computer: Expectation-based sentences in context, Computer Science Department, Research Report 78, Yale University.

Reitman, W. R. 1965. *Cognition and thought: An information processing approach*. New York: Wiley.

Restle, F., *et al.*, eds. 1975. *Cognitive theory*. Vol. 1. Hillsdale, N.J.: Erlbaum.

Riebel, D., and S. Schane, eds. 1969. *Modern studies in English*. Englewood Cliffs, N.J.: Prentice-Hall.

Ringle, M., ed. 1979a. *Philosophical perspectives in artificial intelligence.* Atlantic Highlands, N.J.: Humanities Press.

Ringle, M. 1979b. Philosophy and artificial intelligence. In *Philosophical perspectives in artificial intelligence,* ed. M. Ringle. Atlantic Highlands, N.J.: Humanities Press.

Rips, L., E. Shoben, and E. Smith. 1973. Semantic distance and the verification of semantic relations. *Journal of Verbal Learning and Verbal Behavior* 12: 1–20.

Robins, R. H. 1964. *General linguistics: An introductory survey.* London: Longmans.

Robins, R. H. 1967. *A short history of linguistics.* London: Longmans.

Rosch, E. 1973. On the internal structure of perceptual and semantic categories. In *Cognitive development and the acquisition of language,* ed. T. Moore. New York: Academic Press.

Rosch, E. 1975. Cognitive representation of semantic categories. *Journal of Experimental Psychology: General* 104(3): 192–233.

Rosch, E., *et al.* 1976. Basic objects in natural categories. *Cognitive Psychology* 8: 382–439.

Rosenberg, J., and C. Travis, eds. 1971. *Readings in the philosophy of language.* Englewood Cliffs, N.J.: Prentice-Hall.

Ross, J. 1967. *Constraints on variables in syntax.* Ph.D. diss., Department of Linguistics, Massachusetts Institute of Technology.

Rumelhart, D. 1975. Notes on a schema for stories. In *Representation and understanding,* ed. D. G. Bobrow and A. Collins. New York: Academic Press.

Rustin, R., ed. 1973. *Natural language processing.* Amsterdam: North-Holland.

Sadock, J. 1974. *Towards a linguistic theory of speech acts.* New York: Academic Press.

Sager, N. 1978. Natural language information formating: The automatic conversion of texts to structured data base. In *Advances in computers,* ed. M. C. Yovits, 89–162. New York: Academic Press.

Salomaa, A. 1971. The generative capacity of transformational grammars of Ginsburg and Partee. *Information and Control* 18: 227–232.

Saussure, Ferdinand de. [1916] 1955. *Cours de linguistique general.* 5th ed. Paris: Payot. English translation by W. Bastin, *Course in general linguistics.* New York: Philosophical Library, 1959.

Savin, H., and E. Perchonock. 1965. Grammatical structure and the immediate recall of English sentences. *Journal of Verbal Learning and Verbal Behavior* 4: 348–353.

Sayre, K. M. 1979. The simulation of epistemic acts. In *Philosophical perspectives in artificial intelligence.* Atlantic Highlands, N.J.: Humanities Press.

Scarborough, D. *et al.,* 1977. Frequency and repetition effects in lexical memory. *Journal of Experimental Psychology: Human Perception and Performance* 3: 1–17.

Schaeffer, B., and R. Wallace. 1969. Semantic similarity and the comparison of word meaning. *Journal of Experimental Psychology.* 82: 343–346.

Schank, R. C. 1972. Conceptual dependency: A theory of natural language understanding. *Cognitive Psychology.* 3: 552–631.

Schank, R. C., ed. 1975a. *Conceptual information processing.* Amsterdam: North-Holland.

Schank, R. C. 1975b. The primitive acts of conceptual dependency. In *Theoretical issues in natural language processing.* Cambridge, Mass.:

Schank, R. C. 1977. *Scripts, plans, goals, and understanding: An inquiry into human knowledge structures.* Hillsdale: N.J.: Erlbaum.

Schank, R. C. 1978. Predictive understanding. In *Recent advances in the psychology of language,* ed. R. N. Campbell and P. T. Smith. New York: Plenum Press.

Schank, R. C. 1979. Natural language, philosophy, and intelligence. In *Philosophical perspectives in artificial intelligence,* ed. M. Ringle. Atlantic Highlands, N.J.: Humanities Press.

Schank, R. C. 1980. Language and memory. *Cognitive Science* 4: 243–284.

Schank, R. C., and R. P. Abelson. 1977. *Scripts, plans, goals, and understanding.* Hillsdale, N.J.: Erlbaum.

Schank, R. C., and J. G. Carbonell. 1979. Re: The Gettysburg address: Representing social and political acts. In *Associative networks: Representation and use of knowledge by computers,* ed. N. Findler. New York: Academic Press.

Schank, R. C., and K. M. Colby, eds. 1973. *Computer models of thought and language.* San Francisco: Freeman.

Schank, R. C., N. Goldman, C. Reiger, and C. Reisbeck. 1973. MARGIE: Memory, analysis, response generation and inference in English. *Proceedings of the third international joint conference on artificial intelligence,* Stanford, Calif.

Schopenhauer, A. 1958. *The world as will and representation.* New York: Dover.

Searle, J. 1969. *Speech acts.* Cambridge: Cambridge University Press.

Searle, J. 1975. Indirect speech acts. In *Syntax and semantics,* ed. P. Cole and J. Morgan. New York: Seminar Press.

Sebeok, T., ed. 1974. *Linguistics and adjacent arts and sciences.* Vol. 12 of Current trends in linguistics. The Hague: Mouton.

Shannon, C. E. 1950. Programming a computer for playing chess. *Philosophical Magazine* (Series 7) 41: 256–275.

Shortliffe, E. H. 1976. *MYCIN: Computer-based medical consultations.* New York: Elsevier.

Shepherdson, J. 1959. The reduction of two-way automata to one-way automata. *IBM Journal of Research and Development* 3: 198–200.

Simmons, R. F. 1966. Storage and retrieval of aspects of meaning in directed graph structures. *Communications of the ACM* 9: 211–214.

Simmons, R. F. 1978. Towards a computational theory of textual discourse. Department of Computer Science, TR-NL-35, University of Texas, Austin.

Simmons, R. F., J. F. Burger, and P. E. Long. 1966. An approach toward understanding English questions from text. *Proceedings of the fall joint computer conference.* New York: Spartan.

Simmons, R. F., and A. Correira. 1978. Rule forms for verse, sentences, and story trees. Department of Computer Science, TR-NL-35, University of Texas, Austin.

Simon, H. A. 1976. Artificial intelligence systems that understand. *IJCAI 77, Proceedings.*

Simon, H. A. 1979–1980. Lessons for AI from human problem solving. *Computer Science Research Review,* Carnegie-Mellon University.

Singh, J. 1966. *Great ideas in information theory, language, and cybernetics.* New York: Dover.

Slobin D. 1966. Grammatical transformations and sentence comprehension in childhood and adulthood. *Journal of Verbal Learning and Verbal Behavior* 5: 219–227.

Slobin, D, ed. 1971. *The ontogenesis of grammar.* New York: Academic Press.

Smith, E. 1978. Theories of semantic memory. In *Handbook of learning and cognitive processes,* Vol. 6, ed. W. Estes. Hillsdale, N.J.: Erlbaum.

Smith, E., E. Shoben, and L. Rips. 1974. Structure and process in semantic memory: A featural model for semantic decisions. *Psychological Review* 81: 214–241.

Solomon, R., and D. Howes. 1951. Word frequency, personal value and value duration thresholds. *Psychological Review* 58: 256–270.

Spence, K., and J. Spence, eds. 1968. *Advances in the psychology of learning and motivation research and theory.* Vol. 2. New York: Academic Press.

Sternberg, R. J., ed. 1982. *Handbook of human intelligence.* Cambridge: Cambridge University Press.

Stevens, K., and A. House. 1972. Speech perception. In *Foundations of modern auditory theory,* Vol. 2, ed. J. Tobias, 3–62. New York: Academic Press.

Swinney, D. A. 1979. Lexical access during sentence comprehension: (Re)-consideration of context effects. *Journal of Verbal Learning and Verbal Behavior* 18: 681–689.

Swinney, D. A., and D. T. Hakes. 1976. Effects of prior context upon lexical access during sentence comprehension. *Journal of Verbal Learning and Verbal Behavior* 15: 681–689.

Tanenhaus, M., and J. Carroll. 1975. The clausal processing hierarchy and nouniness. In *Papers from the parasession functionalism*, ed. R. E. Grossman *et al.* Chicago: Chicago Linguistic Society.

Tanenhaus, M., *et al.* 1976. Sentence-picture verification models as theories of sentence comprehension: A critique of Carpenter and Just. *Psychological Review* 83: 310–317.

Tanenhaus, M., J. Leiman, and M. Seidenberg. 1979. Evidence for multiple stages in the processing of ambiguous words in syntactic contexts. *Journal of Verbal Learning and Verbal Behavior* 18: 427–440.

Tarski, A. 1956. *Logic, semantics, and mathematics*. London: Oxford University Press.

Tennant, H. 1980. *Natural language processing: An introduction to an emerging technology.* New York: Petrocelli.

Thompson, F. B. 1968. English for the computer. *Proceedings of the fall joint computer conference*. New York: Spartan.

Thomson, D., and E. Tulving. 1970. Associative encoding and retrieval: Weak and strong cues. *Journal of Experimental Psychology* 86: 255–262.

Thorndyke, P. 1977. Cognitive structures in comprehension and memory of narrative discourse. *Cognitive Psychology* 9: 77–110.

Thorne, J., P. Bratley, and H. Dewar. 1968. The syntactic analysis of English by machine. *Machine intelligence* 3: 281–309.

Tobias, J., ed. 1972. *Foundations of modern auditory theory.* Vol. 2. New York: Academic Press.

Tou, J. T., ed. 1974. *Advances in information systems science.* Vol. 5. New York: Plenum Press.

Tulving, E. 1972. Episodic and semantic memory. In *Organization of memory*, ed. E. Tulving and W. Donaldson. New York: Academic Press.

Tulving, E. 1978. Relation between encoding specificity and levels of processing. In *Levels of processing and human memory*, ed. L. Cermak and F. Craik. Hillsdale, N.J.: Erlbaum.

Tulving, E., and S. Osler. 1968. Effectiveness of retrieval cues in memory for words. *Journal of Experimental Psychology* 19: 593–601.

Tulving, E., and W. Donaldson, eds. 1972. *Organization of memory.* New York: Academic Press.

Turing, A. M. 1936. On computable numbers with an application to the entscheidung-problem. *Proceedings of the London Mathematics Society* 42: 230–265.

Turing, A. M. 1937. A correction. *Proceedings of the London Mathematics Society* 43: 544–546.

Turing, A. M. 1950. Computing machinery and intelligence. *Mind* 59.

Tyler, L., and W. Marslen-Wilson. 1977. The on-line effects of semantic context on syntactic processing. *Journal of Verbal Learning and Verbal Behavior* 16: 683–692.

Wales, R., and E. Walker, eds. 1976. *New approaches to language mechanisms.* Amsterdam: North-Holland.

Walker, E., *et al.* 1968. Grammatical relations and the search of sentences in immediate memory. *Proceedings of the Midwestern Psychological Association.*

Wall, R. 1972. *Introduction to mathematical linguistics.* Englewood Cliffs, N.J.: Prentice-Hall.

Waltz, D. L. 1982. Artificial intelligence. *Scientific American* (October): 118–133.

Wanner, E. 1974. *On remembering, forgetting, and understanding sentences.* The Hague: Mouton.

Wanner, E. 1980. The ATN and the Sausage Machine: Which one is baloney? *Cognition* 8: 209–225.

Wanner, E., and R. Kaplan. 1975. Garden paths in relative clauses. Cambridge, Mass.: Harvard University. Mimeo.

Wanner, E., and M. Maratsos. 1978. An ATN approach to comprehension. In *Linguistic theory and psychological reality*, ed. M. Halle *et al.*, 119–161. Cambridge, Mass.: MIT Press.

Warren, R. 1970. Perceptual restoration of missing speech sounds. *Science* 167: 392–393.

Warren, R., and G. Sherman. 1974. Phonemic restorations based on subsequent context. *Perception and Psychophysics* 16.

Wason, P. 1965. The context of a plausible denial. *Journal of Verbal Learning and Verbal Behavior* 4: 7–11.

Weimer, W., and Palermo, D., eds. 1974. *Cognition and symbolic processes*. Hillsdale, N.J.: Erlbaum.

Weizenbaum, J. 1966. ELIZA. *Communications of ACM* 9: 36–45.

Weizenbaum, J. 1976. *Computer power and human reason*. San Francisco: Freeman.

Wepman, J., and R. Hein, eds. 1963. *Concepts of personality*. Chicago: Aldine.

Wexler, K., and P. Culicover. 1980. *Formal principles of language acquisition*. Cambridge, Mass.: MIT Press.

Wilensky, R. 1978. *Understanding goal-based stories*. Ph.D. diss. Department of Computer Science, Research Report 140, Yale University.

Wilkins, A. 1971. Conjoint frequency, category size, and categorization time. *Journal of Verbal Learning and Verbal Behavior* 10: 382–385.

Wilks, Y. 1973a. An artificial intelligence approach to machine translation. In *Computer models of thought and language,* ed. R. C. Schank and K. M. Colby. San Francisco: Freeman.

Wilks, Y. 1973b. Preference semantics. In *Formal semantics of natural languages,* ed. E. Keenan. Cambridge: Cambridge University Press.

Wilks, Y. 1975. A preferential pattern-seeking semantics for natural language inference. *Artificial Intelligence* 6: 53–74.

Wilks, Y. 1976. Parsing English II. In *Computational semantics,* ed. E. Charniak and Y. Wilks. Amsterdam: North-Holland.

Wilks, Y. 1981. A position note on natural language understanding and artificial intelligence. *Cognition* 10: 53–74.

Winograd, E., and N. Rivers-Bulkeley. 1977. Effects of changing context on remembering faces. *Journal of Experimental Psychology: Human Learning and Memory* 3: 397–405.

Winograd, T. 1972. *Understanding natural language*. New York: Academic Press.

Winograd, T. 1974. Five lectures on artificial intelligence. AI Laboratory Memo No. 246, Stanford University.

Winograd, T. 1975. Frame representations and the declarative/procedural controversy. In *Representation and understanding,* ed. D. B. Bobrow and A. Collins. New York: Academic Press.

Winograd, T. 1976. Artificial intelligence and language comprehension. In *Artificial Intelligence and Language Comprehension*. Washington, D.C.: National Institute of Education.

Winograd, T. 1980. What does it mean to understand language? *Cognitive Science* 4: 209–241.

Winograd, T. 1983. *Language as a cognitive process*. Vol. 1, *Syntax*. Reading, Mass.: Addison-Wesley.

Winston, P. H., ed. 1975. *The psychology of computer vision*. New York: McGraw-Hill.

Winston, P. H. 1977. *Artificial intelligence.* Reading, Mass.: Addison-Wesley.

Winston, P. H. 1984. *Artificial intelligence.* rev. ed. Reading, Mass.: Addison-Wesley.

Winston, P. H., and K. A. Prendergast. 1984. *The AI business: The commercial uses of artificial intelligence.* Cambridge, Mass.: MIT Press.

Woods, W. A. 1967. Semantics for a question-answering system. Report NSF-9. Cambridge, Mass., Harvard University.

Woods, W. A. 1970. Transition network grammars for natural language analysis. *Communications of ACM* 13: 591–606.

Woods, W. A. 1973. An experimental parsing system for transition network grammars. In *Natural language processing,* ed. R. Rustin. Amsterdam: North-Holland.

Woods, W. A. 1978. Semantics and quantification in natural language question answering. In *Advances in computers,* Vol. 17, ed. M. C. Yovits, 2–87. New York: Academic Press.

Woods, W. A., R. Kaplan, and B. Nash-Webber. 1972. The lunar sciences natural language information system: Final report. BBN Report No. 2378. Cambridge, Mass.: Bolt Beranek and Newman.

Wunderlich, D. 1979. *Foundations of linguistics.* Cambridge: Cambridge University Press.

Yovits, M. C., ed. 1978. *Advances in computers.* Vol. 17. New York: Academic Press.

Zampoli, A., ed. 1977. *Linguistic structures processing.* Amsterdam: North-Holland.

Name Index

Subject Index